Computerized Basin Analysis

The Prognosis of Energy and Mineral Resources

COMPUTER APPLICATIONS IN THE EARTH SCIENCES
A series edited by Daniel F. Merriam

Computerized Basin Analysis

The Prognosis of Energy and Mineral Resources

Edited by

Jan Harff

GeoForschungsZentrum Potsdam
Telegrafenberg, Potsdam, Germany

and

Daniel F. Merriam

Kansas Geological Survey
University of Kansas
Lawrence, Kansas

Springer Science+Business Media, LLC

Library of Congress Cataloging-in-Publication Data

Computerized basin analysis : the prognosis of energy and mineral
resources / edited by Jan Harff and Daniel F. Merriam.
       p.   cm. -- (Computer applications in the earth sciences)
   "Based on the proceedings of an international symposium ... held
June 19-22, 1990, in Gustrow, Germany"--T.p. verso.
   Includes bibliographical references and index.
   ISBN 978-0-306-44499-9   ISBN 978-1-4615-2826-5 (eBook)
   DOI 10.1007/978-1-4615-2826-5
   1. Prospecting--Data processing--Congresses. 2. Sedimentation and
deposition--Computer simulation--Congresses.   I. Harff, Jan.
II. Merriam, Daniel Francis. III. Series.
TN270.C674 1993
622'.1--dc20   93-17935
                                                           CIP

Based on the proceedings of an International Symposium on Computerized Basin Analysis: The Prognosis of
Energy and Mineral Resources, held June 19–22, 1990, in Güstrow, Germany

ISBN 978-0-306-44499-9

© 1993 by Springer Science+Business Media New York
Originally published by Plenum Press New York in 1993

PREFACE

This symposium on 'Computerized Basin Analysis for Prognosis of Energy – and Mineral Resources' was organized by Dr. Jan Harff, chairman of the Scientific Committee for the meeting, in Güstrow in what was then East Germany. Sponsors of this meeting were the International Union of Geological Sciences' Commission on Storage, Automatic Processing and Retrieval of Geologic Data (COGEODATA), Academy of Sciences of the German Democratic Republic (GDR), National Oil and Gas Trust of the GDR, and the International Association for Mathematical Geology (IAMG). Main topics of the symposium, held from 19-22 June 1990, were application of computer methods to the exploration and exploitation of oil and gas, coal, and other energy and mineral resources. There were computer demonstrations as well as a one-day field trip to the geothermic heating plant in Waren. The Regional Group for Eastern Europe of COGEODATA also met during the conference.

Fifty-one papers were presented including eight poster sessions by authors from 14 countries. As was to be expected, there was a large percentage of papers from the East Bloc of European countries, especially the GDR, USSR, and the CSSR with a fair representation from the FRG and USA and a smattering from the nine others. Most of the papers were application oriented and related to the mineral industries. There was ample time for exchange of ideas and dissemination of material.

From this program, the Scientific Committee selected an appropriate number of papers for this volume. These written presentations give a good flavor of the tenor of the meeting and the level of sophistication of the subject matter. For those not able to attend then, this collection of papers will give a good representation of Güstrow 1990.

This was a historic international meeting in more ways than one. In addition to being an international scientific meeting of note, major national and international changes were beginning to take place at this time. This was the last international geological meeting to be held in what was then East Germany, before becoming part of the new unified

Germany. The attendants from the East Bloc of countries would now just be listed from the eastern part of Europe. Attendees from the now nonexistent USSR would be from their respective country. Even while the meeting was in the planning stages borders were opened and security loosened so that not only ideas but people freely moved across national borders by the summer of 1990. This was one of, if not, the last meeting sponsored by COGEODATA, which in 1992 has become part of another entity and lost its identity as an IUGS Commission. And lastly, our host the Central Institute for Physics of the Earth in Potsdam ceased to exist just 18 months after the meeting.

We want to thank the many people who helped organize the conference in Güstrow; they did an excellent job. The presentations were excellent and the atmosphere of collegiality was great and extended even to the social functions. Perhaps the highlight of the social activities was the opening of the keg at the Mecklenburgian hunter party in the State Game Reserve near Güstrow! We thank the authors of the papers for providing them; and we thank the many reviewers of the scientific content of those papers.

Visiting Senior Scientist   D.F. Merriam
Kansas Geological Survey
The University of Kansas
Lawrence, KS 66047   USA

CONTENTS

INTRODUCTION

Under the title "Computerized Basin Analysis" an international Symposium was held in Güstrow, (East) Germany on 19-22 June 1990 sponsored by COGEODATA Commission of IUGS and the International Association for Mathematical Geology (IAMG), and organized by the Central Institute for Physics of the Earth, Potsdam together with the Oil-Gas-Geotechnology Gommern Ltd.

The goals of this Symposium were, on the one hand, to demonstrate modern results of computer modeling in the frame of basin analysis and, on the other hand, to bring scientists from the eastern and the western hemisphere together to serve as a platform for the exchange of scientific ideas having been elaborated relatively independently in the East and the West.

About one hundred scientists from fourteen eastern and western countries participated at the meeting and discussed oral, poster, and software presentation delivered on the main topics of

(a) Geological process modeling,
(b) Structure modeling,
(c) Resource analysis and prediction,
(d) Information systems/artificial intelligence.

One of the results of the Symposium was to note that mathematical and computer modeling for geology have reached a new level of sophistication in basin analysis regarding both theoretical background and practical application. The reason is that sedimentary basins can be considered to be relative "closed systems" providing favorable conditions for modeling, data management, information processing, and monitoring. The other is that the practical value of basin modeling for resource prediction and environmental protection is being demonstrated resulting in a concentration of mathematic geological research to this field of work enhancing the elaboration of theory and it's application.

From the presentations, papers were selected by the scientific committee for the publication in this volume which gives a representative overview of the results of the symposium.

One aim of geological process models is to simulate processes associated with basin formation. Realizations can serve as experiments to examine the interaction of variables and compare results to actual geology in stretched space and time-scale modeling, and can help in unraveling geologically complex natural processes. These simulations can be used for a better understanding of the evolution of sedimentary basins and the geological history of individual sedimentary sequences for assessing resource potential. S. E. Medvedev presented a mechanical-mathematical model, its numerical solution and corresponding software for studying the dynamics of the ground surface and the sedimentary layers, interfaces as a function of the changing basement surface at a regional scale. Results from the research on the SEDSIM-Project were outlined by J. W. Harbaugh. He uses a supercomputer (WS) for the three-dimensional simulation of silicilastic sedimentation in small-medium-sized basin structures. An ecological module is included for the simulation of carbonate formation. Numerical experiments in paleohydrogeology for the reconstruction of possible lateral migration paths for pore fluids in the Rotliegend sediments in the Northeast German Basin were undertaken by J. Springer and G. Schwab. They used a PC for their research in a two-dimensional simulation. Burial history, variations of the temperature field, and resulting maturation of hydrocarbon processes in different regions of the Northeast German Basin were analyzed by A. Berthold and K. Menschner by a standard one-dimensional model using new data from this area.

Differential compaction and resulting structural deformation of Triassic sedimentary sequences in the Thuringian Basin, Germany, were investigated by H. Dietrich using a one-dimensional model. R. Ondrak and U. Bayer have developed a model involving application of thermodynamics to simulate the dissolution and precipitation of minerals via temperature and transport processes applied through time during diagenesis along cross sections.

During foreward modeling the parameters of equation systems have to be optimized stepwise. For this purpose, the status of a basin in the geologic past has to be reconstructed by specific independent methods.

Here, reconstruction of a basin's geothermal history plays an important role. P. K. Jensen presented a mathematical procedure for comparing fission tracks annealing temperatures of apatites measurements with basin modeling results of thermal history reconstruction of sedimentary rocks.

On the assumption that a sedimentary basin under study is a closed system, W. W. Hay and C. N. Wold and C.N. Wold, C. Shaw, R. Deconto, and W.W. Hay have developed models for balancing of the mass transfer during sedimentation. The authors have applied the models to sediments of the Gulf of Mexico and the Northeast Atlantic Ocean and their source areas. The models also can be used for paleogeographic reconstruction to assess differential loading of the Earth's crust with respect to isostasy.

Modeling of recent basin structure is based either on stratigraphic records from drilling or on seismic investigation. Well-log imaging and its application to geologic interpretation are explored by L. Huang, D. Richers, and J.E. Robinson through several examples. This paper was not presented during the meeting but was included in this volume for completeness. P. Dowd used geostatistical methods as coregionalization for simulating sequences of beds for structural modeling in sedimentary basins.

For resource assessment, it is effective to develop and to use software systems to examine complex strategies for applying different methods of process modeling to well-log databases. S. Cao, S. Bachu, and A. Lytviak introduced the Quantitative Basin Analysis System (QBAS) used in petroleum exploration via an application to the Western Canada Basin. K. D. Hanemann and K. O. Zeissler used statistical methods and inter-active computer graphics for the interpretation of data from geophysical surveys and well-log data in the frame of lignite exploration in an East German Tertiary basin.

Because of a myriad of possible data types are possible interrelation-ships between them in basin analysis, multivariate methods are required for data integration. D. F. Merriam, B. A. Fuhr, and U.C. Herzfeld developed a method for the integration of geological, geophysical, geo-chemical, and topographic data. A mature oil provence in south-central Kansas was used as a case study of this integrated, automated approach. The theory of the classification of geological objects was developed in the Soviet Union relatively independently from geostatistical theory created in the West. L. W. Watney, J. Harff, J. Davis, J. C. Wong, and G. Bohling tied these theories together developing a regionalized classification method. This method is applied to assessing petroleum resources in Paleozoic sequences in Kansas. Basin subsidence and thermal modeling has been used with the regionalization method to evaluate regional natural gas potential in the Northeast German Basin (J. Harff, P. Hoth, and W. Eiserbeck). Methods of data integration also were used by H. Thiergärtner and J. Rentzsch for metallogenetic research in the Rudolstadt Basin, central Germany. Economic aspects are taken into consideration by D. J. Forman, A. L. Hinde, and A. P. Radlinski in their analysis of exploitation data utilizing particular assumptions about migration fairways in to make assessments of undiscovered hydrocarbon resources in South Australia.

A review of methods of artificial intelligence for basin analysis is given by G. Peschel, H.H. Poppitz, and M. Mokosch. A database comprised of properties from 94 software systems was compiled and is available on diskette.

In general, the basis for successful mathematical and computer modeling has to be an appropriate geological (conceptual) model. That has to be translated firstly into a physical (-chemical) one, then in a mathematical model. Recent information processing and computer technology enables the geologist to simulate geological structures and processes in a sophisticated way. It provides an additional effective tool for geological research, but it cannot, and will not replace, its background and base: the geological thinking, field work, and laboratory measurement.

Chairman of   J. Harff
COGEODATA Working Group on Energy Resource Data
Central Institute for Physics of the Earth
Potsdam, Germany

COMPUTER SIMULATION OF SEDIMENTARY COVER EVOLUTION

S.E. MEDVEDEV

NPO "Sojuznefteotdacha", Ufa, U.S.S.R.

ABSTRACT

A computer program has been developed for the mathematical simulation of the mechanical evolution of a sedimentary basin. It is designed for use in numerical experiments to study the impact of different velocity patterns at the base of the basin on the dynamics of the surface and inner boundaries, as well as for use in reconstructing the formation and evolution of existing basins. A layered basin is considered and the 3D modeling task is reduced to solving of 2D equations describing inner and outer boundaries of basin evolution.

INTRODUCTION

The problem of exact determination of inner boundaries in a layered sedimentary basin is connected closely with the problem of exploration and definition of hydrocarbon reserves.

Periods of evolution involving millions of years are being investigated. This fact allows one to use a model of incompressible high-viscosity fluids for the description of geological structures, as shown by a number of authors (e.g. Rumberg, 1963, Myasnikov and Fadeev, 1980). This model also can be used for the description of the evolution of sedimentary basins (Zanemonetz and others, 1976, Reddy, Stein, and Wicham, 1982).

The theoretical part of this work is prepared using the method of Zanemonetz Svalova (Zanemonetz, Kotylolkin, and Myasnikov, 1974; Zanemonetz, Mikhajlov, and Myasnikov, 1976) and under her direct guidance.

Computerized Basin Analysis, Edited by J. Harff
and D.F. Merriam, Plenum Press, New York, 1993

MODEL EQUATIONS

The mathematical description of a layered sedimentary basin is discussed in this paper. The layers are determined by their rheological characteristics, such as viscosity and density, which are homogeneous inside the layer and differ from one layer to another.

Each layer is assumed to be governed by the 3D Navier-Stokes equations. The dimensionless presentation (in X-axis projection) is:

$$F * St * \frac{dU}{dT} = -\frac{h_k}{L} \frac{\delta \Pi}{\delta X} + \frac{F}{R} \mu_k \left(\frac{\delta^2 U}{\delta X^2} + \frac{\delta^2 U}{\delta Y^2} + \frac{L^2}{h_k^2} \frac{\delta^2 U}{\delta Z^2} \right) \tag{1}$$

where $F = \dfrac{U_0^2}{g * L}$ is the Frude number, $St = \dfrac{L}{t_0 U_0}$ is the Struhal number,

and $R = \dfrac{u_0 L \rho_0}{\mu_0}$ is the Reynolds number.

Dimensionless values of the variables were produced as follows (see Fig. 1):

$X = x/L$, $Y = y/L$, $Z = z/h$, $U = u/u_0$, $V = v/u_0$, $W = w/(u_0 * (h/L))$

$\Pi = p/(\rho_0 * g * h)$, $\mu_k = \mu'_k / \mu_0$, $T = t/t_0$

where L & h_k are the characteristic dimensions of a layer, ρ_0 is the density of layers, μ_0 is the viscosity, u_0 is the characteristic velocity, U,V,W (dimensionless) and u,v,w (dimensional) are cartesian components of velocity, t_0 is the characteristic time of the process, and g is the acceleration of gravity.

The analysis of characteristic values for the present problems gives the following results:

(1) F*S << F/R , so the left term in the equations can be omitted (steady state condition);

(2) F/R << 1 ; and

(3) $\varepsilon = h/L << 1$.

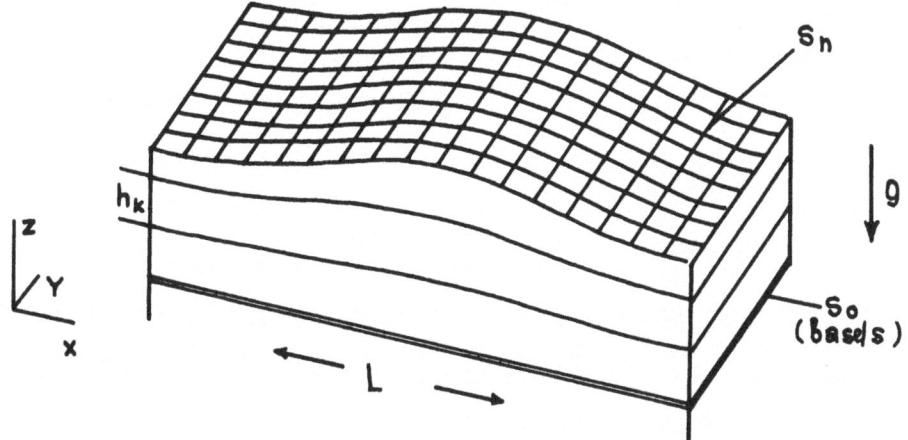

Figure 1. Model of layered sedimentary basin.

Thus, the equation of Navier-Stokes can be reduced to

$$\frac{\delta\Pi}{\delta X} = \mu_k * MS * \left(\varepsilon^2 * \frac{\delta^2 U}{\delta X^2} + \varepsilon^2 * \frac{\delta^2 U}{\delta Y^2} + \frac{\delta^2 U}{\delta Z^2} \right) \tag{2}$$

where $MS = F/R * \varepsilon^{-3}$.

The Y-axis projection of the Navier-Stokes equation is similar to (2). The Z-axis projection (after the transformation similar to (1)-(2)) is:

$$\frac{\delta\Pi}{\delta Z} = -\rho_k + \mu_k * MS * \left(\varepsilon^4 \frac{\delta^2 W}{\delta X^2} + \varepsilon^4 \frac{\delta^2 W}{\delta Y^2} + \varepsilon^4 \frac{\delta^2 W}{\delta Z^2} \right) \tag{3}$$

The value of the MS multiplier should be defined for further investigation. The technique being described in this section can operate with different values (MS<<1, MS>>1, MS=1); and the analytical transformation allows equations to be obtained in the form convenient for solving by computer.

In this work the following example is assumed:

$$MS = 1 \tag{4}$$

This condition (4) limits the application of the program described, but maybe is suitable for the class of problems in question. The different characteristic values can lead to the condition (4), for example:

$$L=100 \text{ km, } h=1 \text{ km , } \varepsilon = 10^{-2}$$

$$u = 1.5 \text{ cm/year, } \mu_0 = 10^{19} \text{ poise, } F/R = 10^{-6} \tag{5}$$

Reddy, Stein, and Wicham (1982) investigated a 13-layered model of the sedimentary basin. The model was based on the United States continental margin of the Gulf of Mexico. Each layer seems to be consistent with the condition (4).

The existence of the small parameter ε allows the use asymptotic techniques. All investigated variables are expanded in a power series in ε:

$$f = f^0 + \varepsilon f^1 + \varepsilon^2 f^2 + \ldots \tag{6}$$

The zeroth approximation of the model equations (three cartesian projections of the Navier-Stokes equation and the continuity equation for incompressible fluids) is:

$$\frac{\delta \Pi^0}{dX} = \mu_k \frac{\delta^2 U^0}{\delta Z^2}$$

$$\frac{\delta \Pi^0}{\delta Y} = \mu_k \frac{\delta^2 V^0}{\delta Z^2}$$

$$\frac{\delta \Pi^0}{\delta Z} = -\rho_k \tag{7}$$

$$\frac{\delta U^0}{\delta X} + \frac{\delta V^0}{\delta Y} + \frac{\delta W^0}{\delta Z} = 0$$

These equations describe the evolution of one layer. For the investigation of the dynamics of n-layered sedimentary basins, these equations are integrated for each layer using interconnected boundary conditions:

continuity of velocity and forces on inner layer boundaries, stress-free surface condition for upper boundary and prescribed flow field at the base of the basin.

The solution gives the expression of the unknown values as functions of the X and Y coordinates, T, and the unknown position of layer boundaries:

$$f = f(X,Y,T,S_k(X,Y)) \tag{8}$$

where S_k is the height of the boundary between the layers k and k+1.

For determination of boundary dynamics these expressions should be put in the condition of impenetrability through the boundaries:

$$St * \frac{\delta S_k}{\delta T} + U * \frac{\delta S_k}{\delta X} + V * \frac{\delta S_k}{\delta Y} + W \mid S_k = 0 \tag{9}$$

for k=0,1,..,n.

As a result the dynamics of the boundaries is described by the system of n parabolic nonlinear equations:

$$S_t * \frac{\delta S_k}{\delta T} = \sum_{i=0}^{n} \left[\left(\frac{\delta^2}{\delta X^2} + \frac{\delta^2}{\delta Y^2} \right) \left(A_i^k \left\{ (S_a - S_b)^4 \right\} \right) + \frac{\delta}{\delta X} B_{ix}^k \left\{ (S_a - S_b)^2 U \right\} + \right.$$
$$\left. \frac{\delta}{\delta Y} B_{iy}^k \left\{ (S_a - S_b)^2 V \right\} \right] + C_k \{W\} \tag{10}$$

for k=1,2,..n; $A_i^k(.)$, $B_{iy}^k(.)$, $B_{iy}^k(.)$ and C_k are linear functions of noted arguments and viscosity and density, a,b = 0,1,..,n ; and the hyperbolic equation for the low boundary description. The equations are not given in full detail because of their large size.

A term such as $(k * \Delta S_n)$, representing an erosion function, is added to the upper boundary equation (Culling, 1960). Terms for modeling of sedimentation also can be added.

Thus, 3D modeling of sedimentary basins is conducted with the help of a system of 2D differential equations. The solution was performed numerically with a computer.

The peculiarity of this system is the complexity of the complete set of equations, caused by the boundary conditions. This complication is caused by the mobility of all boundaries of the basin, especially if the basin's margin is situated in the active area. Thus the model shown in Figure 1 can be used if the area in question is a small part of a larger area where the boundary conditions are determined by isostatic forces. Thus, the boundary conditions for S are set by horizontal forces:

$$\frac{\delta S_k}{\delta n_D} \Big| D = 0 \tag{11}$$

where n_D - normal to the vertical basin surface,
      D - intersection of boundary with vertical basin surface.

When the basin's margin is restricted by some structures, special initial data can be used. In Figure 2 the area in question was extended, such that all layer boundaries come to the surface smoothly and then extended as one boundary. The subvertical lines can have the same form as the adjacent structures. The boundary condition for S was set by zero forces (8).

NUMERICAL MODEL

The finite-difference approximation was used for solving the equations. Central differences were used mainly, but there is a possibility of using exponential or polynomial splines.

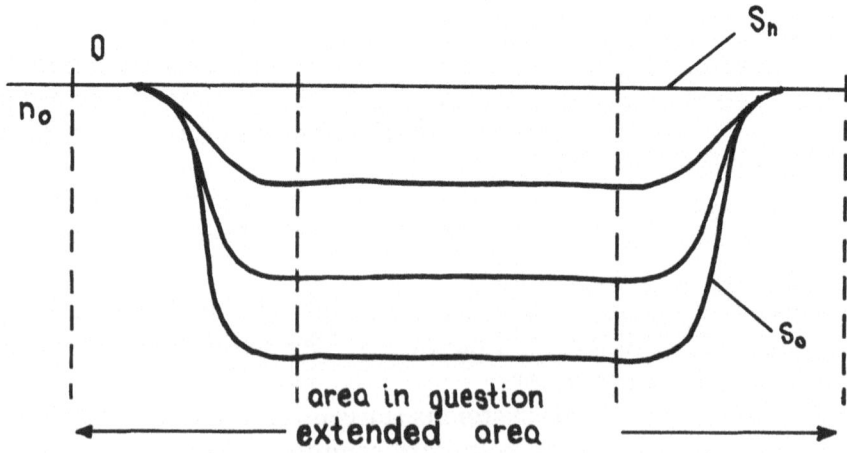

Figure 2. Special initial data.

The lower boundary was calculated by a method of characteristics with approximation order of ($\Delta t \, \Delta x$) (Δt and Δx are the steps in time and space in finite difference network). The other boundaries were calculated with approximation order of($\Delta t \, \Delta x^2$) by iterative alternating direction implicit method (IADI). Three or four iterations are conducted, and accuracy of the solution is checked at each time step.

Analysis has shown that for this scheme to be stable a stable configuration of layers is necessary, that is the density of the upper layer should be less than the next lower layer density:

$$\rho_i \geq \rho_\kappa \quad \text{when} \quad k > i$$

The following observed facts help establish confidence in the accuracy of the solution:
- retention of the symmetry of the solution with the presence of symmetric initial and boundary data;
- conservation of mass, defined by boundaries.

RESULTS

Existing basins are not considered here. Therefore the conditions and results are presented in the dimensionless form. The conditions and results transformed into the dimensional form [through the use of characteristic values introduced in (5)] are given for illustration only.

Trial mathematical tests were conducted on the grid network of 40*40 points for 2, 3, and 4 layers. The time step was taken $\Delta t=0.03$ (dimensional: 0.3 million years), with grid spacing $\Delta x = \Delta y=0.05$ (500 m). Different velocity profiles have been set with max. $U_{max} \leq 1.2(1.8\,\text{cm/year})$. The erosion coefficient varied from 0 to 1.5. The sedimentation process has not been simulated at the present stage.

The investigations were performed for double-layered model with viscosity $\mu_1 = \mu_2 = 1$ (10^{19} poise) and density $\rho_1 = 1.1$ and $\rho_2 = 1.0$ (2.2 g/cm^3 and 2.0 g/cm^3). Initial data: $S_0 = 0, S_1 = 0.5, S_2 = 1, L=1$ (accordingly 0, 0.5 km, 1 km, 100 km).

Figure 3 represents the system compression along the X axis by a flow field on the basis given by the following representation :

$$U = b*\sinh(c*x)$$

with b=-1, c=6, (max(U)=1.5 cm/year) during the period of time t=0.5 (5 million years). Compared with the initial data, Figure 3 shows the rising and the bending of the interlayer boundaries.

Figure 4 represents extension of the basin by a velocity along the X axis:

$$U = b*\sinh(c*x)/(\cosh(c*x))^2$$

with b=-1, c=6, (max(U)=0.75) during the period of time t=0.5 (5 million years).

Figure 4 also shows the comparison of the results with different erosion coefficients (k=0 and k=1.5). Figure 4 shows an active influence of surface effects on the inner dynamics of the basin which is noticeable: this is dificult to explain at present.

Calculations of a sedimentary basin's evolution have been conducted earlier, mainly for 2D models. Of particular interest is the validity of 2D modeling. Two mathematical tests have been conducted. The first test used compression only along the X axis, during the first one-half of the test (until t=0.2):

$$U = -b*\sinh(8x)/(\cosh(8x))^2$$

during the remaining time (until t=0.4) compression was performed only along the axis Y:

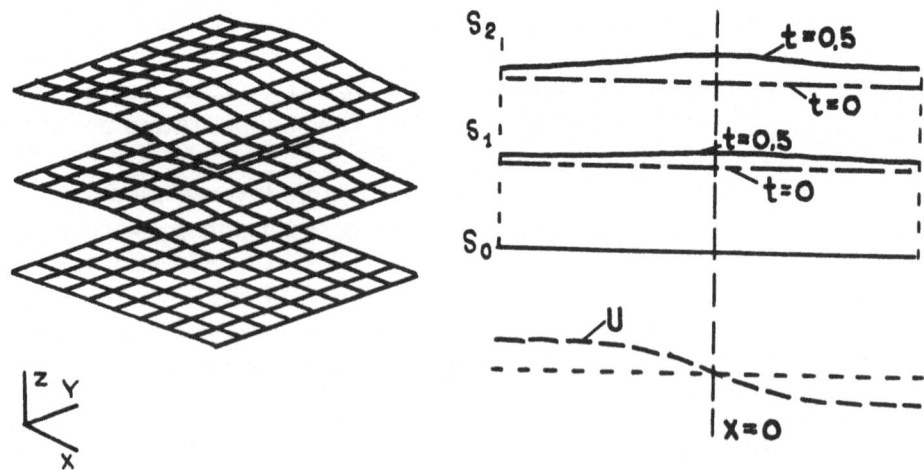

Figure 3. Compression of basin: general 2D view (left), cross section along X axis (compared with initial data) (top right), basis velocity profile (right bottom).

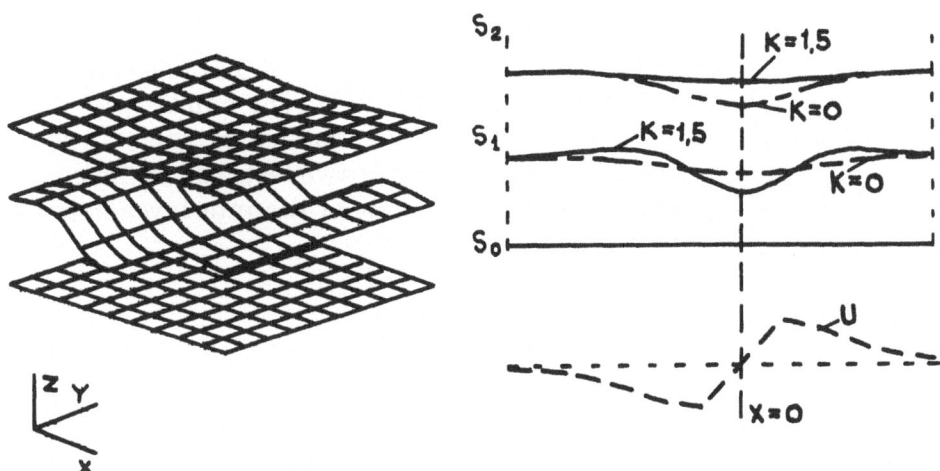

Figure 4. Extension of basin: general 2D view (for k=1.5)(left), cross section along X axis (compared with k=0) (top right), base velocity profile (right bottom).

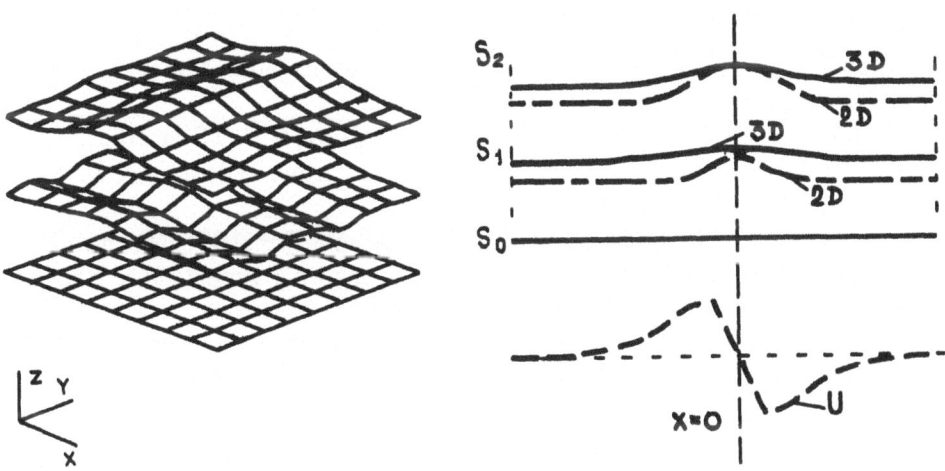

Figure 5. Compression of basin: general 2D view (for 3D-modeling)(left), section along X axis (compared with 2D-model.) (top right), base velocity profile (along X)(right bottom).

$$V = -b*\sinh(8x)/(\cosh(8x))^2$$

(b = 2 in both examples). Thus, two separate 2D modeling phases were conducted.

During the second test the same time (until t=0.4) was used for simultaneous compression along axes X and Y with one-half the previous velocities (b=1). The final transfer of the basin's base is equivalent in both examples. The comparison in Figure 5 of these two tests shows the great difference in the quantitative characteristics of the interlayer boundaries.

REFERENCES

Culling W.E.H. (1960). Analytical theory of erosion: Jour. Geology, v. 68, no. 3, p. 336-344.

Myasnikov, V.P., and Fadeev, V.E., 1980, Hydrodynamical model of Earth and the planets of earth group: Advances in science and engineering (in Russian): Sciencies about Earth Ser., v. 5, 283 p.

Ramberg, H., 1963, Fluid dynamics of viscous buckeling applicable to folding of layered rocks: Am. Assoc. Petroleum Geologists Bull., v. 47, no. 3, p. 484-505.

Reddy, J.N., Stein, R.J., and Wicham, J.S., 1982, Finite-element of folding and faulting: Intern. Jour. Numer. and Analyt. Meth. Geomech., v. 6, p. 425-440.

Zanemonetz, V.B., Kotyolkin, V.D., and Myasnikov, V.P., 1974, About the dynamics of lithospheric motions (in Russian): Izvest. AN SSSR, Fizika Zemli, v. 5, p. 39-53.

Zanemonetz, V.B., Mikhajlov, V.O., and Myasnikov, V.P., 1976, Mechanical model of block folding formation (in Russian): Izvest. AN SSSR, Fizika Zemli, v. 10, p. 13-21.

SIMULATING GEOLOGIC PROCESSES THAT CREATE SEDIMENTARY BASINS: INCORPORATION OF A CARBONATE-ECOLOGY MODEL

John W. Harbaugh
Department of Applied Earth Sciences
Stanford University, Stanford, CA USA

ABSTRACT

Computer process simulation can be used to model sedimentary basin features. The models can be run only forwards and are finite in form; time can be segmented into discrete increments and fundamental laws must be obeyed. A carbonate-ecological model is given as the example of relatively simple procedures that can be used to represent carbonate depositional processes. The carbonate-ecology model will be incorporated as a module with SEDSIM, a large 3D system developed for modeling sediment transport, deposition, and compaction.

INTRODUCTION

A principal issue before "basin modelers" is whether models can be created that operate the way the Earth actually operates. By "model", we denote computer simulation models that progressively change, such that the evolution of a basin can be represented through a span of geologic time, say from 50,000 to a million years or longer. These models incorporate specific geological processes in simplified form, such as representation of the flow of water by differential equations that are solved repeatedly numerically. Furthermore, these flow equations can be linked

Computerized Basin Analysis, Edited by J. Harff
and D.F. Merriam, Plenum Press, New York, 1993

with sediment-transport and other equations to create a mathematical structure that begins to mimic the Earth's processes. By supplying rates and defining the boundaries of the domain over which they are to be solved, we can represent processes operating at specific intensities within geographic segments of a basin through specific segments of geologic time. This is the essence of process simulation modeling in the sedimentary basin context that is focussed on here.

The major question is whether such models are genuinely useful? We acknowledge that any process model must be simplified with respect to the real world, where myriad interdependent processes work continuously at levels of detail that are infinitely fine by comparison with our models. If these models behave with sufficient realism and accord with basic laws, they may be useful in providing insight as to how geologic processes operate to produce features in sedimentary basins that are of major scientific and commercial interest.

CONSTRAINTS UNDER WHICH PROCESS MODELS OPERATE

An assertion usually is made that sedimentary basin process models are needed that will run backwards. These models would be supplied with observational data from outcrops or wells and seismic sections, and then would progressively unfold the past in three-dimensional detail. In addition to the historical revelations thus provided, these models also would supply information about geologic features that exist between wells and seismic lines, so that detailed 3-D interpretations through geologic time would be achieved as an end product. Such a model is too grand to be true. Similar to a perpetual-motion machine, it is impossible for reasons that stem from the second law of thermodynamics, which relates to the irreversibility of actual processes. Water will not run uphill, nor will a process model run backwards. Both can move only forward through time, thus imposing the fundamental constraint of "forwardness".

"Back-stripping" models of sedimentary basins are used widely, in which sediment in a basin progressively is stripped away from a sequence of beds within a basin. Although useful analytical models, they are not process models. Back stripping is not a geological process, for it would contradict fundamental physical laws if regarded as an actual process. However, it is a useful artifice, just as process models are useful artifices.

A second constraint stems from limits on the level of detail that can be represented by a model. Process models are finite in form. A geographic area can be represented by a gridwork of square or rectangular cells arranged in uniform columns and rows. The vertical dimension also can be represented by a finite number of cells, so that a geologic body can be represented by vertical columns of cells defined by the combination of the geographic grid with the vertical cellular increments. Generally the thicknesses of individual cells in the vertical dimension are not uniform.

Time also is segmented into discrete increments, and may be measured in minutes or hours, as well as in tens or thousands of years. For example, fluid flow may be represented in successive time increments or steps that are individually only a few seconds long, whereas other processes may be segmented into increments measured in hundreds of years. On one hand, we may wish to operate simulation models so that fine details are represented in both space and time; on the other, we must compromise because computers are limited in their ability to store information and carry out the arithmetic in representing processes operating through space and time. By contrast, the actual Earth is a continuum in space and time composed of an almost infinity of particles that exist continuously through time. So we always are constrained by the limitation that the bigger the region or the longer the total time span to be represented in the simulation, the coarser the divisions of time and space must be in the simulation. These constraints exist no matter how large or how fast the computers are that implement the mathematical models embodied as programs.

A third major constraint lies in the relationships incorporated in our simulation models. Do we understand the Earth's operations well enough to represent its behavior in dynamic models? Although there are major shortcomings in our understanding of the Earth, a principal objective in simulation is to explore the consequences of assumptions about relationships by observing the models' responses when alternative assumptions are supplied.

Many relationships incorporated in models represent fundamental laws, such as the conservation laws (conservation of mass, energy, and momentum), which are of universal validity (Cross and Harbaugh, 1989). Others are empirical, with varying degrees of applicability. A model may include both fundamental laws and empirical relationships that operate in concert, such as sediment-transport models in which conservation of mass is scrupulously adhered to, but the equations that relate current velocity to suspended sediment load are subject to change, depending on assumed conditions.

A CARBONATE-ECOLOGY MODEL

The remainder of this paper focuses on some relatively simple procedures to represent carbonate depositional processes. At Stanford University we are developing currently an integrated series of three-dimensional process models for representing sedimentary basins, in a project termed the Stanford SEDimentary Basin SIMulation Project, or "SEDSIM". SEDSIM's components include stream flow (Tetzlaff and Harbaugh, 1989), wave activity (Martinez, 1987), sediment transport, basin circulation, compaction, pore-fluid expulsion, and isostatic compensation. Most of these modules or components involve mixtures of relationships, ranging from fundamental laws to highly empirical relationships.

Next, a carbonate-depositional module that is to be incorporated in SEDSIM, is outlined. The concepts in the model are widely applicable, and were developed initially in the 1960's. Figures 1 to 6 are from this early work (Harbaugh and Merriam, 1968; Harbaugh and Bonham-Carter, 1970). Although of relative antiquity in comparison with other geologic process simulation modules, they remain useful. They are empirical by comparison with other SEDSIM modules, and are partially stochastic because they contain elements that are deliberately random. In spite of their empiricism and randomness, they represent carbonate-producing processes that are linked closely with bottom-dwelling marine animals and plants. The model can be termed a "carbonate-ecology" model because it involves ecological relationships in the production of carbonate sediment. Once carbonate sediment has been produced, it is available for transport, deposition, and compaction by virtue of its links with other SEDSIM modules

The carbonate-ecology model simulates interactions between communities of carbonate-secreting organisms through use of triply-dependent, third-order Markov chains with nonstationary transition probabilities (Harbaugh and Bonham-Carter, 1970, p. 125-131). The transition probabilities are recomputed at each time step. The area that the organism communities occupy is divided into a grid with rectangular cells of arbitrary size, as used in SEDSIM. Each cell contains only one community at a time, each community being defined as consisting of one or more species, with communities serving as the basic unit, rather than species or individual organisms. In computer programs representing the model (Harbaugh, 1966; Harbaugh and Wahlstedt, 1967), each community occupying a specific cell is symbolized by an integer (Fig. 1), and the integers are stored in a three-dimensional array in which two of the

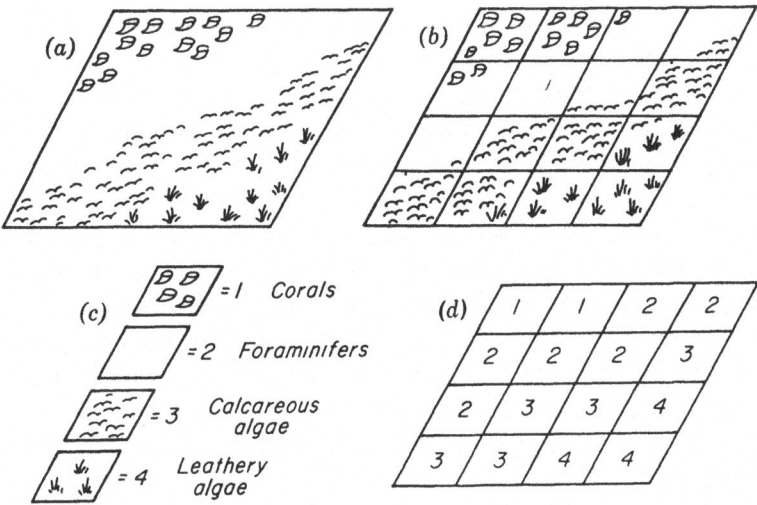

Figure 1. Ways of representing geographic distribution of organism communities: (A) Seafloor is populated by different communities which (B) can be assigned to cells (C) to which integers are assigned as equivalents. (D) Two-dimensional integer array contains essentially same information as graphic symbols, except for loss of detail because of coarseness of cells.

array's dimensions pertain to geographic dimensions, and the third represents successive time increments.

The organism communities are endowed with properties that affect their ability to compete with the other communities. Competition and ecologic succession are simulated by the Markov procedure, which selects communities to occupy cells during successive time increments. The geographic distribution of communities in preceding time increments and their "relative vitality" affects their behavior through time. "Relative vitality" is the relative fitness of each community for the local environment at the specific moment and place, and is determined by the transition probabilities in the triply-dependent Markov chains which are modified progressively at each time increment. Environmental factors, including depth of water and proportion of mud and sand in suspension, as well as the occupants of neighboring cells, affect the transition probability values. The random aspect of selection is simulated by use of pseudorandom numbers

The transition probability values developed in the model are such that, moving forward through time increments of short duration, the most probable community to occupy a cell is the same community that occupied that cell in the time step immediately before. This assumes that other factors in the environment are relatively unchanged, and that adjacent cells do not harbor communities that would overwhelm the cell's existing community. The occupation of the cell by a community that is next in an ideal ecologic succession is the second-most probable event. Occupation by communities that are removed progressively farther in an ideal ecologic succession have progressively lower probabilities. Given sufficient time, a pioneer community will be replaced gradually by other communities, until the ecological climax community is reached. Although there is a tendency for such a succession to move unidirectionally toward the climax, interim reversals may occur as random fluctuations, and major reversals may be produced by major changes in the environment, which can include catastrophic events, such as river floods that bring in large influxes of silt and mud, and hurricanes.

In simulating competition between communities, the model considers the geographic distribution of communities in the three immediately preceding time increments. The geographic stability of a population through successive time increments has strong influence. For example, if a given type of community has occupied a particular cell for three successive time increments, the probability that the same community will occupy the same cell in the forthcoming time increment is greater than it would be if different communities had occupied the cell during the three preceding increments. This effect of this may be likened to inertia, in that long-established communities tend to resist change more than communities whose occupation has been brief. This inertial effect is adjustable in the model by changing the three-step transition probabilities, so that the degree to which communities tend to be stable geographically can be closely regulated.

Communities in cells that are proximate geographically to a specific cell influence the selection of future communities that may occupy the cell, being analogous to propagation by seed dispersal in land plants in which the influence of a community declines with increasing distance. This seeding effect provides a mechanism by which communities can migrate, compete for space, and interact with other communities. The result is to enable a community that is better adapted to a given set of environmental conditions (expressed as high "relative vitality"), to replace progressively a community of lower vitality (Fig. 2A).

If a community has high relative vitality it may spread geographically through time. Reef or banklike deposits created by corals or calcareous algae (Fig. 2B) are examples whose spreading behavior can be readily simulated, even when environmental change is represented. For example, the effect on communities of a river debouching as a plume of muddy water into a shallow clear tropical sea can be simulated by

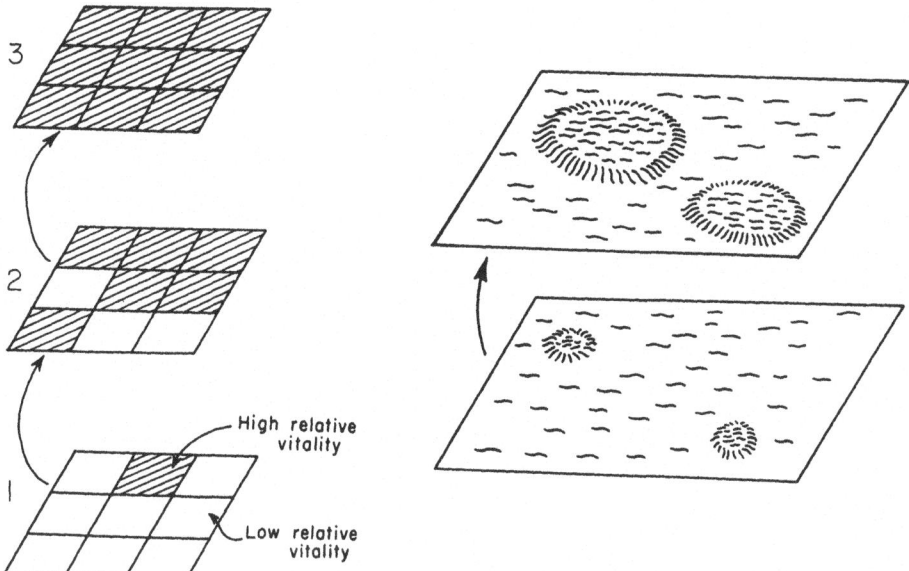

Figure 2. A, Simulation model of replacement of organism community of low relative vitality by one of high vitality through time increments 1, 2, and 3. B, Spread of organism community is analogous to epidemic spread of disease.

changing the relative vitalities of the communities involved. An increase in mud would stimulate a community composed of *Lingula*, a mud-loving brachiopod, and would diminish a community of calcareous algae. The effect on a clam community might be mixed, being favorable to some species and detrimental to others. By contrast, strong wave activity in a tropical sea would stimulate corals and calcareous algae to build rigid frameworks or reefs. Furthermore, changes in water temperature also would have strong affect on relative vitality, as would changes in water depth. The influence of depth for an algal community could be assumed to range between three points (Fig. 3), namely (a) an upper depth limit, above which the community cannot survive, (b) a lower depth limit below which the community cannot survive, and (c) a most favorable depth where the community is best able to compete. The rate of supply of terrigenous sediment could be assumed to have negligible influence on a specific algal community until a specified threshold level is reached (Fig. 4). Above this level, an increase in rate of supply would cause vitality to decrease linearly to zero, where an intolerable level is reached.

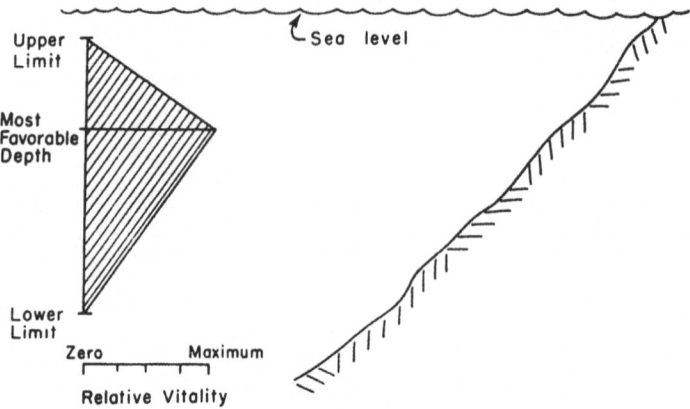

Figure 3. Effect of depth on relative vitality of organism community.

An example type of application of the carbonate-ecology model is outlined in Figures 5 to 7. Figure 5 illustrates actual carbonate banks formed by calcareous algae in Late Pennsylvanian time in southeastern Kansas, in the central United States (Harbaugh, 1960). The banks formed seaward of a debouching fluvial system and delta complex. At intermediate distances, thick carbonate beds of the algal banks pass laterally into shales, siltstones, and sandstones of the delta complex. It is likely that algal communities flourished in shallow water that was

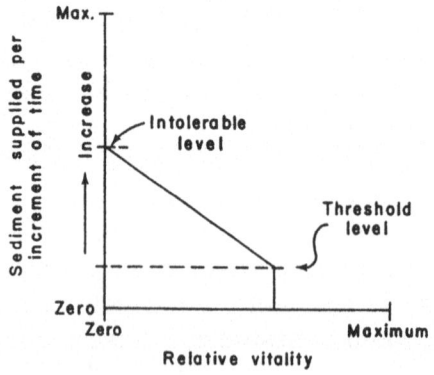

Figure 4. Effect of rate of supply of terrigenous sediment on relative vitality of algal community.

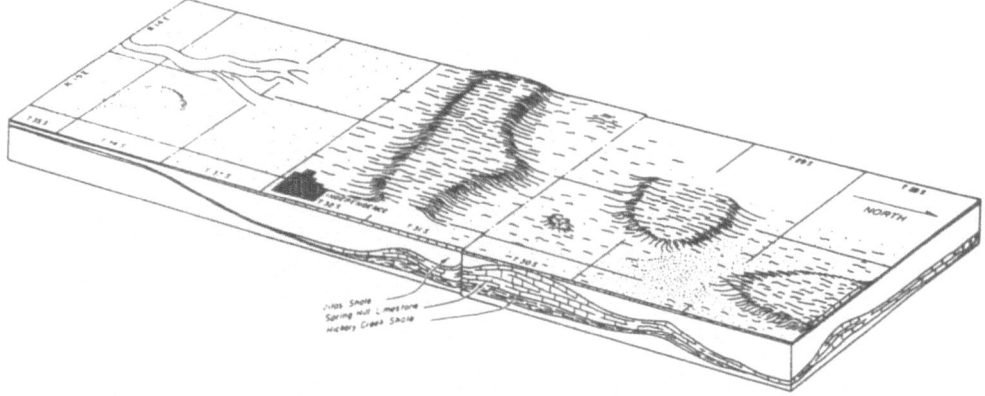

Figure 5. Block diagram showing deposition of algal banks in Late Pennsylvanian time in southeastern Kansas. Block is about 65 km long in north-south direction and 25 km wide in east-west direction. Algal banks have formed seaward of debouching river (from Harbaugh, 1960).

relatively clear, whereas in slightly deeper water nearby, algal communities were rare or nonexistent, the algae being inhibited by water clouded with mud and silt

In an experiment with the original carbonate-ecology model devised in the 1960's, Harbaugh and Bonham-Carter (1970, p. 453-455) simulated a series of deposits (Fig. 6) that represent the Pennsylvanian algal banks and other features shown in Figure 5. The simulation incorporated the assumptions outlined here with respect to depth and supply of terrigenous sediment. However, the original carbonate-ecology model is primitive by comparison with SEDSIM's present capabilities because, for example, the original model contained no provision for transport of carbonate sediment once it was produced.

The new plan is to link the carbonate-ecology model with SEDSIM, and redo the Kansas algal-bank experiments as well as performing new experiments involving a debouching river in which deltaic deposits are subjected to varying degrees of wave activity and are populated by various carbonate organism communities, including frame-building communities. Thus, variations in wave intensity, water depth, and water clarity will affect the three-dimensional bodies of sediments created in the simulations, and should provide detailed experimental examples for comparison with both ancient and modern reef and delta complexes. The work is expected to span two years, beginning in 1991. Modern two-

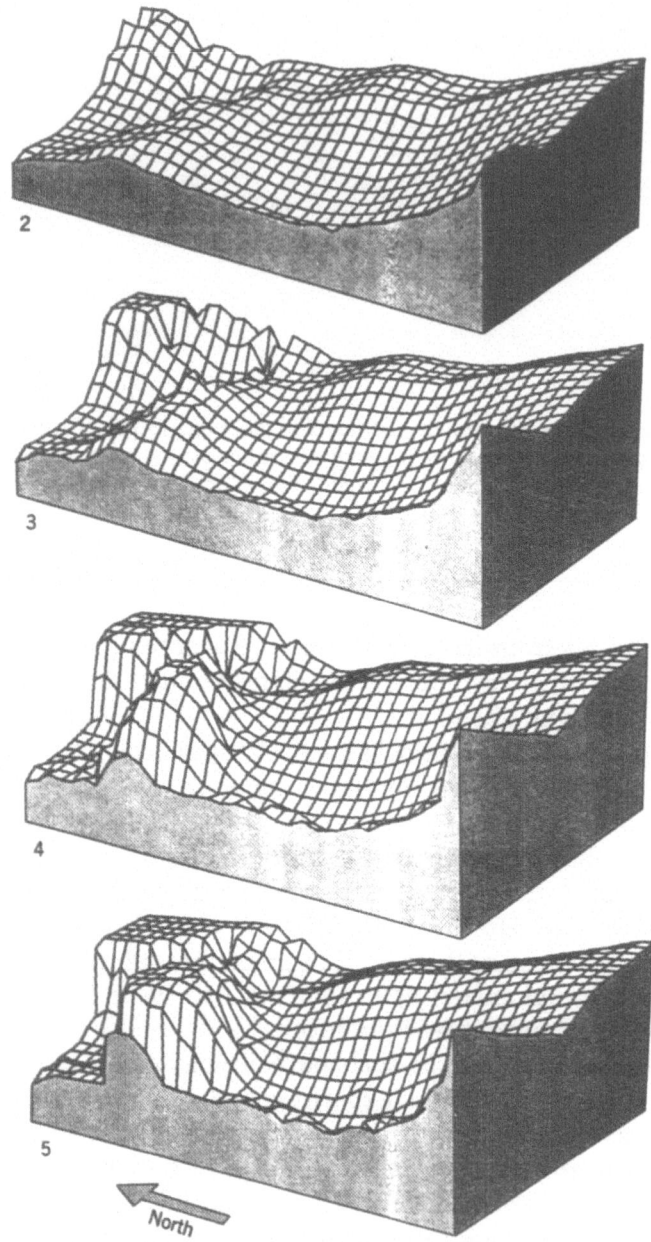

Figure 6. Depositional topography obtained as response during simulation experiment involving area represented in Figure 5. Note that north is toward left in contrast to Figure 5.

dimensional sedimentary process models have been shown to be remarkably successful in linking carbonate depositional processes with sediment-transport processes (Lawrence, Doyle, and Aigner, 1990), and the expanded version of the overall SEDSIM model will extend the linkage to three dimensions.

The general form of the initial experiments to be constructed with the expanded SEDSIM are illustrated in Figure 7, which involves a response obtained in a SEDSIM experiment in which a river debouches into a shallow marine basin. The river's channel is shown in the upper

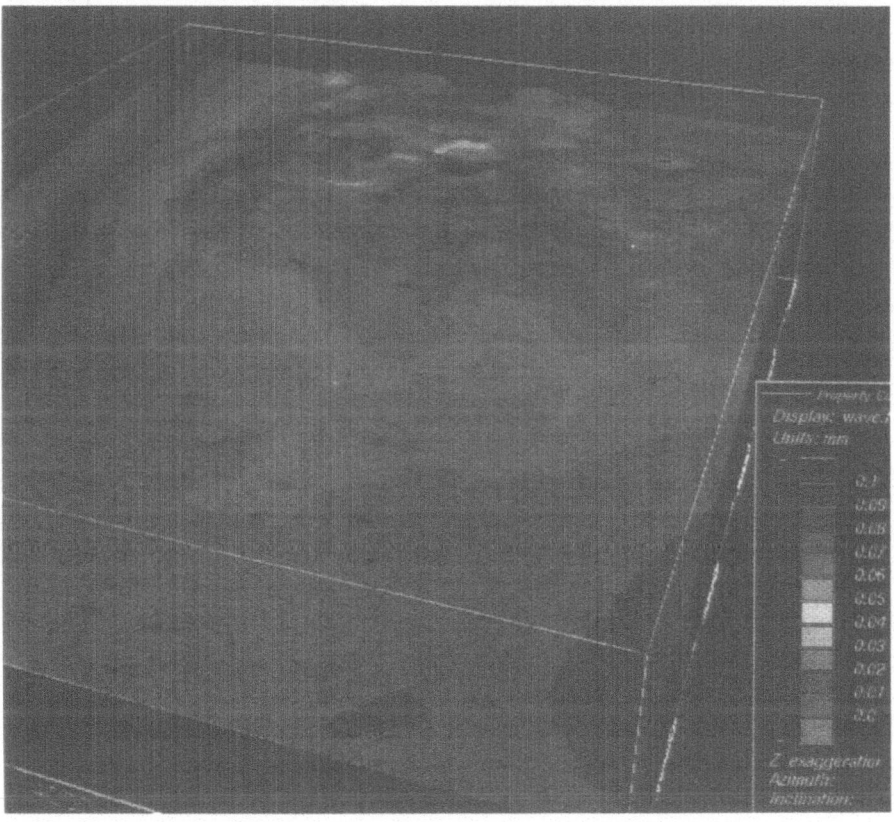

Figure 7. Simulated wave-dominated delta formed where river debouches into shallow sea. Gray-scale variations denote average sizes of clastic grains in mm. Wind direction is from lower left. Display ordinarily is shown in color. Carbonate depositional processes are not represented in experiment, but will be in forthcoming experiment.

part of the 3-D perspective diagram, and deltaic deposits have formed adjacent to the mouth of the river. Strong prevailing winds are presumed to blow from the lower left, so that waves attack the deltaic features, causing coarser sediment to be shunted to the right. Depending upon assumptions about the ecological responses of carbonate-secreting organism communities and other environmental factors, the development of reefs and banks as part of deltaic complexes will be analyzed.

ACKNOWLEDGMENTS

I thank Johannes Wendebourg and Dominik Ulmer of Stanford University for reading the manuscript, Dynamic Graphics Corporation for providing the *Interactive Volume Modeling* ("IVM") software used to generate Figure 7, and Paul Martinez of Stanford for the simulation experiment shown in Figure 7. Furthermore, I acknowledge the cooperation of Professor Reinhard Pflug, of the University of Freiburg, in the overall SEDSIM project, particularly with regard to graphic display procedures, as well as the assistance of members of his group, including Christoph Ramshorn and Herbert Klein. Finally, Brenna Lawrence of Stanford University typed the manuscript; her assistance is gratefully acknowledged.

REFERENCES

Cross, T.A. and Harbaugh, J.W., 1989, Quantitative dynamic stratigraphy: a workshop, a philosophy, a methodology, *in* Quantitative Dynamic Stratigraphy, Cross, T.A., ed.: Prentice-Hall, Englewood Cliffs, New Jersey, 625 p.

Harbaugh, J. W., 1960, Petrology of marine bank limestones of Lansing Group (Pennsylvanian), southeast Kansas: Kansas Geol. Survey Bull. 142, pt 5, p. 189-234.

Harbaugh, J. W., 1966, Mathematical simulation of marine sedimentation with IBM 7090/7094 computers: Kansas Geol. Survey, Computer Contr. 1, 52 p.

Harbaugh, J. W. and Bonham-Carter, G., 1970, Computer simulation in geology: Wiley-Interscience, New York, 575 p.

Harbaugh, J. W., and Merriam, D. F., 1968, Computer applications in stratigraphic analysis: John Wiley & Sons, New York, 282 p.

Harbaugh, J. W., and Wahlstedt, W. J., 1967, FORTRAN IV program for
  mathematical simulation of marine sedimentation with IBM 7040 or
  7094 computers: Kansas Geol. Survey, Computer Contr. 9, 40 p.

Lawrence, D. T., Doyle, M, and Aigner, T., 1990, Stratigraphic simulation
  of sedimentary basins: Concepts and calibration: Am. Assoc. Petroleum Geologists Bull., v. 74, no. 3, p. 273-295.

Martinez, P. A., 1987, Simulation of sediment transport and deposition
  by waves: for simulation of wave-versus fluvial-dominated beach
  environments: unpubl. masters thesis, Stanford Univ., 406 p.

Tetzlaff, D. M., and Harbaugh, J. W., 1989, Simulating clastic sedimentation: Van Nostrand Reinhold, New York, 202 p.

Numerical Simulation of Pore Fluid Movements in the Upper Rotliegend of the North German Depression

J. Springer and G. Schwab
Central Institute for Physics of the Earth
Potsdam, Germany

ABSTRACT

The paper presents a method and some results for the numerical simulation of pore fluid movements, which are connected strongly with migration of hydrocarbons. The mathematical model consists of a two-dimensional initial/boundary value problem with a partial differential equation for the pore pressure in a sedimentary layer. From the given input data (initial condition, hydrodynamic boundary conditions, material parameters) the pore pressure and the Darcy velocity of pore fluids may be computed as functions of space and time. A case study is given from the Upper Rotliegend sediments in a part of the North-German Depression bordered by a system of faults. It is assumed that the overlying Zechstein salt layers hindered the escape of pore fluids, and so an abnormally high pore pressure is produced in the considered sediments. The simulation describes the temporal change of pore pressure and horizontal water flow after some tectonic movements which allowed the fluids to escape along the faults.

INTRODUCTION

The problem treated in this paper arose from the search for hydrocarbons in the central part of the North-German Depression. It is assumed that hydrocarbons generated in the Upper Carboniferous layers

Computerized Basin Analysis, Edited by J. Harff
and D.F. Merriam, Plenum Press, New York, 1993

are dissolved in the pore fluids of the Upper Rotliegend and then transported horizontally to the traps. So the reconstruction of velocities and directions of pore fluid flow becomes an interesting task. Besides this the fluid flow is connected with the pore fluid pressure which influences the compaction process and thus the development of sedimentary structures.

For the reconstruction of geologic processes the numerical simulation becomes more and more an important tool. To perform such a simulation we must perform four main steps:

> - physical understanding
> - mathematical formulation
> - software
> - input data.

PHYSICAL UNDERSTANDING

To get a physical understanding in our examaple we consider a certain complex of sedimentary layers (Fig.1). The compaction and fluid flow in this sedimentary complex is influenced by internal and external conditions (Kluge and Milde, 1975; Welte and Yükler, 1981; Bethke and others, 1988; Tetzlaff and Harbaugh, 1989):

internal: - geometry
- material parameters:
  - permeability (hydraulic conductivity)
  - compressibility (influences the amount of water produced during compaction)

external: - subsidence of basement
- stress exerted by the overburden
- hydrodynamic conditions at the boundaries of the complex (flow in or out)

MATHEMATICAL FORMULATION

The considered process may be described by the spatial and temporal change of several physical quantities (functions of the two horizontal coordinates x and y, the depth z, and the time t):

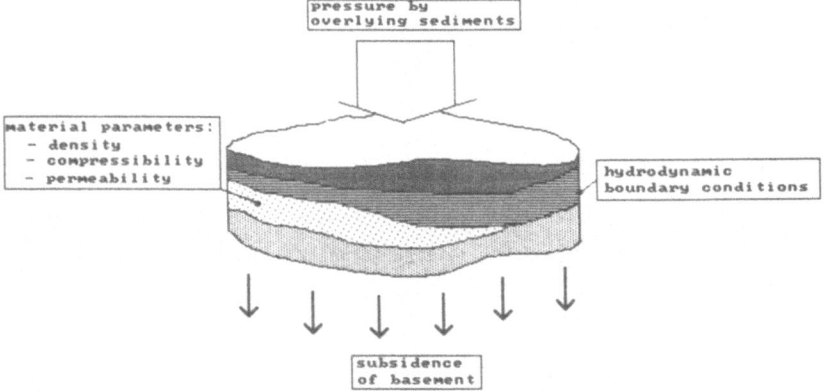

Figure 1. Complex of sedimentary layers with conditions influencing compaction and fluid flow.

p : pore pressure,
h : hydraulic head, h = $p/(g\Theta_w)$ + z,
    where g is the acceleration of gravity and Θ_w the (constant)
    density of water,
ø : porosity.

The hydraulic conductivity K and the compressibility a are assumed to depend on the lithotype and the porosity, the rock density Θ_r is constant. The movements of rock (only vertical) and pore fluid are described by the velocity vectors $\mathbf{v}_r$ = (0 , 0 , v) and $\mathbf{v}_w$.
    The physical laws which control the process are:

– continuity law for the rock

$$\frac{\delta}{\delta t}\left((1\text{-}\phi)\,\Theta_r \right) + \text{div}((1\text{-}\phi)\,\Theta_r\,\mathbf{v}_r) = 0 ,$$

– continuity law for the water

$$\frac{\delta}{\delta t}(\phi\,\Theta_w) + \text{div}(\phi\,\Theta_w\,\mathbf{v}_w) = 0 ,$$

– Darcy's law (flow in porous media)

$$\phi \, (v_w - v_r) \qquad = - K \, grad \, h \, ,$$

– effective stress (Θ_s: average density of sediment)

$$\sigma_{eff} \qquad = \sigma_{tot} - p$$

$$= g \, \Theta_s \, z \, - \, g \, \Theta_w \, (z+h)$$

$$= g \, (\Theta_s - \Theta_w) \, z \, - \, g \, \Theta_w \, h \, ,$$

– deformation law

$$\phi = \phi(\sigma_{eff}) \, , \quad a = - \, \frac{1}{1-\phi} \, \frac{d\phi}{d\sigma_{eff}} \, .$$

The combination of these physical laws results in the next step, the formulation of a mathematical problem. In this situation, if we assume incompressible rock and fluid (constant Θ_r and Θ_w), we get a system of two partial differential equations with initial and boundary conditions:

$$S \, (\, \frac{\delta h}{\delta t} + v \, \frac{\delta h}{\delta z} \,) \quad = div \, (\, K \, grad \, h \,) + Q$$

$$\frac{\delta v}{\delta z} \qquad = div \, (\, K \, grad \, h \,)$$

with

$$S = \alpha \, g \, \Theta_w \qquad \text{storage coefficient}$$

$$Q = \alpha \, g \, (\, \Theta_s - \Theta_w) \qquad \text{source term.}$$

To solve this system in three dimensions requires an immense computational effort. To allow simulation on PCs simpler models are used for different special questions:

- two dimensional (horizontal water flow; Hähne and Kluge, 1976),
- one dimensional (compaction and vertical water flow; Bredehoeft and Hanshaw, 1968; Gibson, England, and Hussey, 1967; Gibson, Schiffmann, and Carbill, 1981; Sharp and Domenico, 1976; Yükler, Cornford, and Welte, 1978; Keith and Rimstidt, 1985),
- compaction model assuming hydrostatic pore pressure (decompaction; Perrier and Quiblier, 1974)

Here the two-dimensional model is considered only.

SOFTWARE

In the two-dimensional situation the system of equations reduces to one parabolic differential equation for the hydraulic head h:

$$S \; \frac{\delta h}{\delta t} = \mathrm{div} \, (\, K \, \mathrm{grad} \, h \,) + Q \, ,$$

where S, K, and Q are computed from the corresponding parameters in the three-dimensional model by depth integration.

To solve an initial/boundary value problem with this differential equation we used the program FEFLOW developed at the Institute for Mechanics (Chemnitz) which uses the Finite Element Method. The program was complemented by the program FLOWGRAF for the graphical presentation of the results.

INPUT DATA

A case study is presented from the North German Basin. The hydrodynamic development in the Upper Rotliegend sandstone layers can be simulated numerically.

The simulation begins at the end of the sedimentation of the considered layers, that is at the begining of the Zechstein (252 million years before present). It is assumed that the pore pressure is hydrostatic at this time, thus the hydraulic head is zero relative to sealevel (initial condition).

The boundaries of the selected area must be of hydrodynamic importance to allow estimation of boundary conditions. Faults are suitable boundaries because the pore fluids may flow along faults to the surface. So we selected an area between 5 faults in North-East Germany (Fig. 2). The boundary conditions are changing in time corresponding to the activities of the faults.

In the considered area the Upper Rotliegend layers are overlain by around 1000 m Zechstein salt and 3000 m of Mesozoic and Cenozoic sediments.

The Figures 3 to 8 show the simulated development of pore pressure (shading) and flow velocity (lines beginning at the grid nodes) in the Upper Rotliegend:

The Rotliegend sediments filled a basin which then rapidly subsided and was covered with the almost impermeable Zechstein salt. This resulted in an increasing pore pressure in the entire area, Figure 3 shows the state at the end of the Zechstein.

At this time the first vertical movements of the faults are assumed, which leads to a pressure decreasing at these faults and a fluid flow because of the pressure gradient (Fig.4).

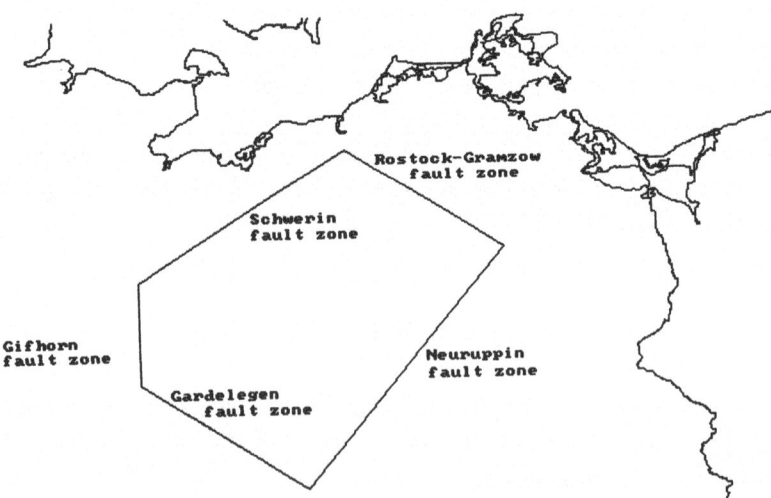

Figure 2. Area of investigation in North German Depression bounded by five fault zones.

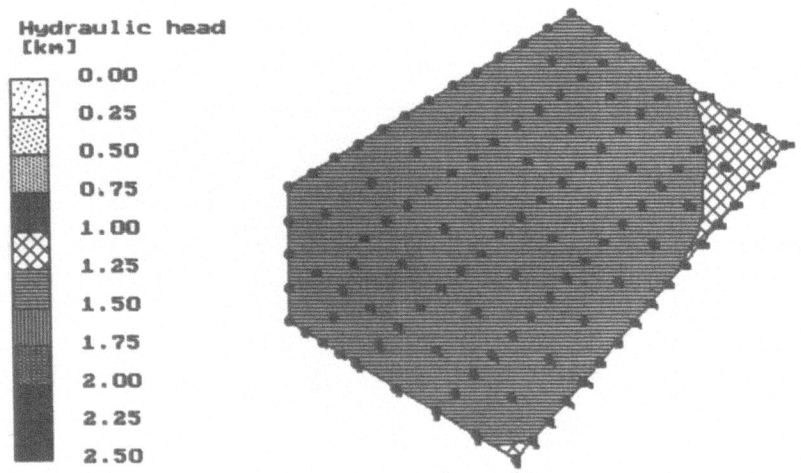

Figure 3. Result of simulation, 247 million years before present (end of Zechstein time).

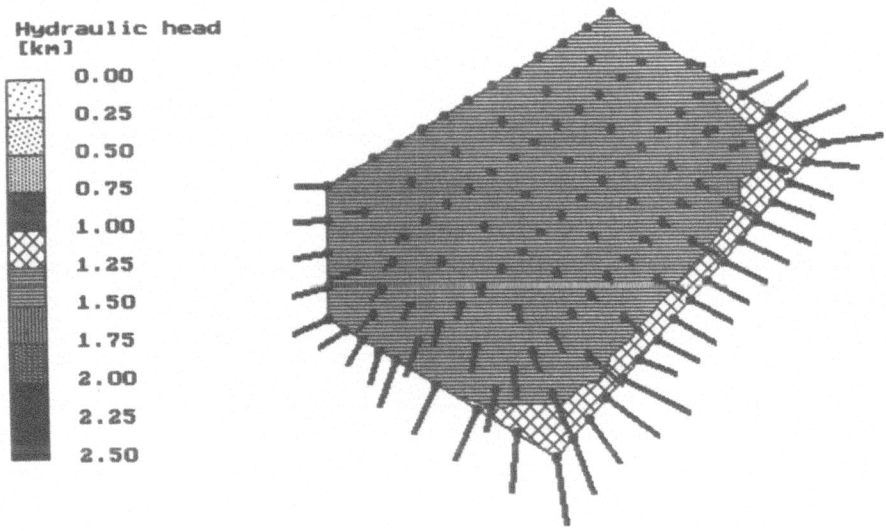

Figure 4. Result of simulation, 246.75 million years before present (beginning of Triassic time).

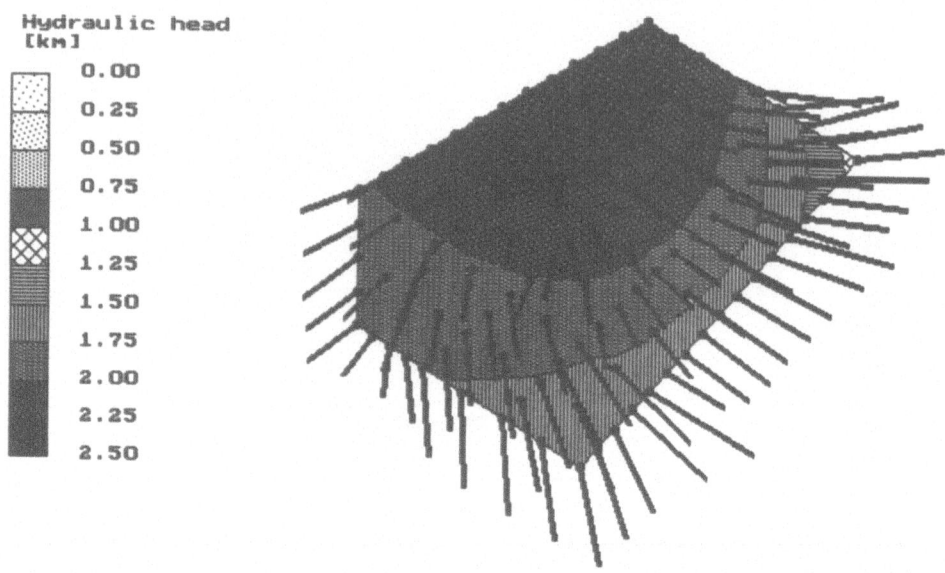

Figure 5. Result of simulation, 242 million years before present (end of Middle Bunter time).

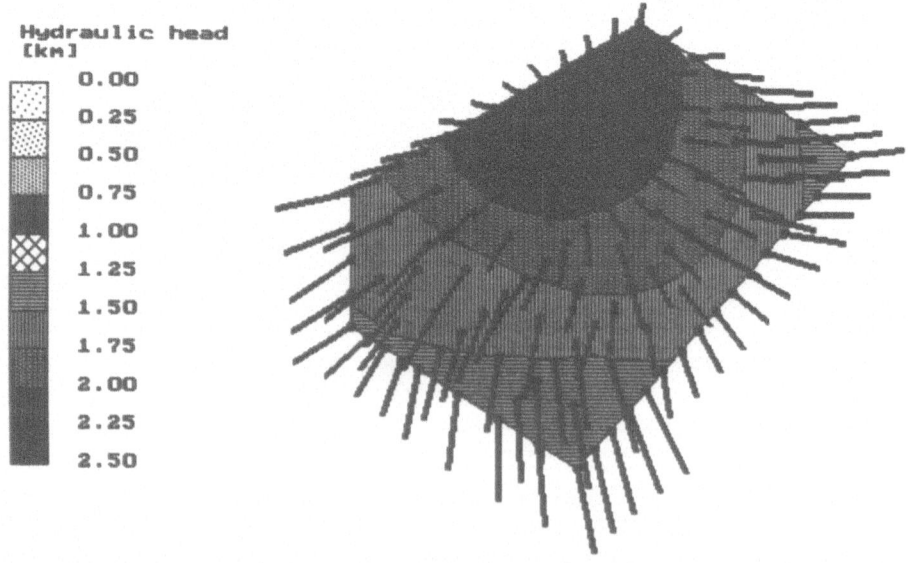

Figure 6. Result of simulation, 241.5 million years before present (beginning of Late Bunter time).

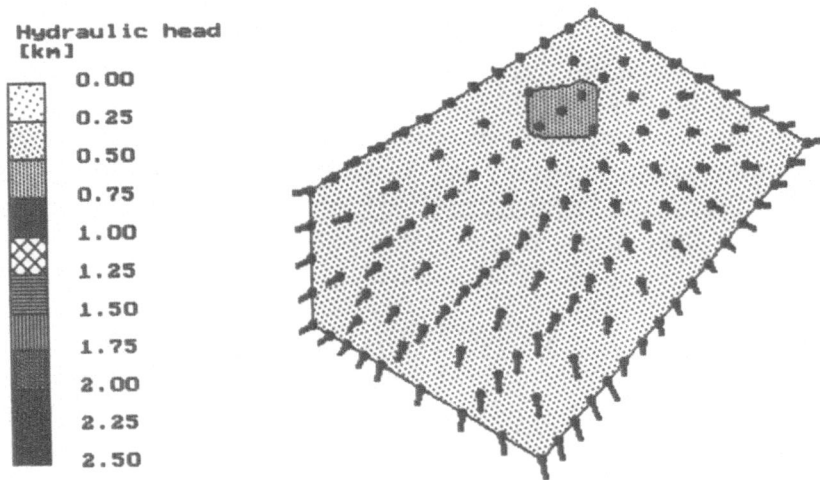

Figure 7. Result of simulation, 232 million years before present (Keuper time).

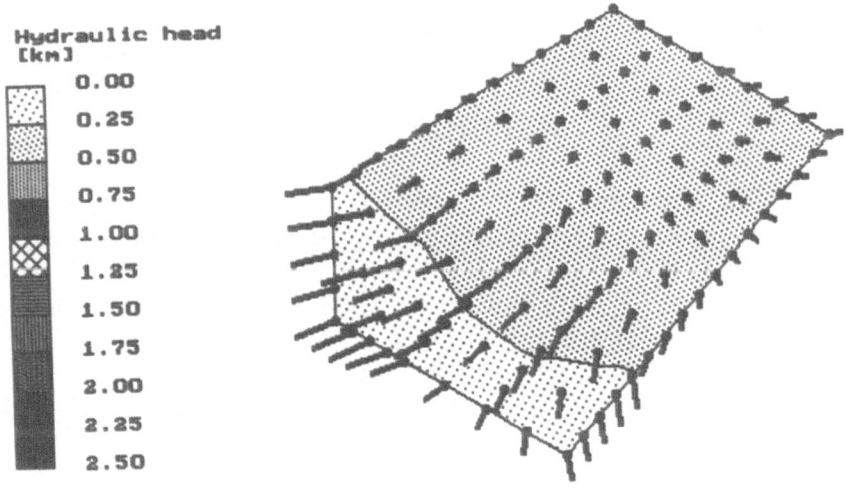

Figure 8. Result of simulation, 231.5 million years before present (Keuper time, after activity of Gifhorn Fault)

Nevertheless the hydraulic head increases in the area until the end of Middle Bunter time (Fig.5).

Now the movement of the faults becomes more pronounced, it is the time of the Hardegsen disconformity. The hydraulic head decreases because of the fluid flow along the faults (Fig.6).

In the Keuper time the hydraulic head already is near zero (Fig. 7).

New movement, especially of the Gifhorn Fault, results only in a short-time flow in the western direction (Fig.8), then the hydraulic head is zero in the whole area.

CONCLUSIONS

The increase of hydraulic head is controlled primarily by the fast subsidence of the layers, but the decrease is controlled by the boundary conditions of the faults. Different facies (larger permeability in the southwest) had only a little influence on the process.

The modeled process, of course, is a simplified model of reality. The numerical simulation allows one to perform experiments with the geologic processes, using different material parameters and boundary conditions. Together with the knowledge of the geologist a refined model may be achieved.

REFERENCES

Bethke, C.M., Harrison, W.J., Upson, C., and Altaner, S.P., 1988, Supercomputer analysis of sedimentary basins: Science v. 239, no. 4837, p. 261-267.

Bredehoeft, J.D., and Hanshaw, B.B., 1968, On the maintenance of anomalous fluid pressures. I. Thick sedimentary sequences: Geol. Soc. America Bull., v. 79, no. 9, p. 1097-1106.

Gibson, R.E., England, G.L., and Hussey, M.J.L., 1967, The theory of one-dimensional consolidation of saturated clays. I. Finite nonlinear consolidation of thin homogeneous layers: Geotechnique, v. 17, p. 281-279.

Gibson, R.E., Schiffmann, R.L., and Carbill, K.W., 1981, The theory of one-dimensional consolidation of saturated clays. II. Finite nonlinear consolidation of thick homogeneous layers: Can. Geotech. Jour., v. 18, p. 280-293.

Hähne, R., and Kluge, W., 1976, Rekonstruktion paläohydrodynamischer Verhältnisse des Buntsandsteins im Mitteleuropäischen Becken mit Hilfe der Elektroanalogie: Z. angew. Geol., v. 22, no. 12, p. 574-583.

Keith, L.A., and Rimstidt, J.D., 1985, A numerical compaction model of overpressuring in shales: Jour. Math. Geology, v. 17, no. 2, p. 115-135.

Kluge, W., and Milde, G., 1975, Theoretische Grundlagen zur Rekonstruktion paläo-hydrodynamischer Verhältnisse: Z. angew. Geol. v. 21, no. 7, p. 315-322.

Perrier, R. and Quiblier, J., 1974, Thickness changes in sedimentary layers during compaction history; methods for quantitative evaluation: Am. Assoc. Petroleum Geologists Bull., v.58, no. 3, p. 507-520.

Sharp, J.M., and Domenico, P.A., 1976, Energy transport in thick sequences of compacting sediment: Geol. Soc. America Bull., v. 87, no. 3, p. 390-400.

Tetzlaff, D.M., and Harbaugh, J.W., 1989, Simulating clastic sedimentation: Van Nostrand Reinhold, New York, 202 p.

Welte, D.H., and Yükler, M.A., 1981, Petroleum origin and accumulation in basin evolution - a quantitative model: Am. Assoc. Petroleum Geologists Bull., v. 65, no. 8, p. 1387-1396.

Yükler, M.A., Cornford, C., and Welte, D.H., 1978, One-dimensional model to simulate geologic, hydrodynamic and thermodynamic development of a sedimentary basin: Geol. Rundschau, bd. 67, Heft 3, p. 960-979.

MODELING OF SUBSIDENCE, TEMPERATURE, AND MATURITY IN THE NORTH GERMAN BASIN

A. Berthold and K. Menschner
University of Leipzig
Leipzig, Germany

ABSTRACT

The temporal variations of the temperature field in several parts of the North German Basin are determined by a one-dimensional PC program simulating burial history, geothermal, and maturity evolution for formation of the basin.

The subsidence model includes sedimentary compaction, erosion events, and halokinetics.

The nonsteady-state heat conduction equation is solved numerically by a finite difference method. The temperature field in the lithosphere is calculated in dependence of thermal activation processes, distribution of heat flow, changes in surface temperature, erosion, and variations of geothermal parameters, and of layer thicknesses (halokinetics) from the Carboniferous to the Cenozoic.

Thermal maturity of organic material and timing of hydrocarbon generation are simulated by different calculation methods on the base of computed temperature distributions. Depth horizons of hydrocarbon generation in the North German Basin can be determined with certain factors of thermal maturity.

INTRODUCTION

Temperature is one of the most important parameters for generation of hydrocarbons in sedimentary basins. Therefore reconstruction of

Computerized Basin Analysis, Edited by J. Harff
and D.F. Merriam, Plenum Press, New York, 1993

geothermal history is an essential part of basin investigations. Some research in this subject during several years at the University of Leipzig, Institute of Geophysics, are represented here.

The research discussed in this paper is focused on the North German Basin. We developed a one-dimensional PC program simulating burial history, geothermal, and maturity evolution during the basin formation to determine the temporal variations of the temperature field in several parts of the basin.

THE COMPUTATION MODEL

The model consists of three parts, that is subsidence model, calculation of temperature distribution, and computation of thermal maturity. Figure 1 shows the scheme of the several sections of the computation model. Called INPUT parameters are needed for calculation of SUBSIDENCE MODEL, NUMERICAL SOLUTION OF HEAT CONDUCTION EQUATION, and MODEL OF THERMAL MATURITY. Computed OUTPUT DATA are represented as graphics or as a data list. The computer program TMOD based on the flowchart allows comfortable interactive working at any time. When comparison of results with recent measuring values is not satisfactory a feedback process is necessary with changed input data. If it has a good fit the last step is the INTERPRETATION.

The Subsidence Model

The stratigraphy, lithology, age, and thickness of sedimentary beds are known by drilling or seismic data and form the basis for reconstructing the fossil subsidence movements in an area. The recent, secondary layer thicknesses are affected by compaction, erosion, and halokinetics and are different from the primary thicknesses. Compaction occurs with the accumulation of porous sediments and strongly influences the petrophysical properties of rocks. One way to determine the dependence of geophysical parameters on depth is by expressing them in terms of exponential porosity dependence (Sclater and Christie, 1980) [Eq. (1)].

$$P(z) = P(0) \exp(-z/b) \qquad\qquad (1)$$

$P(z)$ - porosity of rock in depth z
$P(0)$ - initial porosity at z =0
b   - coefficient

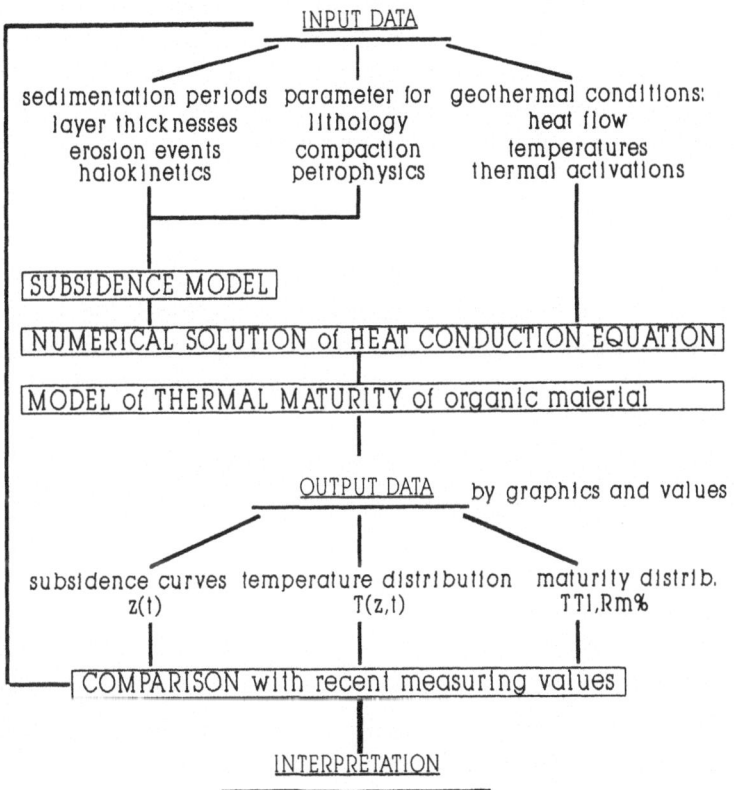

Figure 1. Generalized structure of simulation program.

The thickness, L of uncompacted sediments at the time of deposition is calculated by Equation (2).

$$L_0 = \frac{b}{1-P(0)} \ln \frac{P(0) + (1-P(0))\exp(L_2/b)}{P(0) + (1-P(0))\exp(L_1/b)} \qquad (2)$$

L1, L2 - upper and lower layer boundary after compaction (recent time)

The layer thicknesses that decrease by compaction during subsidence are calculated by this equation. Erosion events and halokinetics also are included. The variations of the geophysical parameters (that is, thermal conductivity, density, specific heat capacity, and heat production) are expressed in relation to the porosity.

The Temperature Model

The temperature field of the Earth changes with time, for example in reality stable geothermal gradients do not exist for larger time distances. They differ in space and time. Hence geothermal modeling in sedimentary basins requires the solution of the nonsteady-state heat conduction equation (Eq. 3).

$$\frac{\delta}{\delta t}\left(\rho c_p T\right) = \frac{\delta}{\delta z}\left(\lambda \frac{\delta T}{\delta z}\right) + A \qquad (3)$$

rc_π - specific volume heat capacity
λ - thermal conductivity
A - heat generation
T - temperature

Equation 3 is solved numerically by an implicit finite difference method adapted to a deformable grid.

The initial temperature-depth distribution is calculated by an analytical formula with the paleoheat flow. The upper boundary condition is a given changeable surface temperature. The lower boundary condition is a given fixed temperature.

Periods of thermal activation can be taken into account as "uprising mantle diapirs."

The Maturity Model

Vitrinite reflectance ($R_m\%$) is one of the most widely used measuring parameters to establish the thermal maturity of organic matter in sedimentary rock. There are several mathematical approaches to model vitrinite reflectance at the base of a computed temperature distribution. One of the first calculation methods was developed by Lopatin (1971) where he calculated the so-called temperature-time-index (TTI) shown in Equation (4)

$$TTI = \sum_{n} \left(\Delta t_n \right) 2^n \tag{4}$$

Δt_n - time a sediment horizon has remained in a temperature interval of $\Delta T = 10°C$

or written as an integral in Equation (5)

$$TTI = \int_{t_1}^{t_2} 2^{(T(t)-105)/10} \, dt \tag{5}$$

In the program the use of both formulae is possible, even so a transformation into the vitrinite reflectance. The factor 2 is changeable in order to increase or decrease the time influence on maturity. A third used possibility to estimate thermal maturity was developed by Middleton (1982) (Eq. 6)

$$R_m (\%)^{\alpha} = b \int_{t_1}^{t_2} \exp \left(cT(t) \right) \, dt \tag{6}$$

a,b,c - modified constants

The maturity evolution of an interesting depth horizon can be determined in order to estimate the timing of hydrocarbon generation qualitatively.

RESULTS OF NUMERICAL CALCULATIONS

Influence of Geological Processes on Temperature

The results of numerical calculations in the North German Basin give an answer to the question how do geological processes influence on geothermal field. Such processes are for example:

- the compaction of sediments during subsidence and its influence
  on geothermal parameters
- erosion events
- halokinetics
- the initial heat flow
- periods of thermal activity during basin evolution and the deposition of sediments with different geothermal properties.

The results are demonstrated on the hand of an example from the
Altmark-District in the North German Basin.

Subsidence

The subsidence-age relationship for the Carboniferous-Cenozoic
interval is the starting point of examination. For every geological section
from the Carboniferous to the Cenozoic a set of geophysical properties
was defined as representative for the predominating rocks. The geophysical
properties density, heat conductivity, heat production, porosity, and heat
capacity for the beds were determined by average values of published
data.

In the example the computed subsidence model is represented in
Figure 2. The full lines represent the layer boundaries. The model starts
350 million years before present. The main subsidence period took place
from the middle Paleozoic to the beginning of the Mesozoic. From the
Jurassic to present the sedimentation rates are low, interrupted by
erosion events.

The compaction leads to decrease of sediment thickness and porosity.
The values for heat conductivity and heat production can be more than
doubled depending on the depth of subsidence.

Temperature

The dotted lines in Figure 2 represent the computed depths of
isotherms from the Carboniferous to the Cenozoic. Here the surface
temperature of the Earth is assumed as constant with 10°C. The main
factors influencing temperature are a thermal activation period and the
low-thermal conductivities of sedimentary rocks near the surface of the
Earth.

The model takes the volcanic activities in the Autunian into account
with an uprising mantle diapir. The related thermal wave leads to
increased temperatures and, therefore, to a melting in the lower crust
depending on melting point of rock. This explains the acid to intermediate

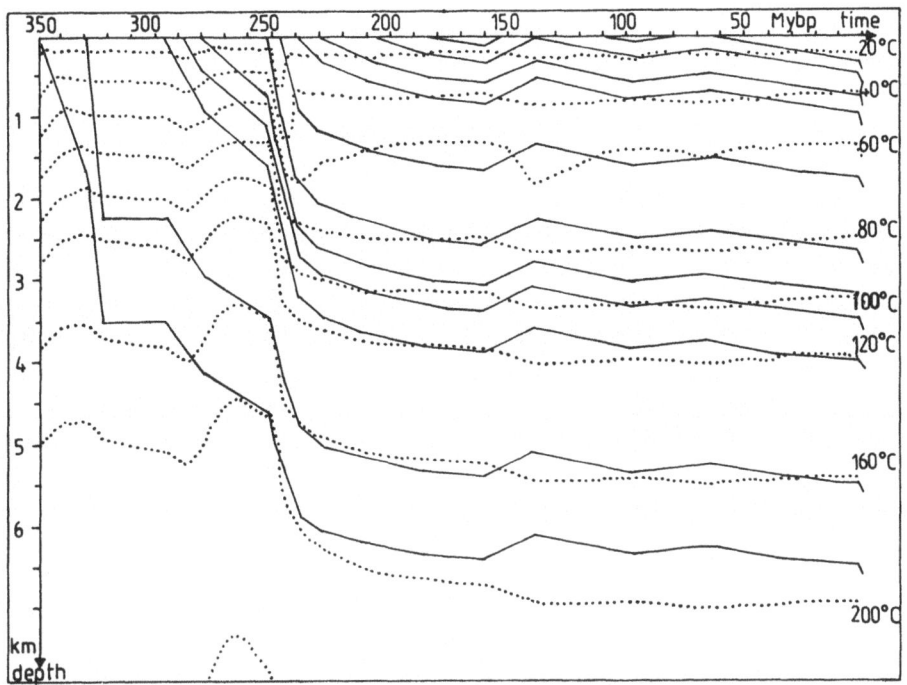

Figure 2. Subsidence model with temperatures. Thermal activation in Autunian (296-288 million years b.p.).   .... = isolines of temperature

character of the volcanic rocks in the area studied. The cooling that sets in after the activation is continued until present thereby indicating the thermal conditions are not stationary.

The decisive influence of thermal conductivity of rock is seen clearly during Zechstein deposition. In the Zechstein time (252 to 247 million years b.p.) the thick accumulation of bedded salt with its high heat conductivity leads to a drop of isotherms in the sedimentary units. In the salt layer a low-temperature gradient is developed. If sedimentary rocks with lower heat conductivity cover the salt layer the isotherms rise upward again.

Erosion effects result by a sinking of isotherms in greater depths because compacted layers with increased thermal conductivity reach the surface of the Earth. In this example the amount of erosion is slight, therefore the isothermal variation is affected little. The maximum temperature change because of erosion is 10° to 20°C and occurs in the salt beds. Here, halokinetics have not occurred, but it can influence strongly the temperature distribution. In the situation of increasing of salt thickness the isotherms will sink in the salt beds and underlying layers as a result of the high thermal conductivity of salt. The contrary effect occurs in the situation of decreasing of salt thickness.

Thermal Maturity

In Figure 3 the computed model of thermal maturity is shown. The dotted lines represent certain equivalent values of vitrinite reflectance characterizing the generation of hydrocarbons. The upper three isolines stand for the onset $(R_m\% = 0.7)$, main phase $(R_m\% = 1.0)$, and the end $(R_m\% = 1.3)$ of the oil-generation window. The fourth line $(R_m\% = 2.2)$ represents the preservation boundary of wet gas. The hypothetical preservation boundary of dry gas $(R_m\% = 4.7)$ is reached at about 6 km depth. With computed depths of these values, the possibility of hydrocarbon presence caused by thermal conditions can be estimated. The temporal maturity evolution of a particular depth horizon can be determined by the timing of hydrocarbon generation.

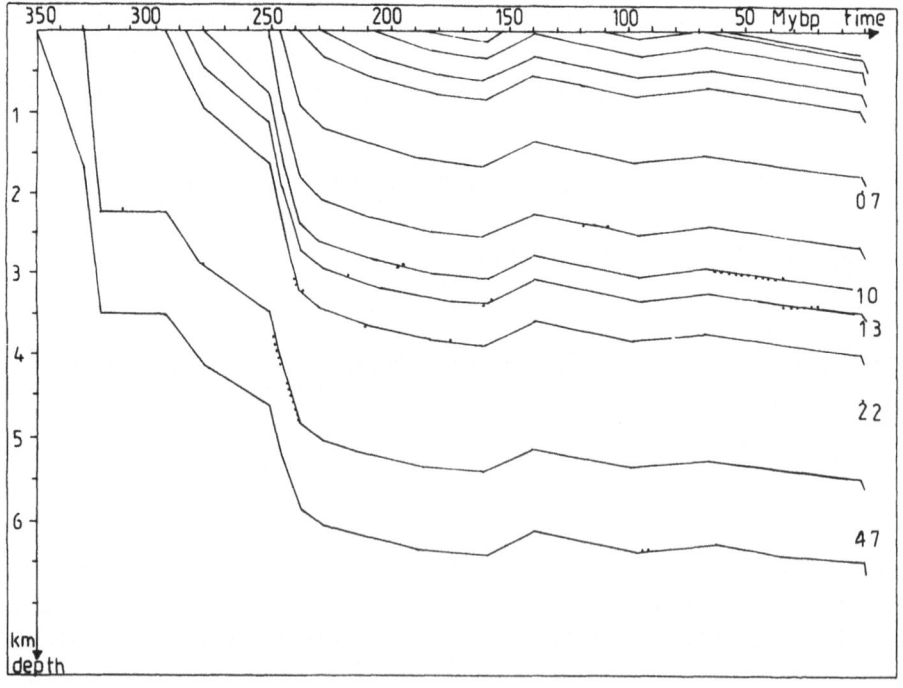

Figure 3. Subsidence model with computed maturity.   .... = isolines of vitrinite reflectance

SUMMARY

In summary, we can say our calculations can be used to determine of burial, geothermal, and maturity history of a sedimentary basin. In the North German Basin there are favorable geothermal conditions for the existence of hydrocarbons in a depth range from 2 to 6 km.

REFERENCES

Lopatin, N.V., 1971, Temperature and geological time as factors of carbonification: USSR Acad. Sci., Izv. Geol. Ser., v.3, p.95-106 (in Russian).

Middleton, M.F., 1982, Tectonic history from vitrinite reflectance: Geophys. Jour. Roy. Astr. Soc., v.68, p.121-132.

Sclater, J.G., and Christie, P.A.F., 1980, Continental stretching: an explanation of the post-mid-cretaceous subsidence of the Central North Sea Basin: Jour. Geophys. Res., v.85, no. B7, p.3711-3740.

DIFFERENTIAL COMPACTION AND STRUCTURAL GENESIS

H. Dietrich
University of Greifswald
Greifswald, Germany

ABSTRACT

An opportunity for differentiating endogenic tectonic activities from compaction-determined facies-depending tectonics is shown. Moreover, different stages of rock diagenesis that are the result of differential compaction lead to facies-dependent structural development.

DIFFERENTIAL DIAGENESIS

Facies differences of sedimentary rocks lead to respectively different thinning by compaction during diagenesis. Figure 1 and Table 1 show results of compaction computations that have been obtained by the exponential function of porosity decrease versus sedimentary depth (Harff and others, 1989; Dietrich, 1990) and by application of values of primary porosity (ϕ_o) and rock-specific compaction constant (b) according to Sclater and Christie (1980).

As shown in Figure 2 the differential compaction of a sand - clay sequence (Figs. 2B and 2C) leads to a dipping of beds and the formation of a facies - derived "diagenetic trap" in the sandstone unit. If the underlying sandstone has been cemented early this process for the superimposed sandstone bed (Fig. 2D) is accentuated even more.

Computerized Basin Analysis, Edited by J. Harff
and D.F. Merriam, Plenum Press, New York, 1993

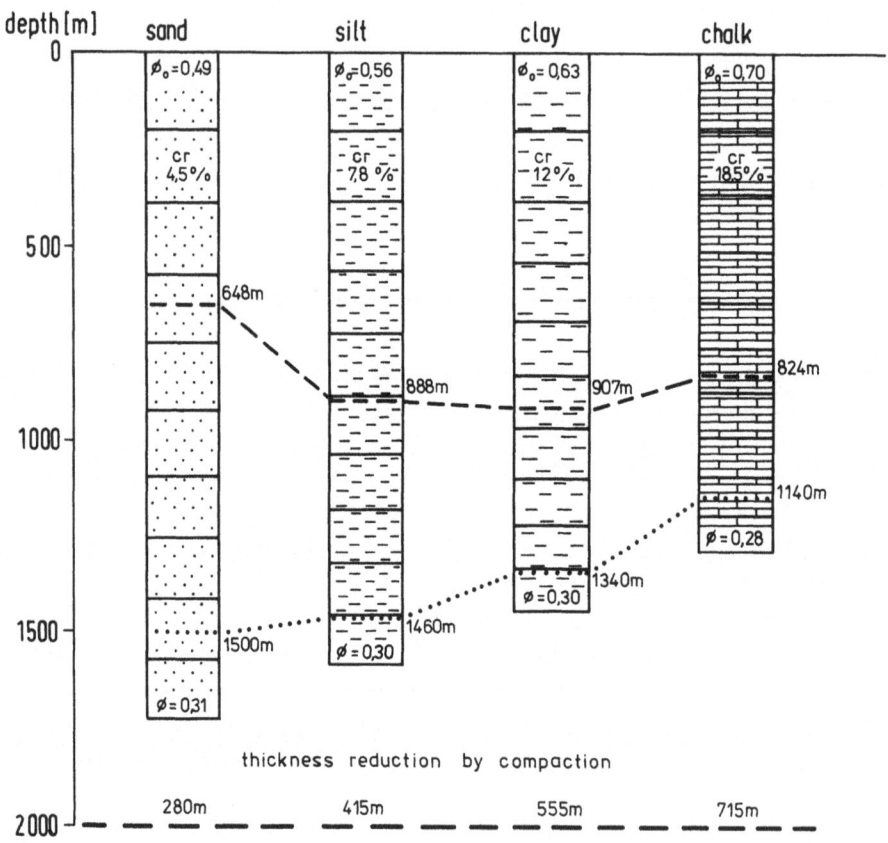

Figure 1. Differential compaction of selected sedimentary rocks (for explanations see Table 1)

Table 1. Compaction values for 10 structural blocks of sediment, each representing an original depositional thickness of 200 m

	Sand	Silt	Clay	Chalk
primary porosity according to Sclater and Christie (1980)	0.49	0.56	0.63	0.70
total thickness reduction for entire 200 meters	280 m	415 m	555 m	715 m
compaction rates between depths of 200 m and 400 m	4.5 %	7.8 %	12 %	18.5 %
Depth in which the compaction rate is below 3 %	648 m	888 m	907 m	824 m
Depth in which the compaction rate amounts to 1,65 % (end of mechanical compaction of sands; Füchtbauer, 1967)	1500 m	1460 m	1340 m	1140 m
Porosity at the base of a sediment thickness of 10 times of 200 m deposited sediment	0.31	0.30	0.30	0.28

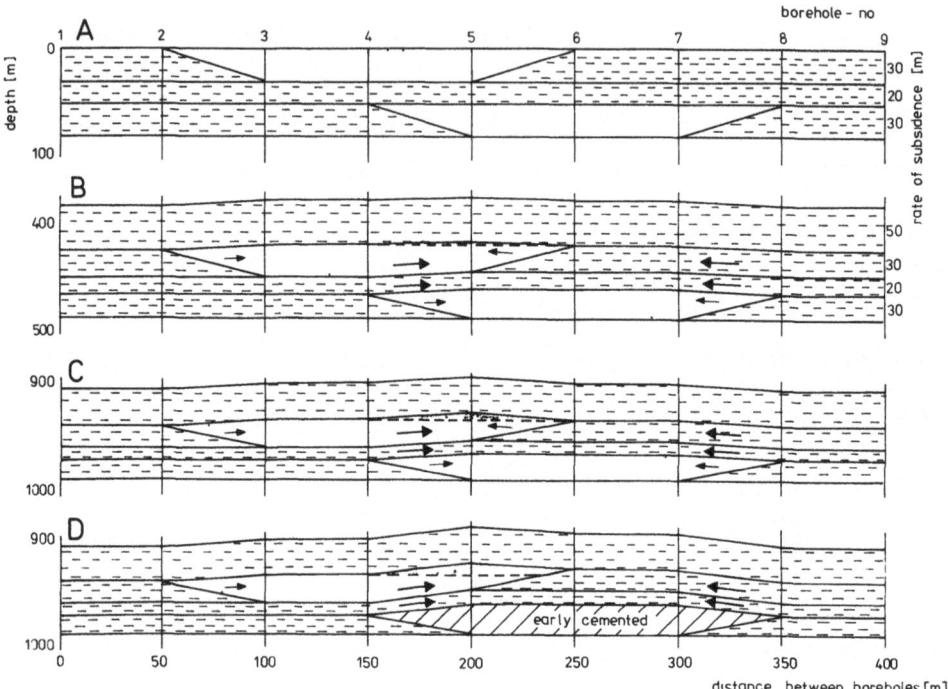

Figure. 2. Compaction of fluviatile channel sands
hatched: clay
pointed: sand
shadowed: facies - derived "diagenetic trap" in sandstone bed
little arrows: flow direction of pore fluid by differential compaction within unit
great arrows: flow direction of pore fluid by differential compaction of underlying
sedimentary units

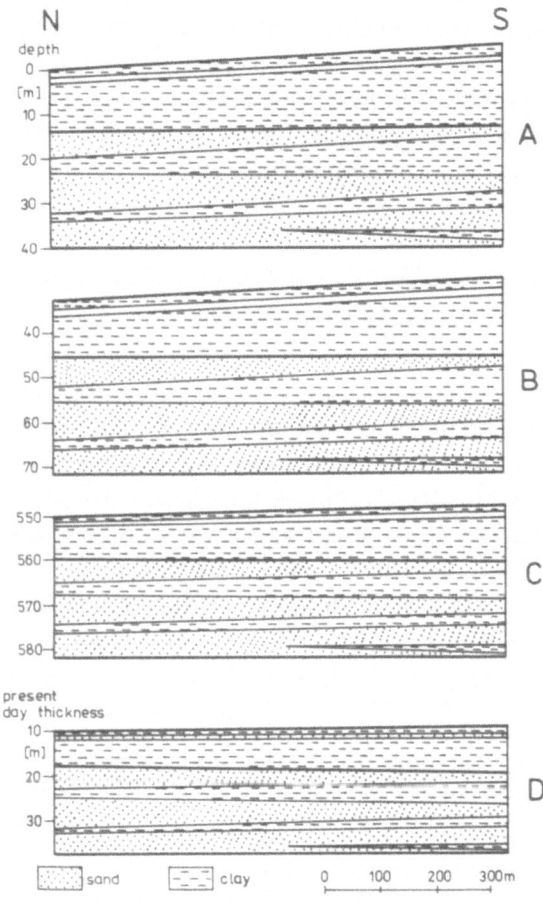

Figure 3. Variation of dipping of beds resulting from differential compaction. A: decompacted; B: post-Solling; C: post-Muschelkalk; D: present day.

A second example illustrates the relationship between two adjoining outcrops in the Northern Thuringian Basin the Detfurth Series (Triassic/ Bunter Sandstone; Fig. 3). The sand - clay distribution corresponds with the real thicknesses.

Observing the bed boundary between the sandy beds at the base and the clay-rich upper part of the Detfurth Series (strong black line) one can recognize that differential compaction leads to a change of dip of this bed boundary with time because of the distinct clay - enrichment in the southern part of the profile. At first, this bed boundary dips weakly to the north (Fig. 3A) and, later, to the south (Fig. 3B, 3C, 3D). Similar examples also exist for carbonates and evaporitic rocks.

TECTONIC ACTIVITIES AND DIFFERENTIAL DIAGENESIS

Besides tectonic processes differential compaction is of great importance concerning patterns of geological structures. However, it is not always easy to distinguish between both processes. In order to recognize (Sersukov, 1971; Schabrodt, Baumann, and Koch, 1986) either synsedimentary tectonic or other processes which may be derived from underlying beds (such as halokinesis, subrosion, or compaction) by thickness differences within sedimentary beds one has to decompact the sequence, especially in the situation of siliciclastic sequences. It is possible to use the difference (D) of the decompacted thicknesses of two outcrops respectively (Dietrich, Kasbohm, and Rechlin, 1990) by use of Equation (1).

$$D = M - N \qquad\qquad (1)$$

Moreover, one can subdivide the sedimentary sequence into numerous sublayers (m_1 to m_k and n_1 to n_k):

$$D_k = \sum_{i=1}^{k} m_i - \sum_{i=1}^{k} n_i \qquad\qquad (2)$$

The difference of thickness changes (ΔD) for a special geological horizon can be determined in order show a tectonic differentiation for this particular horizon:

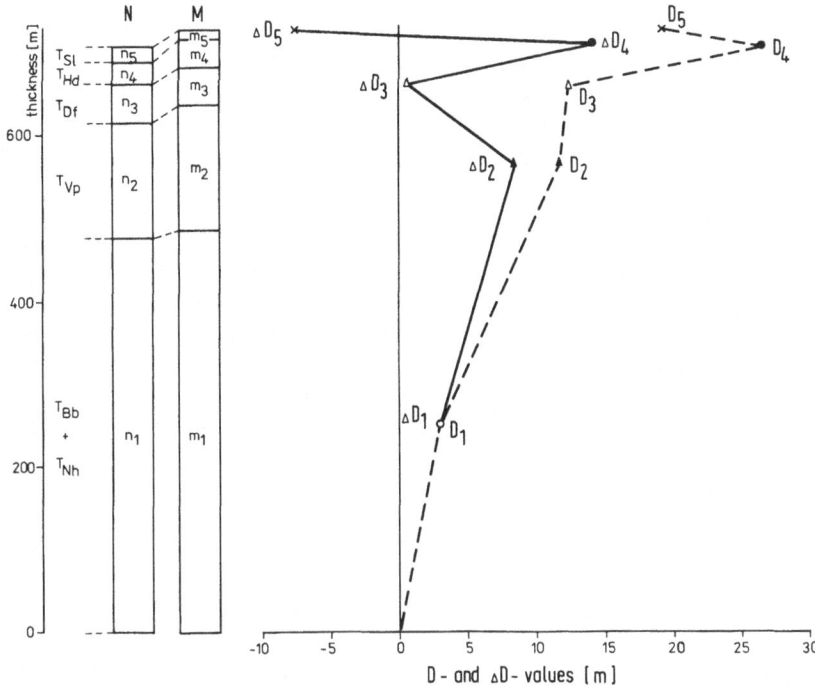

Figure 4. Presentation of example for D- and DD-value determination of Lower and Middle Bunter Sandstone of Thuringia.

$$\Delta D_i = D_i - D_{i-1} = m_i - n_i \qquad (3)$$

Figure 4 shows an example for the determination of D- and ΔD- values.

The aerial distribution of ΔD-values shown in Figure 5 gives an overview of the investigation area for the Schlotheim Graben (Thuringian Basin). In each diagram determination according to Equation (3) the eastern outcrop represents the minuend (M) and the western one represents the subtrahend (N).

At this place a period of tectonic activities has to be mentioned which generally is termed the "Hardegsen-Impulse." Between the outcrops 9 and 3 as well as between 9 and 21 the ΔD-values show great variations for this time interval. This may be the result of tectonic movements in the underground.

Moreover, generally the same conclusion can be made from the relief block diagram of the Zechstein surface of this area which has been

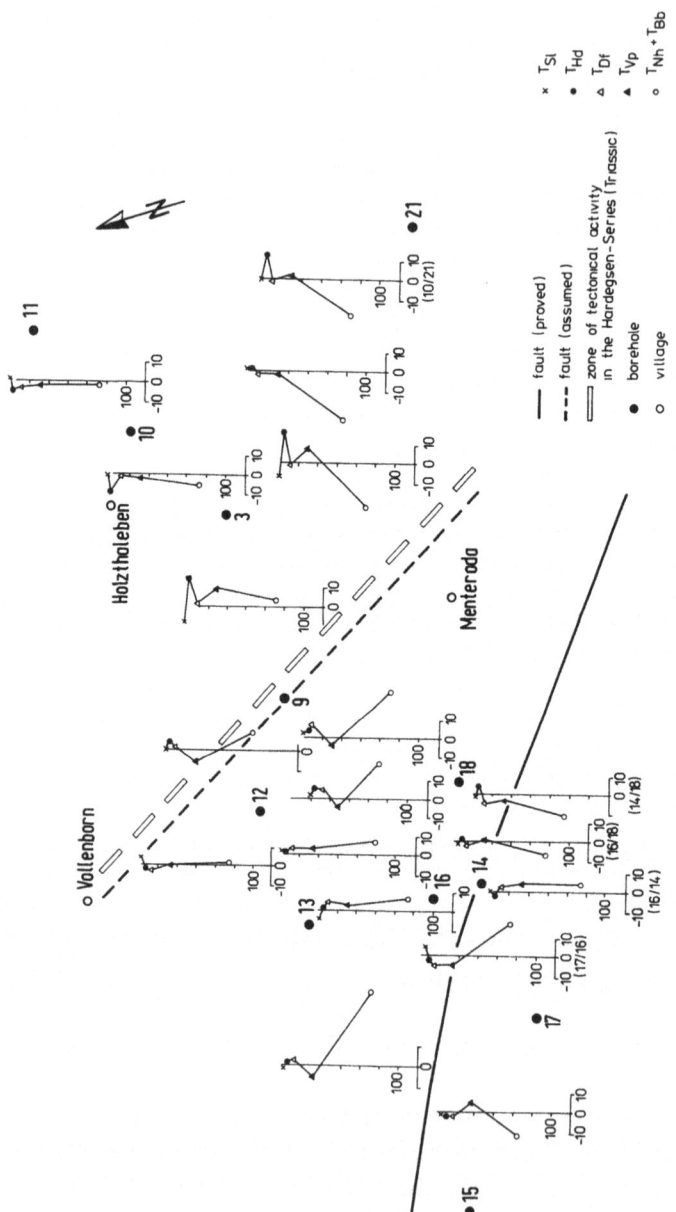

Figure 5. Areal distribution of DD-value diagrams. If positioning between one outcrop and another was impossible, numbers of boreholes were mentioned under diagrams.

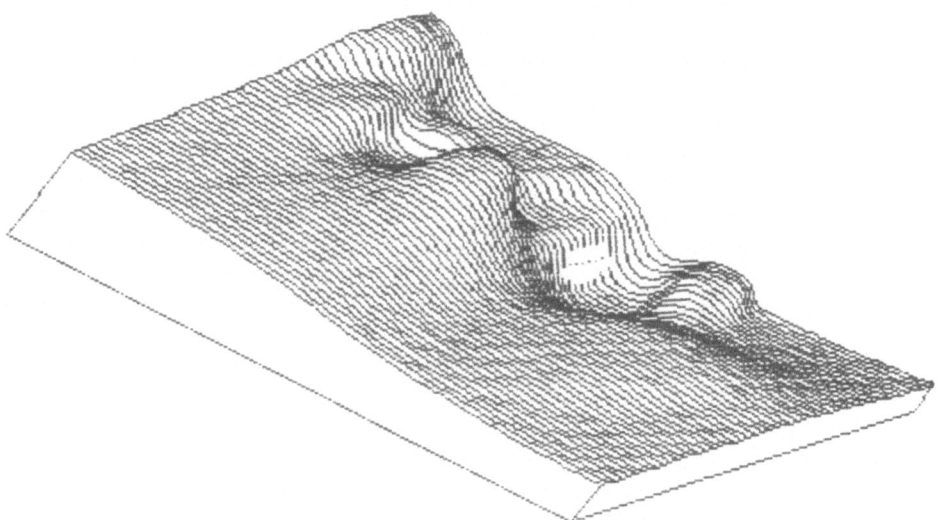

Figure 6. Relief of Zechstein surface after deposition of Hardegsen Series (view from SSE). Subsidence in center of diagram is evident.

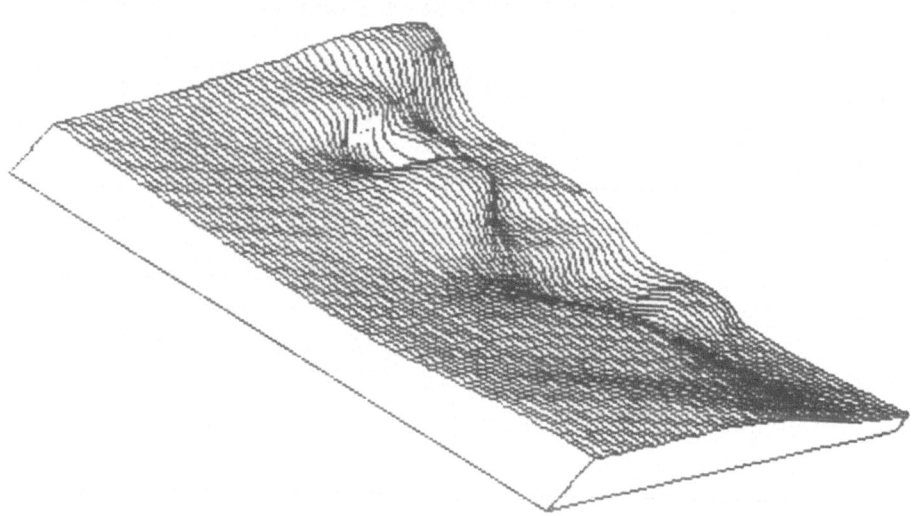

Figure 7. Relief of Zechstein surface after deposition of Detfurth Series (view from SSE).

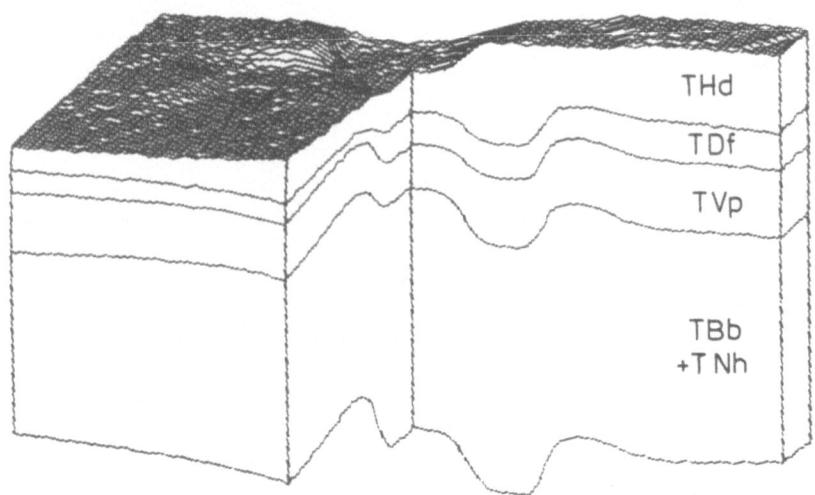

Figure 8. Three-dimensional diagram of study area (time period after deposition of Solling-Series with presentation of Hardegsen surface; view from SSE)

derived from two different periods of geological time. In Figure 6 which shows the Zechstein surface relief of post-Hardegsen time within the center a distinct subsidence can be seen in comparison to pre-Hardegsen time (Fig. 7).

The three-dimensional presentation (Fig. 8) supports the conclusion that the subsidence in the right part of the diagram mainly occurred during Hardegsen-time whereas a continuous elevation of the southern subblock can be assumed in the left part of the diagram.

CONCLUSIONS

In the analysis of the structural development of an area it is necessary to consider the different compaction processes of sediments during diagenesis. This is because of the following reasons:

(1) Lateral facies changes lead to a differential compaction of sedimentary basin sequences which, therefore, has to be considered as a specific structure-forming process.

(2) Accordingly the differential compaction determines the facies of the superimposed deposits, which respectively, underlie a differential diagenesis as well.

(3) In synsedimentary deformed areas intensity and distribution of these tectonic activities in space and time can be correlated much more exactly if the differential compaction is involved.

REFERENCES

Dietrich, H., 1990, Differentielle Kompaktion in siliciclastischen Horizonten: Z. Geol. Wiss., v. 18, no. 9, p. 801-811.

Dietrich, H., Kasbohm, J., and Rechlin, H. 1990, Differentielle Kompaktion und tektonischer Einfluß: Z. angew. Geol., v. 36, no. 11, p. 400-405.

Füchtbauer, H., 1967, Influence of different types of diagenesis on sandstone porosity: Proc. 7th World Petrol. Congr., Mexico, v. 2, p. 353-369.

Harff, J., Schretzenmayr, S., Springer, J., Hoth, P., Eiserbeck, W., and Süssmuth, S., 1989, Mathematisch - numerische Modellierung regionaler bis lokaler Einheiten in sedimentären Becken an einem Beispiel für die Kohlenwasserstofferkundung im Rotliegenden: Z. Geol. Wiss., v. 17, no. 7, p. 747-759.

Schabrodt, T., Baumann, L., and Koch, K., 1986, Untersuchungen zum Zusamenhang zwischen der Beckenentwick-lung und der Halititsedimentation im Werra-Kaligebiet: Z. Geol. Wiss., v. 14, no. 4, p. 427-435.

Sclater, J.G., and Christie, P.A.F., 1980, Continental stretching: an explanation subsidence of the central North Sea basin: Jour. Geophys. Res., v. 85, no. B7, p. 3711-3739.

Sersukov., V.V., 1971, Graficeskij metod obrabotki materialov pri tectoniceskom analize moscnostej: Geol. Razv., no. 4, p. 19-22.

DISSOLUTION AND CEMENTATION IN BASIN SIMULATION

R. Ondrak and U. Bayer
Institute of Petroleum and Organic Geochemistry (ICH-5)
F.R.G.

ABSTRACT

Mineral dissolution and precipitation processes change the storage- and transport-coefficients of sedimentary rocks during geological basin evolution. An algorithm for the simulation of the temperature and transport dependence of diagenetic processes was developed. The algorithm is based on the linearized transport-reaction equation. The differential equation is transformed into an integral-equation which describes solution and precipitation as a sequence of impulses and their associated response functions. The nonlinear interaction between solution, transport, and porosity evolution is solved by iteration. The model was tested on the example of quartz solution/precipitation in a schematic sedimentary basin with a simplified burial history. Flow of pore-water is a boundary problem, and the flow vectors are adjusted for storage- and transport-coefficients changing with time. The results of these calculations show that the processes are controlled primarily by the temperature-gradient-field. Changes of the flow velocity do not significantly influence the resulting porosity patterns, although they influence the magnitude of mass transport. Based on the model, it is possible to estimate the paleogradient field of temperature from observed cementation and dissolution patterns.

Computerized Basin Analysis, Edited by J. Harff
and D.F. Merriam, Plenum Press, New York, 1993

INTRODUCTION

Porosity is an important reservoir property. For this reason much attention has been paid to the diagenetic processes affecting this rock property. Several attempts have been made to quantify cementation and dissolution in sandstones due to quartz redistribution (Füchtbauer, 1978, 1979; Bjorlykke, 1979; Wood and Hewett, 1982; 1984; Leder and Park, 1986). The numerical simulation of the changes in porosity during basin evolution has been approached among others by Bjorlykke (1983), Wood and Hewett (1982), and Leder and Park (1986).

Bjorlykke (1983) made a rough calculation of the porosity reduction as a result of quartz overgrowth by silica derived from dewatering shales. He came to the conclusion, that the solubility of silica in water is too small to account for the observed porosity reduction. Leder and Park (1986) presented a model for the numerical evaluation of porosity reduction in a sandstone during its burial history. Beside some shortcomings of their model (Fang, Visscher, and Davis, 1988; Hanor, 1988), the main restriction is, that the model can simulate only the porosity reduction because of quartz precipitation, whereas it cannot account for porosity increase because of dissolution of quartz. Another theoretical model was presented by Wood and Hewett (1982), who studied the mass transfer and the resulting porosity changes in porous sandstones as a result of fluid convection. All models are similar in that they consider only transport-induced redistribution. In none of the models a reaction term is included, which describes the interaction between the solid phase and the pore-fluid. It usually is assumed that the pore-fluid is in chemical equilibrium with respect to temperature and pressure. Precipitation or dissolution of the silica is induced by changes of the P/T conditions.

In order to overcome the restrictions of the models mentioned, we will develop an algorithm, which allows to model the changes of reservoir properties resulting from both mass transfer resulting from fluid flow and the reaction between the solid and the fluid phase. It is important that mass transfer, cementation, and dissolution of quartz, and the resulting change in porosity is modeled using a single simulator. The algorithms are tested by a rather simple model.

DIAGENETIC PROCESSES

Modeling diagenetic processes changing porosity is a complex task involving various driving mechanisms such as recrystallization, solution, and moving reaction fronts. Table 1 provides a simplified classification scheme of processes and parameters possibly involved in diagenesis. Of course, any transition between the identified classes is possible in nature, however, the classification refers to the different dominant processes and their associated typical length scale (time and distance). The latter one provides an important criteria to select the proper simulation technique for a specific problem.

Table 1. Simplified classification of diagenetic processes suitable for numeric modeling. See text for discussion

PROCESSES INFLUENCING POROSITY

RECRYSTALLIZATION	SOLUTION		MOVING FRONTS
Solid-Solid Reactions	Solid-Fluid Reactions		Solid-Fluid Reactions
Exsolution	Incongruent	Congruent	Reactant Flux
	(Incomplete)	(Complete)	
	Dissolution	Dissolution	

Chemical and Physical Factors controlling the system:

P & T	P, T & chemical composition of the solution: e.g. pH, Eh, pCO_2	Density Differences, chemical composition of the fluid, T, P

Physical Processes:

Solid-State Diffusion	Diffusion and "Mixing"	Flow/Diffusion velocity ratio

Typical Scale:

mm	m	km

Recrystallization of minerals is a solid-solid reaction. Minerals stable under one set of temperature and pressure conditions can become unstable if these parameters are altered. The process of recrystallization takes place by solid-state diffusion on a typical scale of not more than millimeters. It therefore is a locally restricted process, which will not be considered in our model. Examples for this process are the recrystallization of amorphous silica to quartz, of aragonite to calcite, and Mg-calcite to dolomite and calcite.

Solution of minerals involves reactions between the solid and the fluid phase. Two processes have to be distinguished in this context: solution can take place either by congruent or incongruent dissolution. In congruent dissolution the mineral dissolves completely, whereas in the situation of incongruent dissolution only a part of the mineral goes into

solution and an insoluble part remains in place. For this reason incongruent dissolution can be regarded as transitional between solution and
recrystallization. Examples for congruent dissolution are the solution of
salt, quartz, and calcite in water and for incongruent dissolution the
solution of feldspar in water with the resulting formation of clay minerals. The process of solution is controlled by the temperature and pressure
conditions of the system, which control the thermodynamic conditions for
the increasing or decreasing solubility of minerals. In addition the
chemical composition of the solutions, for example the pH, Eh, pCO_2 also
plays an important role in controlling the system. The solution of
minerals can be controlled by diffusion of the solute away from the surface
or by fluid flow and resulting mixing of solutions with chemical composition, which are not in equilibrium with the surroundings. The typical
scale through which this process takes place are meters. For high volume
fluxes larger dimensions are possible; this would be transitional towards
a moving front system.

Moving front systems also are characterized by solid-fluid reactions.
In this situation a reactant flux causes a change in the mineral composition of the solids. Usually the flux is driven by density differences in the
pore-fluids or a hydraulic head. This fluid flow results in a replacement
of the original pore-fluid, and thus in a change of the chemical composition
of the pore-fluid. The formation of moving fronts and the width of the front
depends mainly on the flow and diffusion velocity ratio. In a flow
dominated system a moving front will develop more readily and will be
more distinct. Although in a diffusion dominated system, a moving front
may form or the front will be wider, and not recognized as a front. The
typical scale of such systems is in the order of kilometers. Examples for
such systems are regional dolomitization, uranium roll fronts, and on a
smaller scale redox reactions based on fluctuations in the groundwater
table.

In our model we will study the effects of dissolution and precipitation
of minerals in combination with fluid flow on a basin wide scale. Therefore,
a combination of a simple congruent dissolution reaction with a large-
scale moving front system is used. The only reactant phase in the aqueous
system will be silica providing a simple but interesting reaction-transport
system.

Quartz Solubility

Dissolution of quartz in water and the resulting formation of
monosilicic acid (H_4SiO_4) is described sufficiently by the following equation:

$$SiO_2(s) + 2 H_2O <==> H_4SiO_4 \qquad (1)$$

This notation implies, that the molecule going into solution is monosilicic acid and polymerization is negligible (Alexander, Heston, and Iler, 1954; van Lier, DeBruyn, and Overbeek, 1965; Fournier and Potter, 1982). Depending on the pH, monosilicic acid can dissociate further into disilicic acid ($H_3SiO_4^-$), which, in turn, can dissociate into trisilicic acid ($H_2SiO_4^{2-}$). The dissociation processes can be described by reaction (2) and (3):

$$H_4SiO_4 <==> H_3SiO_4^- + H^+$$
$$\qquad (2)$$

$$H_3SiO_4^- <==> H_2SiO_4^{2-} + H^+ \qquad (3)$$

With the assumption that the concentration of monosilicic acid remains constant, the dissociated species contribute to the total amount of silica in solution. It can be calculated as follows:

$$\Sigma SiO_2(aq) = [H_4SiO_4] + [H_3SiO_4^-] + [H_2SiO_4^{2-}]$$

Under these considerations the solubility of quartz in water is determined by one reaction which is independent of the pH of the solution and two reactions which are dependent on pH. For this reason it must be evaluated under which pH conditions the reactions (2) and (3) contribute significantly to the total silica in solution. The pH dependence of the formation of disilicic acid can be calculated according to the following equation:

$$[H_3SiO_4^-] = (K_{eq}*K_1)/[H^+]$$

Where $[H_3SiO_4^-]$ stands for the concentration of disilicic acid, $[H^+]$ is the concentration of the H^+ ion, K_{eq} and K_1 are the equilibrium constants for reaction (1) and (2). In the same manner the formation of trisilicic acid is calculated:

$$[H_2SiO_4^{2-}] = (K_{eq}*K_1*K_2)/[H^+]^2$$

with $[H_2SiO_4^{2-}]$ standing for the concentration of trisilicic acid and K_2 being the equilibrium constant for reaction (3). For standard state conditions (25° C and 1 atm pressure) the following values for the constants Keq, K_1, and K_2 are given (Blatt, Middleton, and Murray, 1982; Rhyzenko, 1967; van Lier, De Bruyn, and Overbeek, 1965):

$$\log K_{eq} = -3.77 \text{ to } -4.0$$
$$\log K_1 = -9.46 \text{ to } -9.82$$
$$\log K_2 = -11.7 \text{ to } -12.56$$

The magnitude of the dissociation constants K_1 and K_2 already indicates, that the dissociated species of monosilicic acid will contribute to the total silica in solution only at pH ranges higher than 8-9. The relative amount of each species contributing to the total silica in solution (ΣSiO_2 (aq)) as a function of pH is calculated and plotted in Figure 1 for standard state conditions (25°C, 1 atm). For higher temperatures the dissociation constants K_1 and K_2 are less well determined, but the published data indicate, that the general picture changes only slightly. For this reason it can be assumed safely, that the dissociated species can be neglected for pH ranges between 5 and 9. This is important, because most natural ground and formation waters fall between pH values of 5 - 8.5 (Baas-Becking, Kaplin, and Moore, 1960; Collins, 1975). Hence, in the remainder of the this paper 'silica' stands for the monosilicic acid which is soluble in water, disregarding the minor effects of dissociation of silica.

Beside pH the solubility of silica depends on the temperature and pressure conditions within the solution. The T and P dependence has been studied intensively by several authors (Siever, 1962; Morey, Fournier, and Rowe, 1962; van Lier, De Bruyn, and Overbeek, 1965; Crerar and Anderson, 1971; Walther and Helgeson, 1977; Fournier and Potter, 1982). Most experiments have been performed under changing temperatures and the associated vapor pressures. Only few experiments have been carried out under pressures exceeding the vapor pressure. These experiments show that the increase in solubility of silica in water resulting in higher pressures is small. The increased solubility is interpreted as the result of the higher density of water because of the higher pressure and thus a higher solution capacity of water with respect to silica. This interpretation does not account for increased solubility as a result of pressure solution at grain boundaries. Pressure solution at grain boundaries can not be accounted for by these experiments, because they have been carried out with pure quartz whereas pressure solution and stylolitization is observed only at clay rimmed grains. Furthermore

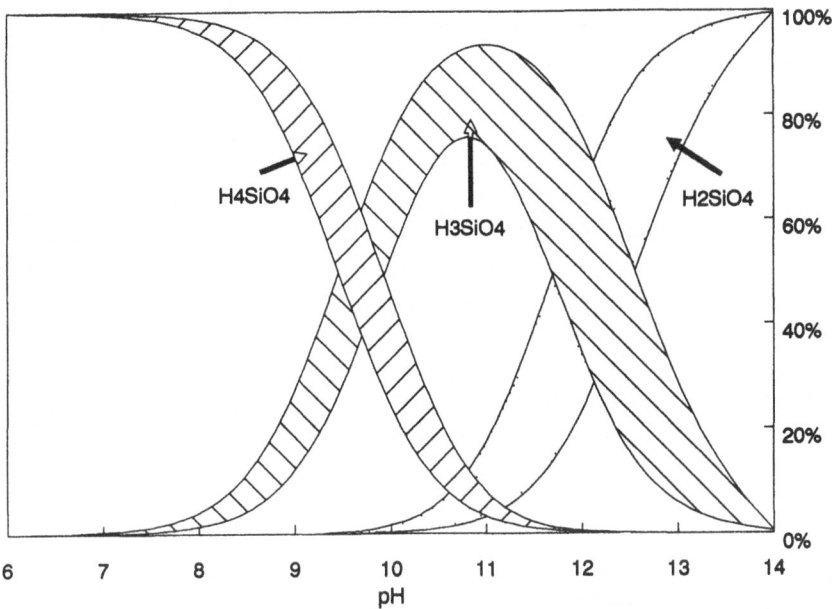

Figure 1. Stability field of silica species with respect to pH of solution. Computed using dissociation constants published in van Lier, De Bruyn, and Overbeek, (1965); Rhyzenko, (1967); Blatt, Middleton, and Murray (1982). Field of interest is limited to pH 6-9, where system is dominated by H_4SiO_4

pressure solution probably takes place in a diffusion dominated system on a small scale. As we assume a pure quartz sandstone in our model, and the increase in solubility of silica in water resulting from increasing pressure is small, we have disregarded the influence of pressure.

As mentioned, the experimentally determined effect of temperature on the solubility of silica has been studied by several authors. Based on the experimental results several empirical equations for the temperature dependence have been derived. The results for a few of these equations are plotted in Figure 2. The curves lie closely together up to temperatures of 200° C and but deviate at higher temperatures. For studies of diagenetic processes it can be assumed in most situations that the temperatures involved will not exceed 200° C. For this reason it is sufficient to use the simple empirical equation derived by van Lier, De Bruyn, and Overbeek (1965):

$$\log K = 4.929 - (1162.0/T) \qquad \text{(log K in ppm)}$$

As illustrated in Figure 2, the solubility of silica in water increases exponentially with temperature. This indicates that the reaction of quartz with water is temperature sensitive, which will be of major importance for the model computations.

Porosity calculation. The dissolution or precipitation of quartz in a porous sandstone affects porosity. The porosity increases or decreases proportionally to the specific molar volume (Vsp) of the dissolved or precipitated quartz. In a closed system changes in porosity can be calculated from the product of the change of concentration (C) in the fluid and the specific molar volume of quartz:

$$\frac{d\phi}{dt} = \frac{d(\phi Cl)}{dt} * Vsp \tag{4}$$

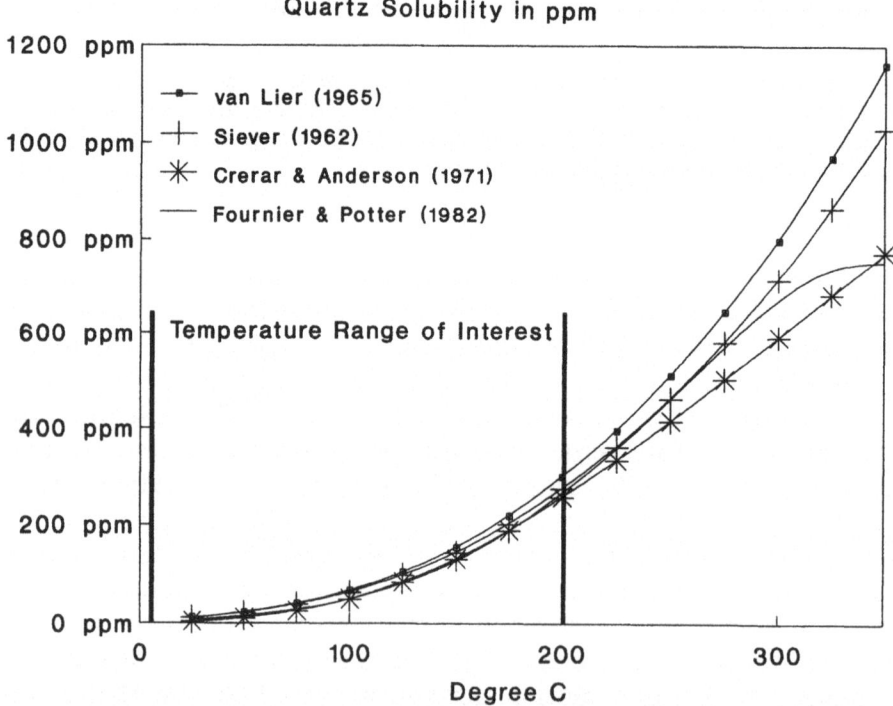

Figure 2. Empirically derived silica solubilities as function of temperature after different authors. In temperature range of interest (0-200° C) functions are similar, therefore, simple equation of van Lier, De Bruyn, and Overbeek (1965) has been used for simulation purposes.

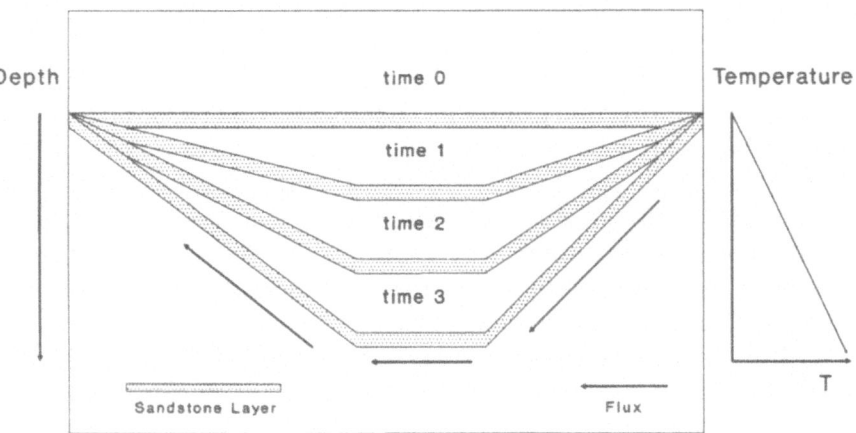

Figure 3. Temporal subsidence of thin sandstone layer with variable subsidence rates along cross section. Constant geothermal gradient is assumed. See text for discussion.

In an open system with additional fluid flow, the changes in concentration depend also on the mixing rate of the fluids providing additional complications as discussed. Nevertheless, Equation (4) provides the basis for calculating the effective changes in porosity.

A Quantitative Model of Quartz Diagenesis

To derive a quantitative model for the evolution of secondary porosity a geological model of the basin history is needed as well as a mathematical model for the processes involved. In this study we focus on the general concepts of quartz redistribution. Therefore we will try to simplify the complex structures to their essentials, in order to highlight the basic factors controlling simple dissolution and precipitation processes within a flow regime and to illustrate the basic principles involved in the associated process of pattern formation.

The Geological Model. The basic features of the geological model are illustrated in Figure 3. A confined sandstone layer or aquifer, continuous throughout the area of interest, subsides as sediment load increases with time. The subsidence rate may differ along the cross section through the basin as illustrated schematically in Figure 3. The temperature field is considered stationary in time and increases linearly with depth, that is the geothermal gradient is constant throughout the basin. Within the sandstone layer the temperature increases proportionally to subsidence or to the depth location. Thus, the temperature gradient within the sandstone layer will vary depending on the slope of the layer with respect to the constant geothermal gradient. The thickness of the sandstone layer is taken to be small with respect to its horizontal extent along the cross section, for example [m] for the thickness and [km] for the horizontal extension. The basin geometry therefore is reduced to a 1-D system with the sandstone layer represented by a simple line. The model does not take into account the compaction as the result of increased sediment load or to porosity changes caused by dissolution.

Within the sandstone we assume a pore-fluid flux which, by definition, will be from the right to the left in all the calculations and figures. The causes of the flux are not modeled. The flux is predetermined as boundary condition and may be related either to a hydraulic head outside the right boundary of the figures, or to a large scale convective flow through the sandstone, driven by density differences derived from the geothermal temperature field. The two concepts provide slightly different solutions as will be discussed next. The illustrations are based upon the flow model driven by density differences. This concept eliminates the flow potential as an additional variable for a given subsidence history. The flow is controlled entirely by the evolving permeability because of the dissolution or precipitation processes. Because the sandstone layer is modeled as a line, the entire flux through the confined layer is controlled by the smallest permeability occurring along the cross section. This is required by the mass-balance of the pore-fluid, which entails, that the amount of fluid entering the cross section from the right side equals the amount leaving the system at the left end.

The dissolution model can be used to calculate the changes in porosity. However in order to handle the changes of the flow velocity with time we need to calculate the new permeability resulting from dissolution or precipitation processes. For the purpose of this study the new permeability is calculated as a function of the smallest porosity by use of the Kozeny-Karman equation (see e.g. von Engelhardt, 1960).

The Mathematical Model. For the derivation of a suitable mathematical model for quartz redistribution we start from a classical transport-reaction equation (Lichtner, 1988). It consists of the flux and diffusion term describing transport and the reaction term defined by a dissolution term alone, as given by Equation (1) of Table 2. The reaction term has been modified with regard to other formulations given in the literature (Berner, 1980; Leder and Park, 1986). This meant introducing

Table 2. Mathematical models of solution-transport equation for SiO_2

Flux	Reaction	Diffusion

$$\frac{\partial \phi C}{\partial t} = \nabla (vC) + k(1-\phi)\phi(C_{eq}-C) + D\nabla^2 C \qquad (1)$$

Pointwise collagation without diffusion

$$\frac{dC_i}{dt} = v^* \left\{ C_i - \left[C_{i-1} + \frac{dC_{i-1}}{dt} \right] \right\} + \Delta C_i \qquad (2)$$

ignoring $\frac{dC_{i-1}}{dt}$, integrate partially

$$C_i = C_o \, e^{-v^*(t)t} + C_{i-1}\left[1 - e^{-v^*(t)t}\right] + \int_{t_1}^{t_2} \Delta C(t_o) \, e^{-v^*(t-t_o)t} \, dt_o \qquad (3)$$

mixing term   source
= impulse response function

where
ϕ = porosity
k = reaction constant
D = diffusion coefficient
C = concentration
C_{eq} = equilibrium concentration
ΔC = source term $\equiv$ reaction term in eq. (1)
v = Darcy velocity
v^* = effective velocity

the porosity dependent product term $(1-\phi)\phi$, which dictates that dissolution ceases if all solids are dissolved, and equally so if no pore space, and therefore, no pore fluid is left because of cementation. If either one of these terms is missing, the differential equation itself becomes unstable either for dissolution or cementation.

Based on our simplified basin geometry the differential Equation (1) is reduced to a 1-D equation along the cross section with the only complication that the effective distance between two controlling points has to be calculated as a line integral along the sandstone layer and thus

changes with time as a function of subsidence. It will be shown later that
the diffusion term is negligible within our scenario. Discretizing the
remaining right-hand side of the equation, therefore, leads to a set of
ordinary differential equations for the numerical model. This set of
equations can be simplified further by the assumption, that the fluxes
and porosities remain constant for a single time step. The nonlinear
interaction between these parameters and the concentration of the solute
is reestablished by an iterative process applied to each time interval.

In order to analyze the basic properties of the model the equations
can be linearized further by the assumption that the input concentration
is constant during a time interval. The boundary value then reduces to an
initial value problem for any time step. Equation (2) of Table 2 can be
integrated formally, yielding Equation(3) of Table 2, providing a stable
explicit scheme to model the evolution of the solute. However, the final
step is not the only solution, other numerical approximations may be
applied to Equation (2) of Table 2. Nevertheless we will to use Equation
(3) of Table 2 in order to elucidate the processes involved. Equation (3) of
Table 2 consists of a mixing term describing the dislocation of the pore
fluid in connection with dispersion or mixing, and a source term considering
the dissolution/precipitation around the collagation point during a time
interval. The mixing term is given as a weighted average of the initial
concentration of two neighboring points with the original concentration
at the point 'i' declining exponentially as a function of the effective flow
velocity. The change in concentration is continuously substituted by the
same amount of pore-fluid having the concentration of the neighboring
point.

The source term is given as an impulse-response function or as the
convolution of the reaction term with the loss function due to the flow
term. Its evaluation provides the changes in concentration which together
with Equation (2) of Table 2 are used to calculate the local porosity
alterations. The reaction term is approximated by linear interpolation
through the time interval with regard to the known change in temperature and is subsequently corrected for the change in concentration,
porosity and flux by an iteration procedure. The linearized model [Eq. (3)
of Table 2] provides a straightforward explicit scheme with respect to the
spatial distribution of the points as far as the flow is undirectional along
the cross-section as we assumed previously (Fig. 3). With regard to the
large time steps used in geological modeling however, an implicit scheme
should be used to evaluate the temporal evolution.

Equation (3) of Table 2 accounts for the important physical assumption that the diagenetic processes possess a time average, that is that

short time disturbances and fluctuations of the flow regime do not alter the final results in a substantial way. This assumption is an essential short coming of the concept because any local alteration by diagenesis necessarily influences the permeability, porosity and the flow rate. The concept derived here therefore can account only for large-scale diagenetic trends, and alterations of the physical properties have to be considered as a large volume average.

COMPUTED EXAMPLES

The previously derived concepts and their consequences for diagenetic pattern formation in large-scale systems will be illustrated by a single model with subsequent sensitivity analysis. Applying the previously discussed restrictions, the model can account only for regional trends which can be approximated by spatial averages of the physical properties. In this section we will introduce a basic model and study the qualitative and quantitative changes in the diagenetic pattern which arise from the variation of some basic parameters.

Basic Model. The general concept of the geological basin model is illustrated in a schematic fashion in Figure 3. Figure 4 shows the burial history of the sandstone used in the numerical model. The sandstone consists of a single layer; its position with increasing burial along the idealized cross section is shown for several stages in Figure 4. The time intervals between the positions are 10 Ma; the following calculations are based on time steps of 1 Ma for a period of 100 Ma. The sandstone layer moves through a stationary temperature field (grad T = 27° C/km). The temperature in the sandstone therefore increases with depth according to the constant geothermal gradient. On the left edge, at location 0 [km] the temperature is constant (surface temperature: 4° C) throughout the entire time because this point does not subside. It reaches a maximum at location 190 [km] and declines again to the right. The flux of water is entering the sandstone layer on the right leaving it again at 0 [km] on the left. The initial porosity of the entire sandstone layer is 35%.

The temperature gradient within the sandstone layer varies with time because of the differential subsidence. Therefore, the solubility of silica changes laterally and temporally as illustrated in Figure 5. Because the time intervals used are large (1 Ma), the concentrations satisfy chemical equilibrium conditions. This is expressed by the fact that the concentration profiles mirror the stages of the subsidence curves. The boundary conditions are: the entering fluid is in chemical equilibrium with the solids and the temperature at the left side is constant (4° C).

The associated local changes in porosity as a result of dissolution or cementation are illustrated in Figure 6 in the same manner as the burial history, that is as temporal stages along the cross section. The basin is divided into a dissolution dominated region at the right side and a cementation dominated region at the left, the cross-over zone being located at the point of maximum subsidence and maximum solubility of silica. The fluids enter the sandstone layer from the right and dissolve quartz as temperature increases and the solubility of silica increases. The upward moving and cooling fluids on the left precipitate quartz because of the decreasing solubility of silica. The amount of quartz dissolved or precipitated correlates with the slope of the sandstone layer and thus with the temperature gradient within the sandstone layer. However with increasing time the flow velocity declines because of the increasing pore closure by cementation, and therefore, reduced permeability. The disso-

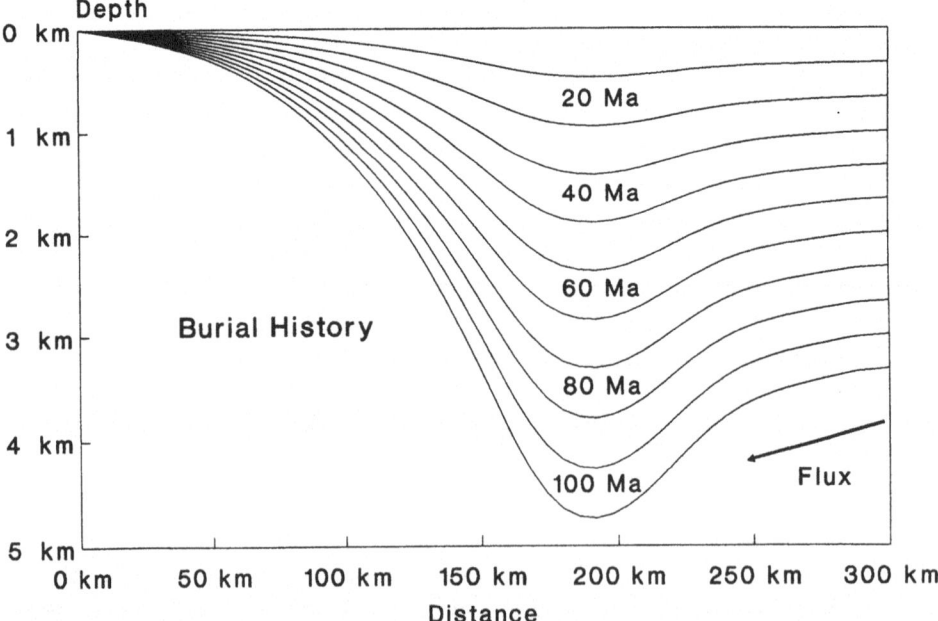

Figure 4. Burial history of thin sandstone layer along cross section through hypothetical basin. Lines indicate position and geometry of layer at different times. Flux within sandstone is from right to left.

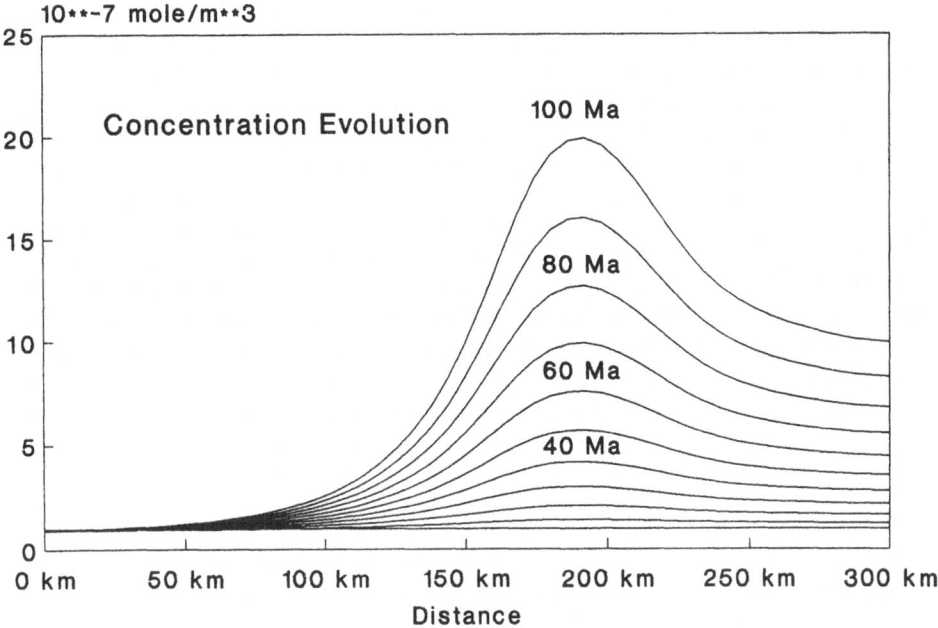

Figure 5. Concentration profiles of dissolved silica in sandstone layer for different times as indicated (cf. Fig. 4).

lution and cementation pattern, therefore approaches a stationary state as indicated by the convergence of the subsequent porosity curves (Fig. 6).

The computational model of Figure 6 exhibits a nonlinear porosity evolution caused by the assumed density driven flow which requires an inverse density gradient within the sandstone layer (i.e. the density of the pore-fluid decreases with depth). This gradient is built up as a result of the depth dependent temperature evolution along the sandstone layer during the differential subsidence of the sandstone along the cross section. Therefore, the fluxes are extremely small during the early phase of subsidence and increase as the temperature gradient and, therefore, the density differences along the layer increase. This additional effect can be eliminated by defining a constant hydraulic head for the entire time period. Figure 7 illustrates the qualitative difference in the time dependent porosity evolution at the point of maximum porosity decrease in the basin (location 155 [km]). In the situation of a constant hydraulic head the porosity declines almost exponentially with time.

Sensitivity Analysis. The model depends on a variety of parameters affecting the qualitative and quantitative results. However, not all parameters are independent in terms of the discrete approximation. The flux occurring in the mathematical model, for example, depends on the permeability and the pressure gradient, and in the discreted version it also becomes a function of the time step and the distance between the grid points. Any variation of the flux, therefore, can be interpreted alternatively as a variation of one or several of these parameters leading to the same final flux value. Although there may be small differences in the long-term effect because of the nonlinear coupling of the processes, some principal changes in the pattern formation can be studied by the variation of permeability alone. Figure 8 illustrates the changes in porosity evolution at high permeabilities or fluxes. The main effect is that the dissolution and cementation fronts converge with time, that is the cementation front migrates against the flux direction. Thus a moving front is evolving with time. In addition the extremes widen and the area of cementation increases in the flow direction especially during the early stages when the porosities are high.

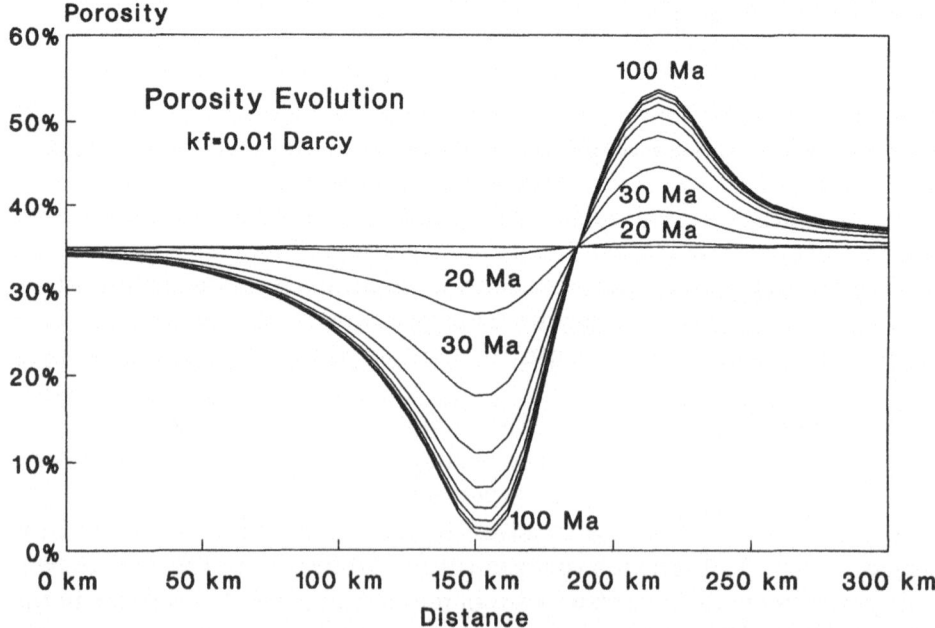

Figure 6. Porosity evolution with time for basin model of Figure 4. Initial porosity is 35%.

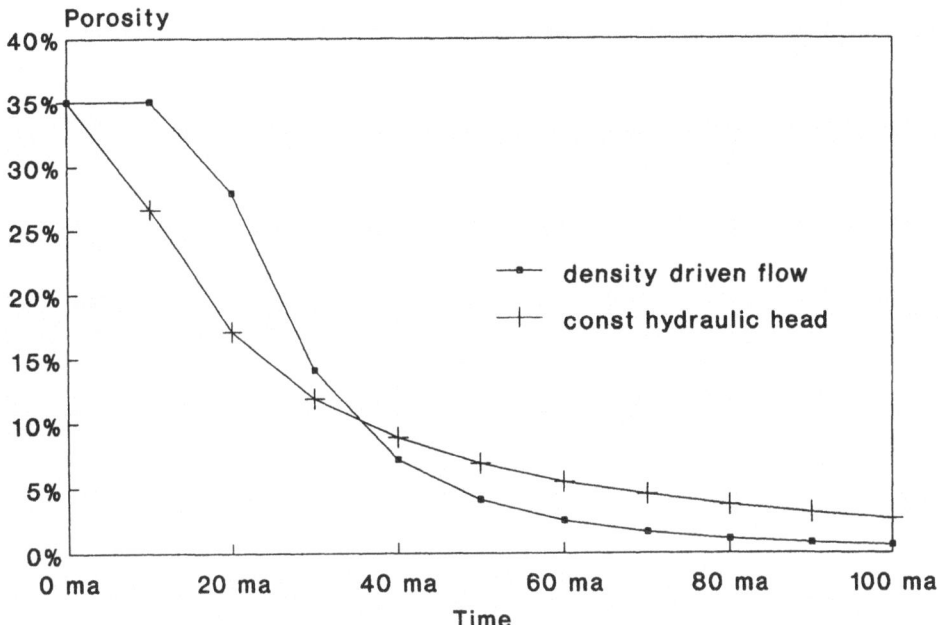

Figure 7. Time dependent porosity evolution in sandstone layer at point 155 [km] for two different flow mechanisms. See text for discussion.

It also can be observed (Fig. 6) that the nonlinear effect of the density driven flux decreases and the porosity pattern approaches the patterns related to a flux derived from a constant hydraulic head. The system evolves from a solubility dominated one into a flow dominated one.

The Effect of Diffusion. In the derivation of the model we neglected the diffusion of the dissolved species. Figure 9 illustrates a purely diffusive transport which has been simulated using realistic diffusion constants of the order $3*10^{-6}$ cm²/s at 20° C (Giles, 1987) and unrealistically high diffusion coefficients of $1.5*20^{-2}$ cm²/s. In both situations the effect on the porosity evolution is negligible being in the order of 0.001 % of the original porosity. This result illustrates that diffusion can be neglected in terms of this model.

Nevertheless, the patterns shown in Figure 9 are interesting in the general framework of our concept, at least qualitatively. Although Figure 9A represents again a solubility dominated system we enter the domain of a thermo-diffusive system in Figure 9B which potentially may become of interest for soluble substances of high diffusibility. The temperature field in the confined sandstone layer causes a field of equilibrium solution

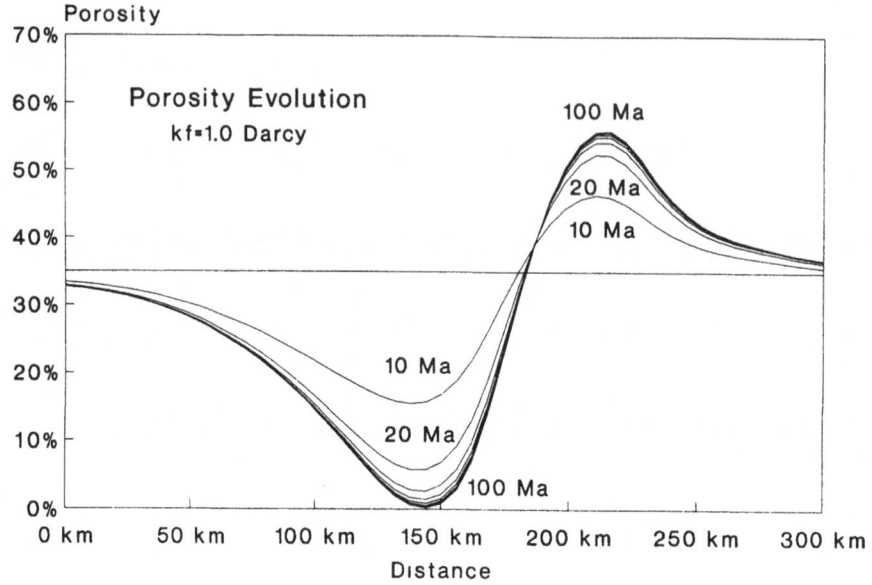

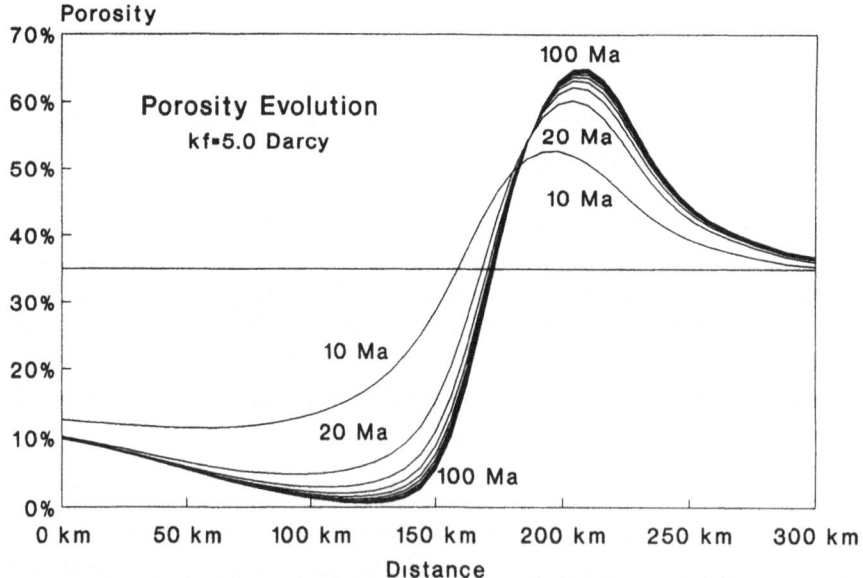

Figure 8. Variations in flux magnitude are simulated by different initial
permeabilities (Fig. 8A: 1 Darcy; Fig. 8B: 5 Darcy) of sandstone layer. Principal
patterns of porosity evolution remain same, only extremes widen and cementation
proceeds faster.

almost identical with Figure 9A and, therefore diffusive fluxes occur to both sides of the maximum concentration. However, as diffusion proceeds towards the shallower parts of the sandstone the temperature and, therefore, the equilibrium concentration decreases. This causes precipitation of the excess material in solution. After the formation of a nucleus the cementation front grows into both directions of the nucleus. On a temporal sequence the transition zone between dissolution and cementation is characterized by dissolution followed by subsequent cementation, thus creating a local diagenetic sequence.

Except for the scale, which is several magnitudes smaller in diffusion dominated systems, the thermo-diffusive effects are similar to the flow dominated systems as would be expected for a wider class of gradient driven systems. Because diffusion has been approximated by a standard finite difference method, we may take these results as a verification of our flow model.

Structural discontinuities. In our model we introduced a strict coupling between subsidence and temperature to account for the general geologic experience. Consequently we expect, that the model reacts sensitively to variations of the geometry of the evolving basin. We shall elucidate this by a simple modification of our original subsidence model given in Figure 4. The smooth approximation of the geometry is broken into linear pieces as illustrated in Figure 10A. The total number of grid points used for the calculation exceeds the number of key points (7) used for the definition of the basin geometry. The additional grid points are defined by linear interpolation. The modified subsidence model therefore is smooth in terms of the depth location and the resulting temperature between the key points, with pronounced discontinuities at the key points. The resulting porosity evolution is illustrated in Figure 10B. The geometric discontinuities are repeated as sharp breaks in the porosity evolution, whereas the porosity evolves more smoothly between the discontinuities. The smoothness of the porosity functions depends only on the number of excess computation points and is approximated by a function of the type $\phi(x) = a(1 - e - \text{ß}x)$ on the straight intervals which can be related to a stationary state derived from the principle of mass conservation (De Groot, 1960). It is obvious, that discontinuities in the temperature field and its gradients will influence the local diagenetic pattern. A similar statement holds for structural discontinuities as far as they translate into disturbances of the temperature field within the sandstone layer. For simulation and modeling purposes we conclude that the reconstruction of the evolving basin geometry is an essential, if not the most important factor controlling the resulting diagenetic patterns.

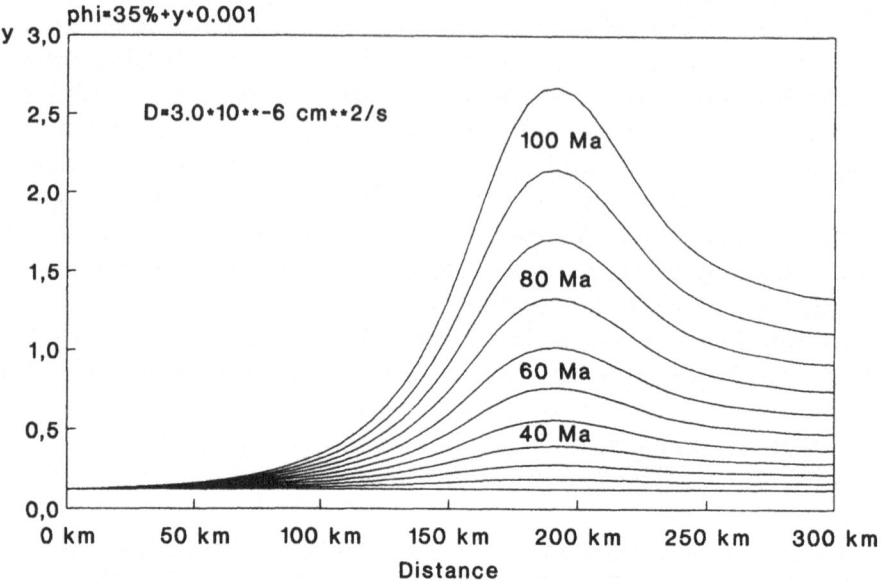

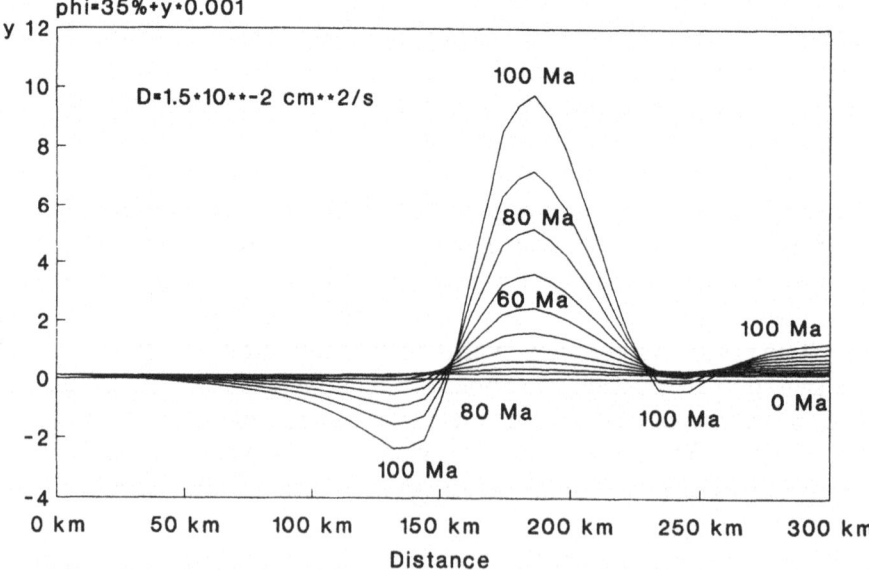

Figure 9. Porosity development of purely thermo-diffusive system using realistic diffusion coefficient for Figure 9A and unrealistic high coefficient for Figure 9B. At unrealistic high diffusion rates (Fig. 9B) dissolution and cementation resemble qualitatively previously discussed patterns derived from directed flux. However, cementation occurs on both sides of thermally induced concentration source. Note that changes derived from diffusion are quantitatively negligible compared to those associated with fluid flow. Y-axis is strongly exaggerated to make results visible ($\varnothing$ = 35% ± 0.01% -0.001%).

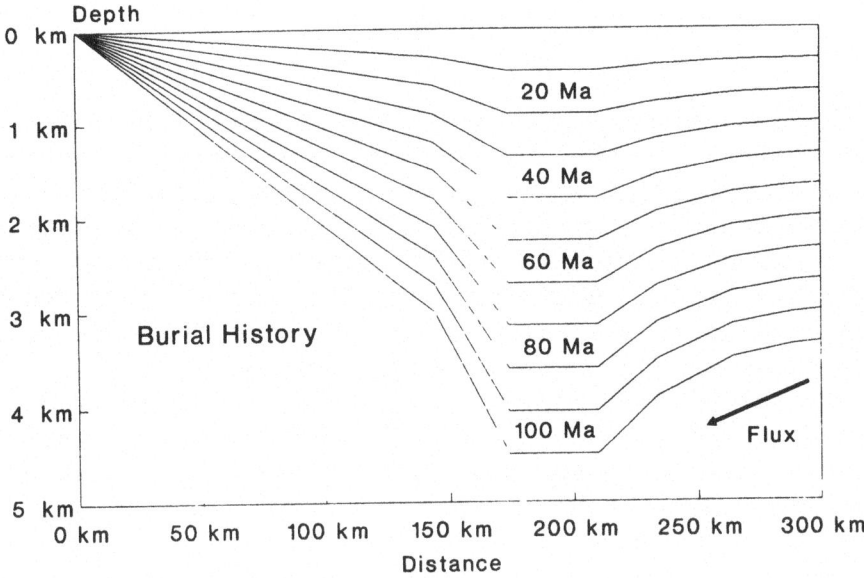

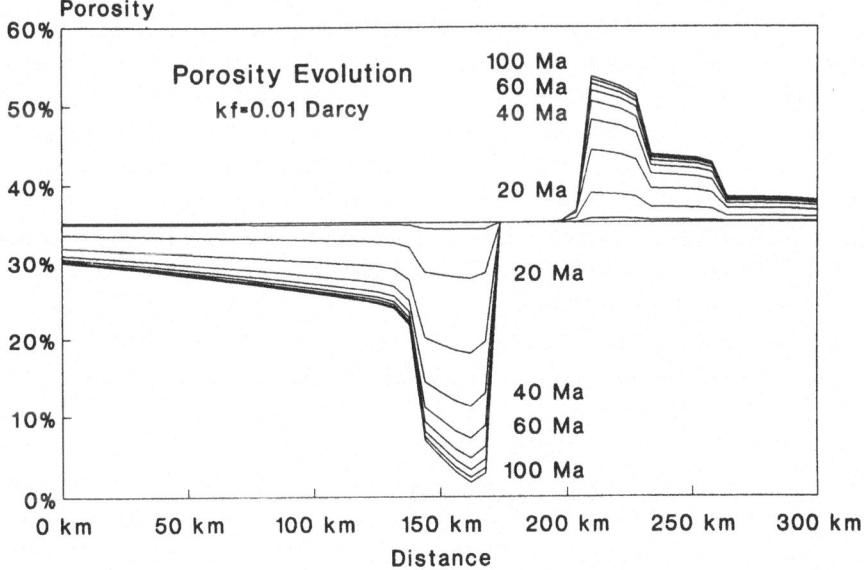

Figure 10. Modified geometric approximation of burial history shown in Figure 4. Geometry is defined by few key points, additional collagation points are defined by linear interpolation between key points. Discontinuities of temperature gradient in sandstone layer cause discontinuities in porosity evolution of the layer.

CONCLUSIONS

Based on the modified transport-reaction equation, we developed an algorithm for the numerical simulation of dissolution and precipitation in basin modeling. The algorithm was tested for quartz solubility on a simple, theoretical geologic model. With this model the sensitivity of the different variables controlling the system was investigated, that is the temperature gradient within the layer, the geometry of the basin, diffusive transport versus transport by fluid flow, flow velocity, the porosity evolution, and the associated permeability.

The regional porosity evolution as a result of quartz precipitation or dissolution is most sensitive to variations of the temperature gradient within the permeable layer. In a stationary geothermal temperature field, the variations of the temperature gradient along the layer are determined by the depth of burial of this layer. For this reason the geometric approximation of the subsiding layer will influence the computed distribution pattern of secondary porosity and cementation. For the numerical simulation of the porosity evolution due to quartz dissolution or precipitation it is most important to have precise knowledge of the burial and temperature history of the basin. Only when these variables are known, can the simulation produce reliable results. The model also can be used for a completely different purpose. Observed diagenetic patterns of temperature sensitive reactions may be used to reconstruct past temperature gradients within a layer. In this way other models for the temperature calculations can be compared against the results of this model. A subsequent combination of the results of the diagenetic simulation model and the different temperature indicators from organic geochemistry would allow a thorough interpretation of all available geological information.

The mechanism of fluid flow, which is one of the most important factors in this system, was not examined in depth in our model. In our simulation we defined simply a fluid flow from the right to the left of the basin, which is driven by density differences similar to thermal convection. As the only modification, we studied a fluid flow resulting from a constant hydraulic head. The differences in the results of either assumption are minor. Of greater importance is the direction and the magnitude of the flow. The magnitude of the flow will determine how fast a front system will evolve, and whether it will form at all. The flow direction determines the temperature evolution of the fluid moving through the sandstone layer, and therefore, the location of dissolution or precipitation. Beside the geometry of the basin and temperature, the flow direction is therefore of vital importance for the numerical simulation.

For more sophisticated numerical simulation of diagenetic processes it will be necessary to include other mineral reactions in the model, in order to account for the usually more complex mineral composition of reservoir rocks. But even then simplifications must be done for the modeling, for example it will not be possible to consider all minor species. In addition a better chemical knowledge of the wide variety of diagenetic processes is necessary.

ACKNOWLEDGMENTS

We want to thank D.H. Welte, H.S. Poelchau, and D. Baker for their helpful discussions.

REFERENCES

Alexander, G.B., Heston, W.M., and Iler, R.K., 1954, The solubility of amorphous silica in water: Jour. Phys. Chem., v. 58, no. 6, p. 453-455.

Baas-Becking, L.G., Kaplan, I.R., and Moore, D., 1960, Limits of the natural environment in terms of pH and oxidation-reduction potentials: Jour. Geology, v. 68, no. 3, p. 243-248.

Berner, R.A., 1980, Early diagenesis - a theoretical approach: Princeton University Press, Princeton, 241 p.

Bjorlykke, K., 1983, Diagenetic reactions in sandstones, in Parker A., and Sellwood, B.W., ed., Sediment diagenesis: NATO ASI Series C: Mathematical and Physical Sciences v. 115, Reidel Publ. Co., Dordrecht, p. 169-213.

Blatt, H., Middleton, G., and Murray, R., 1982, Origin of sedimentary rocks: Prentice Hall, Englewood Cliffs, New Jersey, 782 p.

Collins, A.G., 1975, Geochemistry of oilfield waters: Elsevier Sci. Publ. Co., Amsterdam, 496 p.

Crerar, D.A., and Anderson, G.M., 1971, Solubility and solvation reactions of quartz in dilute hydrothermal systems: Chem. Geology, v. 8, no. 1/2, p. 107-122.

De Groot, S.R., 1960, Thermodynamik irreversibler Prozesse: Bibliographisches Institut, Mannheim, 216 p.

Engelhardt, W. v., 1960, Der Porenraum der Sedimente: Springer Verlag, Berlin, 207 p.

Fang, J.H., Visscher, P.B., and Davis A.M.J., 1988, Porosity reduction in sandstone by quartz overgrowth: discussion: Am. Assoc. Petroleum Geologists Bull., v. 72, no. 12, p. 1515-1517.

Fournier, R.O., and R.W.II, Potter, 1982, An equation correlating the solubility of quartz in water from 25 to 900°C at pressures up to 10,000 bars: Geochim. Cosmochim. Acta, v. 46, no. 10, p. 1969-1973.

Füchtbauer, H., 1978, Zur Herkunft des Quarzzements. Abschätzung der Quarzauflösung in Silt- und Sandsteinen: Geol. Rundschau, Bd. 67, Heft. 3, p. 991-1008.

Füchtbauer, H., 1979, Die Sandsteine im Spiegel der neueren Literatur: Geol. Rundschau, Bd. 68, Heft. 3, p. 1125-1151.

Giles, M.R., 1987, Mass transfer and problems of secondary porosity creation in deeply buried hydrocarbon reservoirs: Marine and Petroleum Geology, v. 4, no. 3, p. 188-204.

Hanor, J.S., 1988, Porosity reduction in sandstone by quartz overgrowth: discussion: Am. Assoc. Petroleum Geologists Bull., v. 72, no. 12, p. 1518-1519.

Leder, F., and Park W.C., 1986, Porosity reduction in sandstone by quartz overgrowth: Am. Assoc. Petroleum Geologists Bull., v. 70, no. 11, p. 1713-1728.

Lichtner, P.C., 1988, The quasi-stationary state approximation to coupled mass transport and fluid-rock interaction in a porous medium: Geochim. Cosmochim. Acta, v. 52, no. 1, p. 143-165.

Morey, G.W., Fournier, R.O., and Rowe J.J., 1962, The solubility of quartz in water in the temperature interval from 25 to 300°C: Geochim. Cosmochim. Acta, v. 26, no. 10, p. 1029-1043.

Rhyzenko, B.N., 1967, Determination of hydrolysis of sodium silicate and calculation of dissociation constants of orthosilicic acid at elevated temperatures: Geochem. Intern., v. 4, no. 1, p. 99-107.

Siever, R., 1962, Silica solubility, 0-200°C, and the diagenesis of siliceous sediments: Jour. Geology, v. 70, no. 2, p. 127-150.

van Lier, J.A., De Bruyn, P.L., and Overbeek J.Th., 1965, The solubility of quartz: Jour. Phys. Chem., v. 64, no. 12, p. 1675-1682.

Walther, J.V., and Helgeson H.C., 1977, Calculation of the thermodynamic properties of aqueous silica and the solubility of quartz and its polymorphs at high pressures and temperatures: Am. Jour. Sci., v. 277, no. 12, p. 1315-1351.

Wood, J.R., and Hewett T.A., 1982, Fluid convection and mass transfer in porous sandstones - a theoretical model: Geochim. Cosmochim. Acta, v. 46, no. 10, p. 1707-1713.

THE USE OF FISSION TRACK
MEASUREMENTS IN BASIN MODELING

Peter Klint Jensen
RisR National Laboratory, Denmark

Torben Bidstrup
Geological Survey of Denmark, Denmark

Kirsten Hansen
Institute for Petrology, Copenhagen University, Denmark

Helmar Kunzendorf
RisR National Laboratory, Denmark

ABSTRACT

Forward modeling of fission track annealing of apatites is used together with basin modeling to estimate past heat flow and amount of erosion. For a known uranium concentration the track length histogram and the surface track density, hold information on the thermal history of the minerals. An equation is developed which is used to determine the number of horizontal (relative to prismatic faces) tracks in the histogram that has been generated after deposition. The post-sedimentary thermal history of the sample is calculated by a 1D-basin model with an initial estimate on the paleoheat flow. The thermal history then is used to calculate the track length histogram including distribution of track lengths caused by distribution of energy of fission products and anisot-

Computerized Basin Analysis, Edited by J. Harff
and D.F. Merriam, Plenum Press, New York, 1993

ropy. Adjustments of the paleoheat flow or the erosion depth are performed to obtain a better match between the calculated and measured track length histograms (post-sedimentary stage). The modeling is applied to the Danish North Sea well D-1. Fission track measurements on apatite are available at three depths.

INTRODUCTION

During the last decade basin modeling has become an important tool in hydrocarbon exploration. The ultimate goal is to predict hydrocarbon accumulations. The basic elements of basin modeling are calculation of the history of (1) subsidence of sedimentary sequences, (2) formation temperature, (3) source rock maturity, (4) generation of hydrocarbons, and (5) migration and accumulation of hydrocarbons. General principles of basin modeling were discussed in, for example, Yükler, Cornford, and Welte (1978) and Ungerer and others (1984). The necessary data for calculation of the subsidence history are formation thickness, age datings, and data for compaction. The paleoformation temperature is calculated from knowledge of paleosurface temperature, paleoheat flow, and thermal properties of rocks. Hydrocarbon generation is simulated on the basis of the paleoformation temperature and a kinetic model for degradation of kerogen. The paleosurface temperature may be estimated from measurements of oxygen isotopes (Buchardt, 1978). An estimate on the paleoheat flow is obtained from geodynamic models (e.g., Mckenzie, 1978; Vejbæk, 1989) and from thermal indicators such as vitrinite reflectance (Middleton, 1982), biomarkers (Lerche, 1988), and fission tracks in apatite (Wagner, 1968; Gleadow, Duddy, and Lovering, 1983; Huntsberger and Lerche, 1987; Green and others, 1989, Bertagnolli, Keil, and Pahl, 1983; Duddy, Green, and Laslett, 1988).

In this paper paleoheat flow is estimated by forward modeling of fission tracks combined with basin modeling. Determination of the paleoheat flow from vitrinite reflectance and biomarkers is discussed also to stress the differences and analogies relatively to the fission track interpretations. Advantages of forward modeling of fission tracks also are discussed.

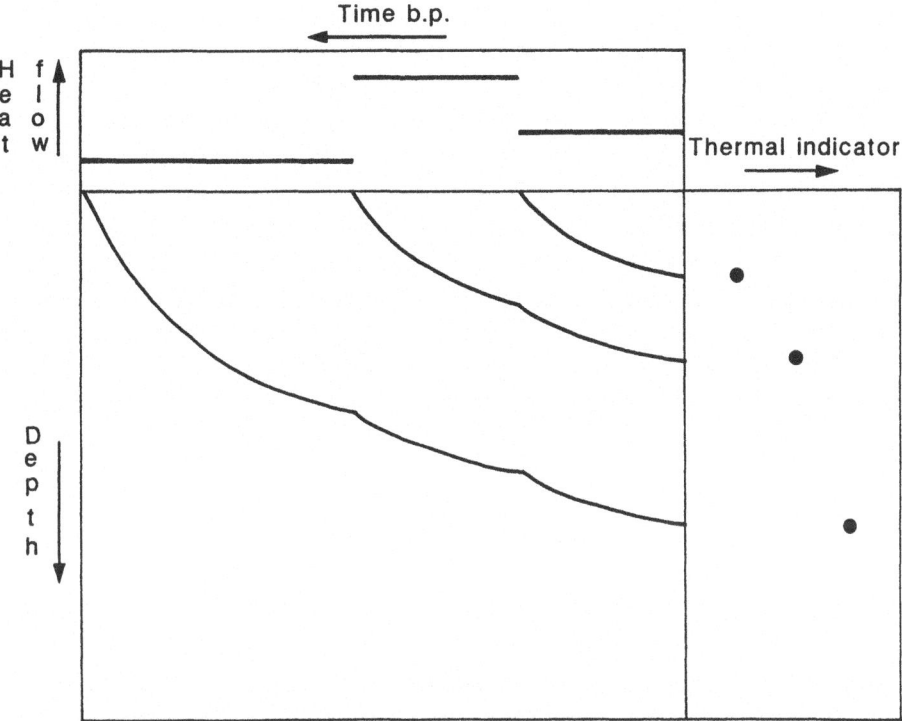

Figure 1. Illustration of relationship between transformation of thermal indicators and determination of paleoheat flow in basin modeling.

PALEOHEAT FLOW FROM VITRINITE
REFLECTANCE AND BIOMARKERS

Vitrinite reflectance and biomarker transformation values are measured down the well as illustrated in Figure 1. The subsidence for each of the measured samples are calculated by basin modeling. The paleoheat flow is assumed to be a step function. The thermal indicators are transformed as an integral function of temperature and time. Having a model for the transformation, the heat flow can be estimated by iterative adjustments until the calculated thermal indicator values are in accordance with the measured values.

The iteration starts with the most shallow measurement and the most recent step of the heat flow function. The burial and the temperature history for the particular sample is calculated by basin modeling. The most recent step of the paleoheat flow is adjusted to obtain a match between simulated and measured values. The iteration continues with the next deeper measurement and the next older heat flow step. The process stops with the deepest measured thermal indicator. Estimation of paleoheat flow from thermal indicators such as vitrinite reflectance and biomarkers unfortunately is highly uncertain resulting from (1) the uncertainty of measurements or (2) migration, and (3) unreliable kinetic alteration models. These problems may lead to unrealistic estimates of the past heat flow. A more practical solution is to start the inversion with an initial estimate of the past heat flow based on a geodynamic model. This estimate thereafter is modified by manual adjustments, having in mind the given idealized inversion principles. Lerche (1988) used tomography for automatic determination of a simple preassumed paleoheat flow function.

MODELING OF FISSION TRACKS

Fission tracks are linear crystal defects caused by the spontaneous fission of ^{238}U (Silk and Barnes, 1959; Young, 1958). The length of the tracks are shortened as a function of temperature and time (e.g. Wagner, 1968; Gleadow, Duddy, and Lovering, 1983).The potential tracks are oriented randomly and are less than 18 μm long and with a diameter of around 50 A in apatite. Tracks which are connected to the mineral surface, are made visible under a light microscope by etching with hydrofluoric acid. The length distribution of the wholly included near horizontal tracks parallel to prismatic faces then are measured in a number of apatite grains. Long tracks are more likely to cut a path to the surface, and being etched, than short tracks. On the contrary, short tracks are more likely to be accepted as being near horizontal. These two biases more or less cancel each other (Jensen, Hansen, and Kunzendorf, 1991). To obtain the paleotemperature from the horizontal track length distribution additional information about the age of the oldest (shortest) track must be obtained either from different radiometric or sedimento-logical methods or from surface track density and uranium concentration (Jensen, Hansen, and Kunzendorf, 1991). The two methods are discussed in the following.

(1) The age of the oldest (shortest) track of the histogram is esti-mated, for example for a tuff where the minerals have cooled rapidly at the moment of sedimentation. The age is determined from radiometric or paleontological methods. Assumptions on the age of the oldest track is

necessary in the forward modeling examples of Green and others (1989) and Huntsberger and Lerche (1987).

(2) Knowing the surface track density and the concentration of ^{238}U the histogram column heights can be converted to time according to Jensen, Hansen, and Kunzendorf (1990). An age of the oldest track (and any other track) then can be calculated. Keil, Pahl, and Bertagnolli (1987) used the frequency density of the horizontal projection of surface tracks together with the uranium concentration to convert to time.

Using age determinations as described in (1) limits the number of examples which can be be studied. For instance, the age of the sedimentation in general does not coincide with the age of the oldest track. The oldest tracks thus may have been generated before sedimentation of reworked apatites, or the oldest track may have been generated after deposition due to annealing after the sedimentation. These problems can be overcome by using the principle described in (2). In this paper, apatites with tracks older than the deposition age are used.

FORWARD MODELING OF FISSION TRACKS

Because basin modeling only is concerned with the time after deposition, the subset of the measured tracks generated within this period, has to be identified. In Jensen, Hansen, and Kunzendorf (1991) it is shown that the number of tracks in each of the histogram columns are related to time, namely the time it takes to generate the tracks belonging to each of the columns. The number of measured tracks in a column is dependent on the amount of apatite included in the measurements, as well as on the uranium content. The equation relating the randomly oriented tracks to time is (Jensen, Hansen, and Kunzendorf, 1991):

$$\Delta t_j = \frac{\omega_j}{\left(\lambda_f * c_{238}\right)} \qquad (1)$$

where Δt_j is the time it takes to generate the tracks belonging to column j of ω_j the track length distribution density (unit: track / cm^3), λ_f is the decay constant of fission of ^{238}U, and c_{238} is the concentration (unit: number of atoms/cm^3).

The uranium concentration c_{238} is determined by measuring the density of induced surface tracks in a mica close to the apatite samples and close to a glass standard with known uranium concentration.

The uranium concentration c_{238} (number of atoms/cm³) is

$$c_{238} = \frac{U_a \, v_a}{M_u} \tag{2}$$

where U_a is the uranium concentration (ppm), v_a is the density of apatite, and M_u is the atomic mass of uranium. ω_j in Equation (1) is related to the surface track density (track/cm²) of the measured histogram:

$$\omega_j = \frac{2 * \rho_s * (n_h)_j}{\sum \left((n_h)_j * (l_m)_j \right)} \tag{3}$$

where ρ_s is the surface density (track/cm²) of spontaneous tracks, $(n_h)_j$ is the measured number of near horizontal tracks in the length interval $[(l_m)_j - \Delta L/2 , (l_m)_j + \Delta L/2]$. $(l_m)_j$ is the mean length corresponding to column no. j. Equations (1) and (2) are used to calculate the time equivalent of the track length histogram (Fig. 2).

If t_k is the time of generation of the oldest track in column no. k, then

$$t_k = \sum_{j=1}^{k} t_t \tag{4}$$

where Δt_j represents the time intervals equivalent to the histogram columns $(n_h)_j$. The number of tracks n_d generated after deposition is related to the number of tracks

$\sum_{j=1}^{k} (n_h)_j$ generated because the oldest track in column no. k

as the time t_d to t_k:

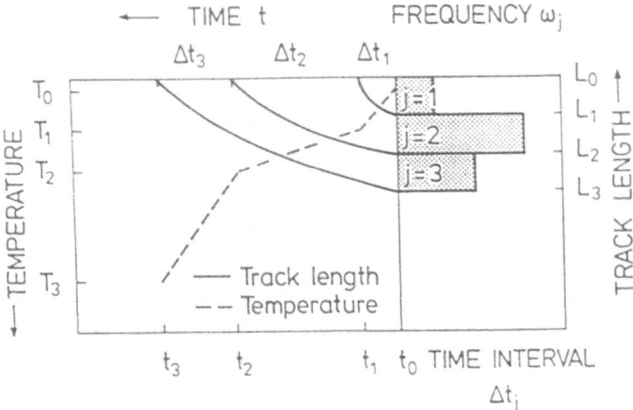

Figure 2. Track length histogram (right) is related to time intervals. Piece by piece linear paleotemperature is determined for each histogram column by iterations until calculated number of tracks in column is equal to measured value. Figure is taken from Jensen, Hansen, and Kunzendorf (1990).

$$\frac{n_d}{\sum_{j=1}^{k} (n_h)_j} = \frac{t_d}{t_k}$$ (5)

where t_d is the time of deposition. Combining the Equations (1) - (5) leads to an expression for the number of horizontal tracks generated after deposition :

$$n_d = t_d \frac{\lambda_f \, c_{238} \sum \left((n_h)_j * (l_m)_j * (l_m)_j \right)}{2\rho_s}$$ (6)

The forward modeling program calculates the reduction in mean confined track length as a function of temperature and time using the annealing law introduced by Green and others (1989). The track length distribution is calculated given the deposition time t_d, and the number of tracks generated after deposition n_d [Eq. (6)], and the thermal history of the sample. For simplicity the tracks are assumed to be generated at equal time intervals.

The discussed annealing model does not include the initial distribution of track lengths, and it is assumed that the initial track length is constant and the later annealing is isotropic. However, anisotropy is important in nature (Laslett and others, 1982). Further, a range also is observed through the measurements of the length histogram of freshly induced tracks. Convolving the simulated histogram of initially constant length and isotropic tracks with the histogram of the measured induced fresh tracks includes the range of track lengths. As an example the measured histogram of tracks from Fish Canyon (e.g., Hansen and others, 1991) is simulated leading to a match assuming a sample temperature equal to 17° C for a period of 30 million years. The fact that the initial track lengths are a distribution indicates that the relation between the number of tracks occurring in a histogram column of the measured tracks can not be related strictly to time (Eq. 1). However, the equation can be used to calculate the number of tracks generated in a given time interval and therefore the number of tracks generated after the deposition can be calculated.

A basin model is run to give an estimate on the thermal history of the sample. For this purpose the initial heat flow history is estimated, for example on the basis of geodynamics. The calculated temperature history from basin modeling for the sample then is used as input for the forward modeling program simulating the corresponding track length histogram. This is compared with the measured histogram and adjustments of the input paleoheat flow are made until there is reasonable agreement with the observations. The match is obtained starting with the longest tracks (most recent time). The tracks generated before sedimentation are not matched in this study.

CASE STUDY

The Danish well D-1 is used as a calculation example. The well is situated in the western part of the Norwegian - Danish Basin close to the margin. The location is not too far from the Central Trough and therefore a preveous estimate of the heat flow history of the region is used as a first estimate. This is based on calibrations with vitrinite reflectance using the Yükler 1D basin model.

The well penetrates a small salt structure, and reaches the Lower Permian. The stratigraphic section is given (Fig. 3): a Pleistocene section of 200 m consists of mainly clay, sand and gravel; a 950 m Tertiary section of gray shale and sand (part of the Tertiary is missing as a result of

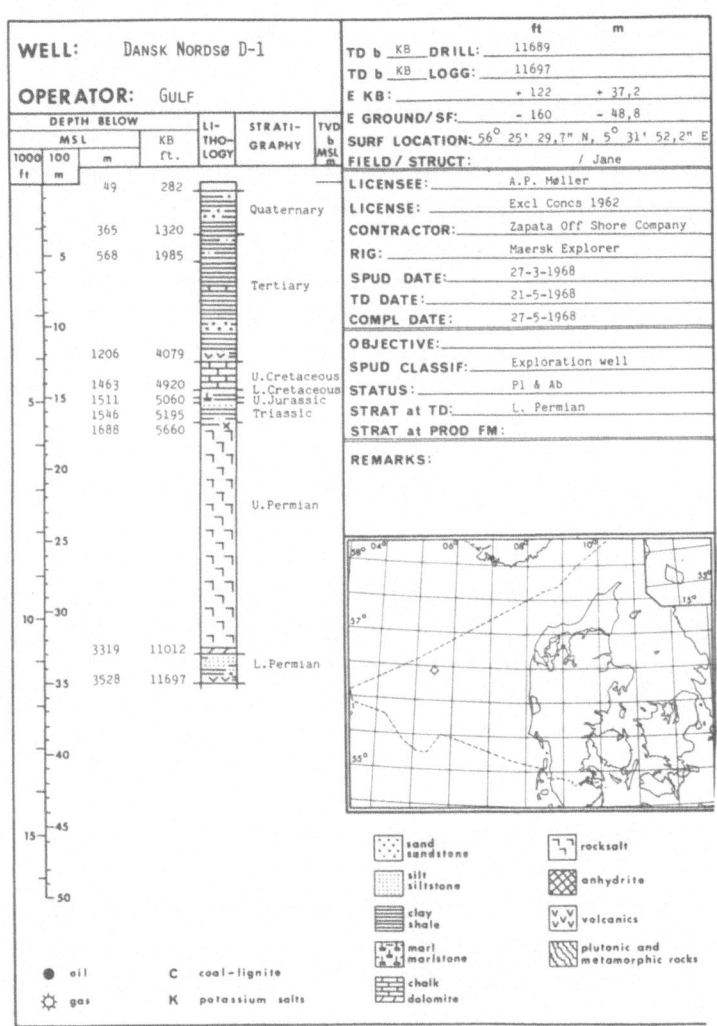

Figure 3. Location of D-1 well and stratigraphy. Figure is modified after Geological Survey of Denmark (1981).

faulting); a Danian to Upper Cretaceous section of 260 m thickness consists of mainly carbonate sediments (limestone, chalk); a Lower Cretaceous to Upper Jurassic section of 80 m is composed of gray shale and sand and a Lower Triassic section of 140 m of red shale, sand, and evaporates seismic data confirm that a marked erosion has taken place between the Lower Triassic and Upper Jurassic; a Zechstein section of 1600 m consists of salt and anhydrite with thin carbonates (the present-day Zechstein section is thicker than originally due to salt flow); a pre-Zechstein section of 200 m of shale and sand ends in volcanics.

The basin model used is the Yükler 1D model (Yükler, Cornford, and Welte, 1978). The most important input data are described here. The stratigraphic column is divided into a number of events characterized by deposition or erosion. Each event has a certain thickness (negative if it is erosion) and duration. The events are characterized further by a lithotype. The lithotypes are used in the program for calculating permeabilities, heat conductivities, and compaction coefficients. The paleowater depth is the water depth at deposition. The paleosurface temperature is the temperature at the sediment/water interface. The paleoheat flow, Figures 4 and 5, is obtained after repeated simulations of the fission track histogram.

The fission tracks measured were from three samples taken at depths of 1000 m, 1615 m (above the salt), and 3322 m (below the salt). Preparation of samples and the measuring technique and results are discussed in Hansen and others (1991). The measured histogram is shown in Figures 6 to 8 together with the calculated histograms.

The following data are used in the Equations (1)-(6):

$$M_u = 3.95 \ 10^{-25} \text{ kg}, v_a = 3.2 \ 10^3 \text{ kg m}^{-3}, \lambda_f = 8.57 \ 10^{-17} \text{yr}^{-1}, \quad L = 0.805 \ \mu\text{m}.$$

Pertinent values for the three samples are given in Table 1.

The best histogram match is obtained with the paleoheat flow shown in Figures 4 and 5. Only the right part of the histogram, corresponding to the post-sedimentation stage, are matched. The heat flow for the samples above the salt structure is increased in agreement with the focussing effect of the salt (Jensen, 1990). The Lower Cretaceous - Upper Jurassic erosion is assumed to be 550 m based on seismic interpretations in the area. With only one sample beneath the erosion surface it is not possible to determine both erosion thickness and heat flow at the same time. The heat flow within this period therefore has to be assumed. A good match of the simulated and measured histograms obtained with the erosion thickness equal to 550 m. In Figure 9 an example is shown with

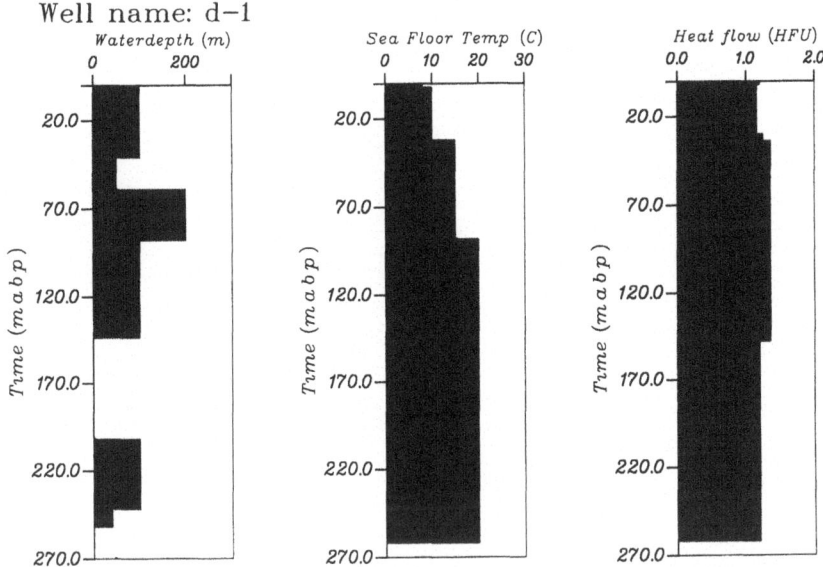

Figure 4. Assumed water depth and seafloor temperature together with calculated paleoheat flow determined by measured fission tracks shown in Figures 6 and 7.

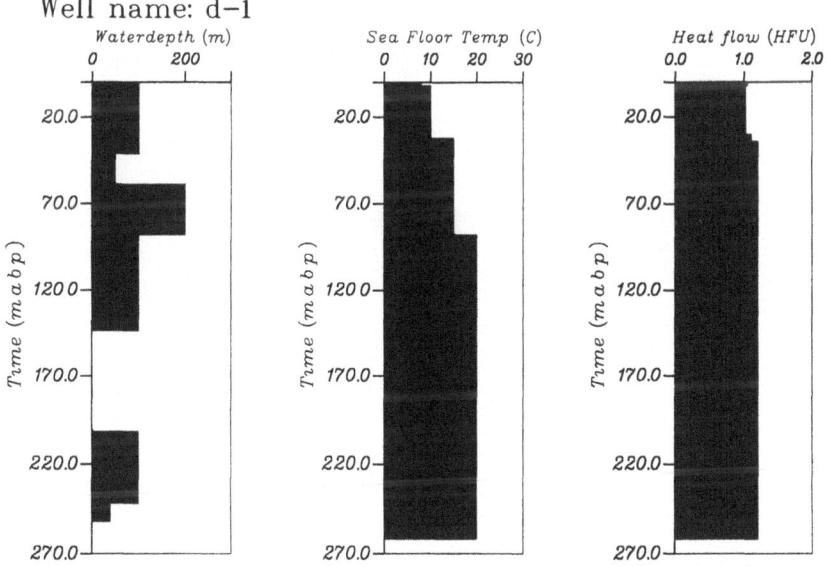

Figure 5. Assumed water depth and seafloor temperature together with calculated paleoheat flow determined by measured fission tracks shown in Figure 8.

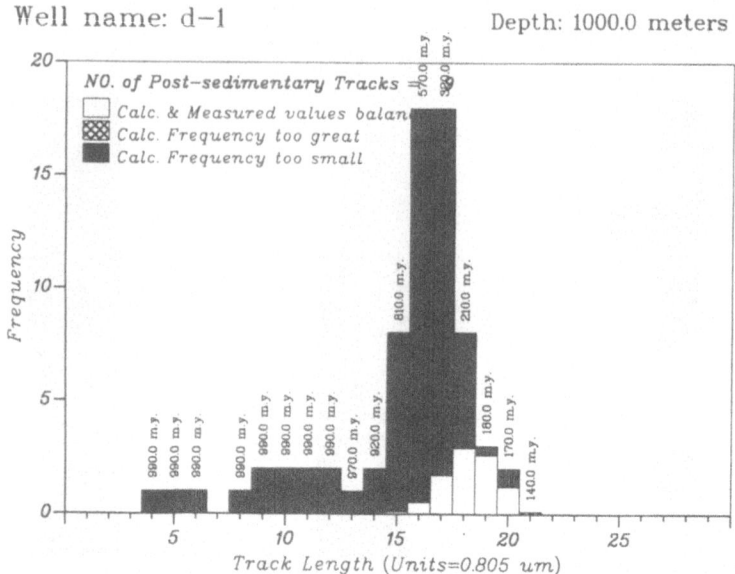

Figure 6. Calculated and measured fission track histograms.Most of measured tracks are generated before sedimentation. Only tracks generated after deposition are simulated.

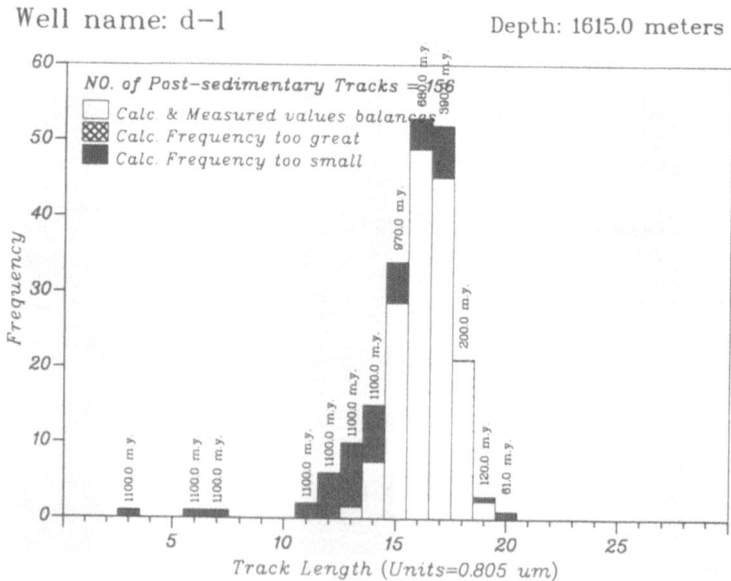

Figure 7. Calculated and measured fission track histograms. Samples are taken above salt structure. Match is obtained using heat flow history shown in Figure 4.

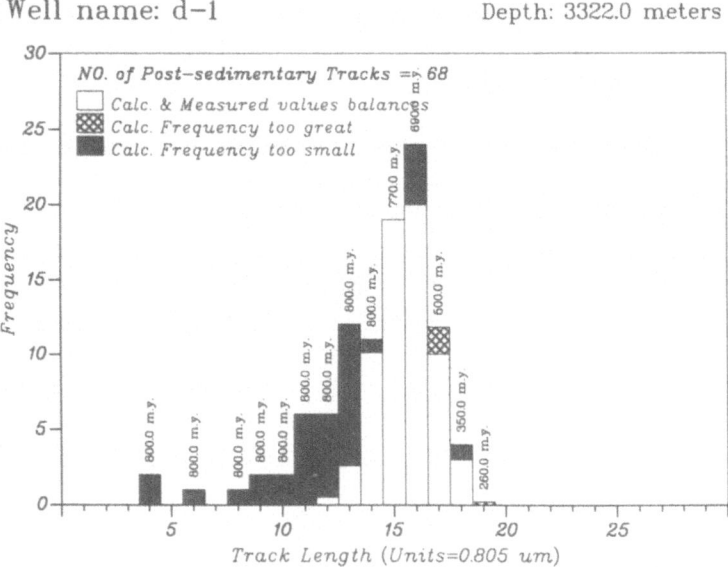

Figure 8. Calculated and measured fission track histograms. Samples are taken below salt structure. Match is obtained using paleoheat flow shown in Figure 5. Corresponding thermal history calculated by basin modeling is shown in Figure 9.

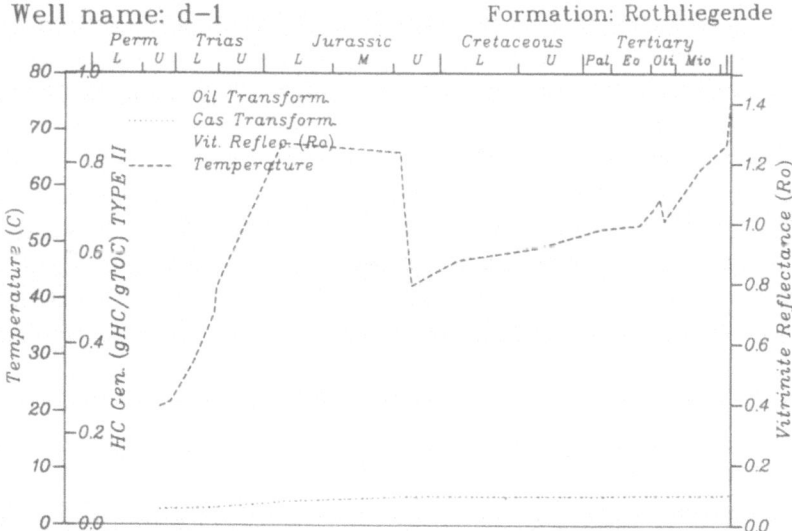

Figure 9. Calculated temperature history for apatite sample taken below salt structure. Temperature history is calculated by Yükler 1D basin model and is determined by comparison of calculated and measured fission track length histogram. Also shown is calculated maturity (vitrinite reflectance) and potential hydrocarbon generation.

the calculated temperature history, vitrinite reflectance, and hydrocarbon generation (for type II kerogen). It is seen that the present-day values represent an unmature situation.

Table 1. Constants used for calculation of number of tracks generated after deposition [Eq. (5)] for three samples from D-1 well. The measured surface track densities are from Hansen (1990).

Constant	Sample Depth (m)		
	1000	1615	3322
U_a (ppm)	7.35	11.31	7.19
ρ_s (track/cm^2)	$7.8\ 10^5$	$14.24\ 10^5$	$9.95\ 10^5$
t_d (mill. yr)	32	232	261

DISCUSSION AND CONCLUSIONS

In this study the range of tracks caused by distribution of initial fission energy, anisotropy of initial length distribution and etching rate has been taken into account by convolving the calculated histogram of mean track lengths with the measured histogram of fresh induced tracks. The theoretical foundation for Equation (6) assumes isotropy of tracks. However, measurements have shown that angular anisotropy is only pronounced for mean track lengths less than about 11 μm (Green and others, 1986). In the same paper it is shown that the standard diviation of confined track length distribution show no significant variation for mean track length longer than about 11 μm. Therefore it is reasonable to use the same filter for all calculated mean track lengths greater than about 11 μm.

An additional problem is the effect of chlorine on the stability of fission tracks in apatites (Green and others, 1985, 1986) which has not been taken into account.

Green and others (1989) showed that the information about the thermal history of a single sample of apatites is best stored in the track histogram during constant uplift, whereas the information is stored only partly during burial. Thus sampling at several depths of the well is nescessary to obtain further information, and the thermal history may be delineated in the same way as for vitrinite reflectance and biomarkers.

The track length unit of the histogram is 0.805 μm which for practical purposes is the division size of the scale bar inserted into the ocular.Further, the number of tracks included in the histogram is limited by the number of suitable apatite grains available and by the fact that only the near horizontal tracks are selected. One track count in the histogram, Figure 8, represents about 4 million years. Because some of the histogram columns contain a larger number of tracks the time resolution is less than this, in the worst situation being about 100 million year. This clearly is unsatisfactory for basin modeling. In general, the information level of a single sample of apatite may be higher than for a single sample of vitrinite or biomarkers. Fission track measurements therefore are an important supplement to other thermal indicators.

Nevertheless the fission track measurements did provide valuable information on the thermal history. The geological model used for the basin modeling of the D-1 well is rather simple, because the time resolution of the fission track method is rather limited. Thus, the post depositional history is only represented by 7 columns in the histogram on Figure 8. The salt flow was not modeled. The model gave a better fit with the measured data using a higher heat flow above the salt structure than beneath it. The seismic evidence of an erosion of 550 m is in agreement with the modeling of the fission tracks.

ACKNOWLEDGMENT

This study has been financed by the Danish Ministry of Energy (EFP-88) and RisR National Laboratory. Thanks are given to Anders Mathiesen for programming the plotting code used to plot Figures 4 to 9.

REFERENCES

Bertagnolli, E., Keil, R., and Pahl, M.,1983, Thermal history and length distribution of fission tracks in apatite, Part 1: Nucl. Tracks., v. 7, no. 4., p. 163-177.

Buchardt, B., 1978, Oxygen isotope palaeotemperatures from Tertiary Period in the North Sea areas: Nature, v. 275, no.5675, p. 121-123.

Duddy, I.R., Green, P.F., and Laslett, G.M.,1988, Thermal annealing of fission tracks in apatite 3. Variable temperature behavior: Chemical Geology, v.73, no. 1, p. 25-38.

Geological Survey of Denmark,1981, Well Data Summary Sheets, volume 1: Geological Survey Denmark, 166 p.

Gleadow, A.J.W., Duddy, I.R., and Lovering, J.F.,1983, Fission track analysis: a new tool for the evaluation of thermal histories and hydrocarbon potential: Aust. Pet. Eng. Assoc. Jour., v. 23, p. 93-102.

Green, P.F., Duddy, I.R., Gleadow, A.J.W., Tingate, P.R., and Laslett, G.M.,1985, Fission track annealing in apatite: track length measurements and the form of the Arrhenius plot: Nucl. Tracks, v. 10, no.3, p. 323-328.

Green, P.F., Duddy, I.R., Gleadow, A.J.W., Tingate, P.R., and Laslett, G.M.,1986, Thermal annealing of fission tracks in apatite, 1. a qualitative description: Chemical Geology, v. 59, no. 4, p. 237-253.

Green, P.F., Duddy, I.R., Laslett, G.M., Hegarty, K.A., Gleadow, A.J.W., and Lovering, J.F., 1989, Thermal annealing of fission tracks in apatite 4. Quantitative modelling technique and extension to geological timescales: Chemical Geology, v. 79, no. 4, p. 155-182.

Hansen, K., 1990, The Fennoscandian border zone; Thermal and tectonic history of a tuffaceous sandstone from Bornholm: submitted to Terra Nova.

Huntsberger, T.L., and Lerche, I., 1987, Determination of paleoheat flux from fission scar tracks in apatite: Jour. Petroleum Geology, v. 10, no. 4, p. 365-394.

Jensen, P.K., 1990, Analysis of the temperature field around salt diapirs: Geothermics, v. 19, no. 3, p.273-283.

Jensen, P.K., Hansen, K., and Kunzendorf, H., 1990, A numerical model for the thermal history of rocks based on confined horizontal fission tracks: submitted to Nucl. Tracks.

Keil, R., Pahl, M., and Bertagnolli, E., 1987, Thermal history and length distribution of fission tracks: Part II: Nucl. Tracks Radiat. Meas., v. 13, no.1., p. 25-34.

Laslett, G.M., Kendall, W.S., Gleadow, A.J.W., and Duddy, I.R., 1982, Bias in measurement of fission-track length distributions: Nucl. Tracks, v. 6, no. 213, p.79-85.

Lerche, I.,1988, Inversion of multiple thermal indicators: Quantitative methods of determining paleoheat flux and geological parameters. I. Theoretical development for paleo heat flux: Jour. Math. Geology, v. 20, no.1, p. 1-36.

Mckenzie, D.,1978, Some remarks on the development of sedimentary basins: Earth and Planetary Sci. Lett., v. 40, no. 1, p.25-32.

Middleton, M.F.,1982, The subsidence and thermal history of the Bass Basin, South Eastern Australia: Tectonophysics, v. 87, no. 1-4, p. 383-397.

Silk, E.C.H., and Barnes, R.S., 1959, Examination of fission fragment tracks with an electron microscope: Phil. Mag. v.4, p. 970-971.

Ungerer, P., Bessis, F., Chenet, P.Y., Durand, B., Nogaret, E., Chiarelli, A., Oudin, J.L., and Perren, J.F., 1984, Geological and geochemical models in oil exploration; principles and practical examples: Am. Assoc. Petroleum Geologists Bull., v. 35, no. 1, p. 53-77.

Vejbæk, O.V.,1989, Effects on asthenospheric heat flow in basin modelling exemplified with the Danish Basin: Earth Planet. Sci. Lett., v. 95, no. 1-2, p. 97-114.

Wagner, G.A., 1968, Fission track dating of apatites: Earth Planet. Sci. Lett., v. 4, no. 5, p. 411-415.

Young, D.A.,1958, Etching radiation damage in lithium fluoride: Nature, v. 182, no.4632, p. 375-377.

Yükler, M.A., Cornford, C., and Welte, D., 1978, One dimensional model to simulate geologic hydrodynamic and thermodynamic development of a sedimentary basin: Geol. Rundschau, B. 67, Heft 3, p. 960-979.

MASS-BALANCED RECONSTRUCTION OF PALEOGEOLOGY

William W. Hay and Christopher N. Wold
Cooperative Institute for Research in Environmental Sciences (CIRES)
University of Colorado, Boulder, CO

ABSTRACT

Mass-balanced paleogeographic reconstruction can be extended into reconstruction of the paleogeology, using information inherent in the mass/age distribution of the sediments.

The reconstruction method removes sediments from their site of deposition and replaces them as overburden onto the sites from which they were eroded. The mass/age and spatial distribution of the sediments contain the information required to determine the age and lithology of the rocks that made up the overburden. Replacing the overburden onto the source areas in terms of age and lithology defines the geologic history of a region in much greater detail than previously thought possible.

INTRODUCTION

Mass-balanced paleogeographic reconstruction can be extended from a simple replacement of sedimentary mass of unknown age and lithology onto source areas to reconstruction of the paleogeology. The information necessary to reconstruct the paleogeology is inherent in the mass/age distribution of the sediments. It is requisite that the region under consideration forms a closed system with respect to the detrital sediments, but it is recognized that it is open with respect to biogenic or chemically precipitated sediments. Knowledge of the paleogeology is essential to the paleogeographic reconstruction processes, because the potential source areas are defined as those areas that were not sites of deposition during a given interval. In the technique for constructing mass-balanced paleogeographic maps described by Hay, Shaw, and Wold (1989), the geologist making the reconstructions must make knowledge-able guesses of the age of the materials being replaced. The technique suggested here uses the information inherent in the mass/age distribution to suggest the age and lithology of the materials replaced. It constructs a paleogeologic framework consistent with the geologic history of the region interpreted from the sedimentary rocks existing today. Because of space limitations, the discussion and examples given here are restricted to replacement of the sediments by age. The replacement of sediments by both age and lithology is discussed at the end of the paper.

DEFINING THE PROBLEM

In another paper in this volume, Wold and others (in press) described how the mass-balanced reconstruction of paleotopography requires determination of the amount of overburden eroded from each site. The nature of the overburden was not defined in terms of age or lithology. The problem now becomes one of determining the age and lithologies of the sediment filling the overburden space at each site, that is, the age and nature of the sedimentary materials before their erosion. To simplify the discussion and the example presented here we consider only a single "lithology," detrital sediment. The key to replacement by age lies in knowing the amount of sediment of each age eroded to yield the mass of sediment being replaced.

DETERMINING ANCIENT REGIONAL
SEDIMENT FLUXES

A simple method for estimating ancient sediment fluxes has been described briefly by Hay, Shaw, and Wold (1989), Hay and Wold (1990), and Wold and others (in press) in this volume. The technique has been described in detail by Wold and Hay (1990). It yields estimates of ancient flux rates that closely approximate the real, ancient flux rates, assuming a closed system in which all younger sediments form from cannibalistic destruction of older sediments.

An alternative method, which gives exact answers under the assumption of a constant sedimentary mass, involves calculating the value of the proportional decay parameter, b, for each time increment. This method also makes use of the fact that the mass/age distribution of sediment in a closed system has the form of an exponential decay

$$y = Ae^{bt} \tag{1}$$

where y is the mass of sedimentary rock of age t existing at present, A is the 0-age intercept of the decay curve (the 0-age mass flux), and b is the proportion of the total sedimentary mass recycled during each age increment.

Using the terminology of Hay, Shaw, and Wold (1989), Hay and Wold, (1990), and Wold and Hay (1990), and assuming a closed system, the mass of sediment deposited during the interval from age i to age j that would be in existence if there were no fluctuations in the rate of sediment cycling, $M^{\#}_{ij}$, is

$$M^{\#}_{ij} = M^{\#}_{p}\, e^{b(i+j)/2} \tag{2}$$

where $M^{\#}_{p}$ is the average long-term sediment flux (the intercept of the exponential decay curve with the 0-age axis), b is the proportion of the sediment mass recycled per unit time, and $(i + j)/2$ is the age of the mid-point of the time interval. However, mass/age distributions invariably indicate that the rate of sediment cycling has differed with time. This indicates that b and $M^{\#}_{p}$ are in fact not constants, but covary with time, being related through the total sedimentary mass, TSM.

$$TSM = \int_{0}^{t_{max}} A\, e^{(bt)dt} \tag{3}$$

over the length of the mass/age distribution 0 to t_{max}. From this, it follows that the proportional decay constant, b, for the youngest interval, 0 to k, is

$$b = \frac{M^{*}_{0k}}{TSM} \tag{4}$$

where M^{*0k} is the sediment mass for the time interval from the present, 0, to age k. By definition it is both the amount of sediment deposited and the sediment flux during the interval from age 0 to age k. TSM = Total Sedimentary Mass, the total of all masses in the mass/age distribution. The estimate of the amount of sediment deposited during some interval i to j that existed at the end of that interval, M^{*}_{ij}, can be determined only sequentially, working backwards one interval at a time from the present. The process is iterative and hence much more time consuming than the method described by Hay, Shaw, and Wold (1989), Hay and Wold (1990), and Wold and Hay (1990). A new mass/age distribution is generated with each iteration, with the mass in each interval increased according to the inverse of the exponential decay. However, this method requires that the number of intervals in the distribution (t_{max}) remain constant. To achieve this, the difference between TSM and the sum of all younger masses is assigned to a new oldest interval created with each iteration.

Figure 1 shows the present mass/age distribution for the northwestern Gulf of Mexico and its present drainage basin. The data have been normalized from classic geologic units of unequal length (i.e. Miocene, Early Cretaceous, etc.) to 10 my increments as described in Wold and Hay (1990). Figure 1 also shows the normalized mass/age distribution for the same area for 10 Ma. The mass/age distribution for 10 Ma assumes that the value of b for the interval 0-10 Ma is the ratio of the mass of sediment of age 0-10 Ma, 3.11×10^{21}g to the total sedimentary mass, TSM, 53.97×10^{21}g, that is $b_{0-10\ Ma} = 0.0576$. Because of the large mass of Pliocene/Pleistocene sediment delivered to the northwestern Gulf of Mexico margin the value of b for this interval is larger than the average value of b calculated for 10 my increments for the entire Phanerozoic (0.0017). Figure 1 also shows the amount of sediment contributed by older strata to produce the mass of sediment that has accumulated during the past 10 my. Figure 2 is a detailed histogram of the amounts of older sediment contributed to the 0-10 Ma sediment mass.

The mass/age distribution is critical to the paleotopographic and paleogeologic reconstructions. Errors in the mass/age distribution of sediment will result in erroneous interpretation of ancient conditions. However, because the model requires internal consistency, the interpretation of ancient conditions should conform to the knowledge available for making the reconstructions. If the mass/age distribution is incorrect, the paleotopographic and paleogeologic reconstructions will be consistent with it, but incorrect.

10^{21} g/10 my

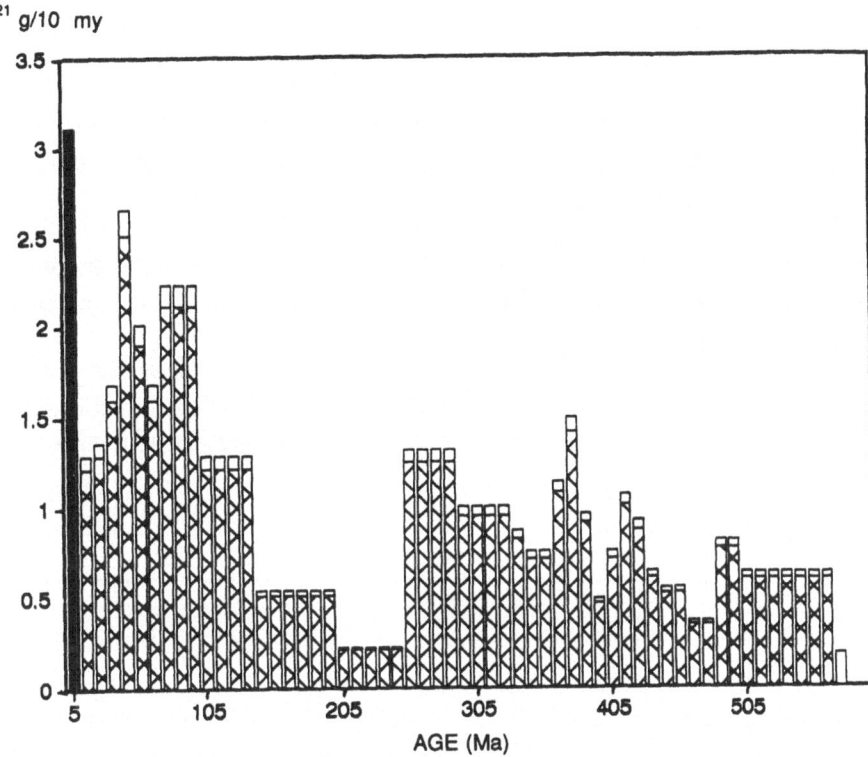

Figure 1. Mass/age distribution for northwestern Gulf of Mexico and its present drainage basin, normalized to 10 my increments. Solid bar represents mass that has accumulated in last 10 million years (0-10 Ma). Bars with cross hatching represent older masses of Phanerozoic sediment. Open bars represent masses of sediment eroded from each of older increments to form 0-10 Ma sediment mass. Total height of bars from 10-570 Ma represents mass/age distribution as it would have appeared at 10 Ma. Open bar at right end of diagram represents mass of 0-10 Ma sediment derived from erosion of Precambrian rocks. Mass/age distribution for 10 Ma assumes that value of b for interval 0-10 Ma is ratio of mass of sediment of age 0-10 Ma, 3.11 x 10^{21}g to total sedimentary mass, TSM, 53.97 x 10^{21}g, that is, b0-10 Ma = 0.0576.

REDISTRIBUTING THE SEDIMENT ONTO THE SOURCE AREA BY AGE

The contribution of material of each older age to the 0-10 Ma sediment mass is known (Fig. 2). The redistribution of this material onto the source area by age requires that the surficial geology be converted from its usual form of presentation as units of classic geologic age to a numerical geological map that shows the distribution of sediments in terms of time increments of equal length corresponding to those of the normalized mass/age distribution. Figure 3 shows a numerical geologic map of the northwestern Gulf of Mexico and its present drainage basin in terms of 10 my increments for the past 60 my and a single category for all older strata. It should be noted that usually there is a problem in preparing a numerical map of the present-day surficial geology because most geologic maps do not show Pleistocene glacial deposits even though they may be thick.

Backstripping the sediment younger than 10 Ma shows the potential source areas that may have contributed to the 0-10 Ma sediment mass (Fig. 4). This is a numerical subcrop map. It is closely related to the classic

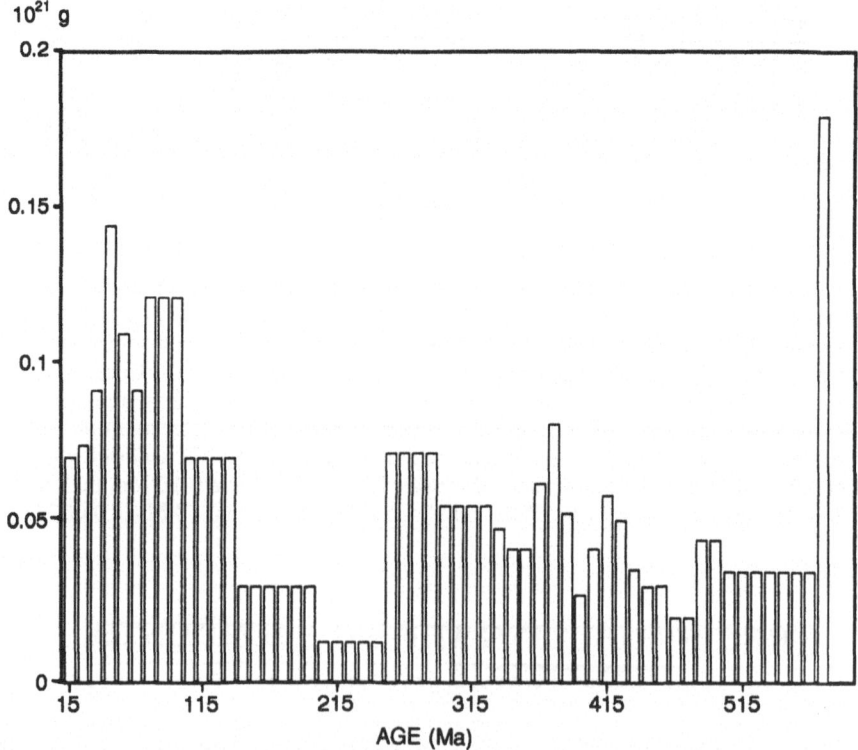

Figure 2. Masses of sediment contributed by each older 10 Ma stratigraphic increment to 0-10 Ma   sediment mass. Bar at right end of distribution is mass contributed from Precambrian rocks.

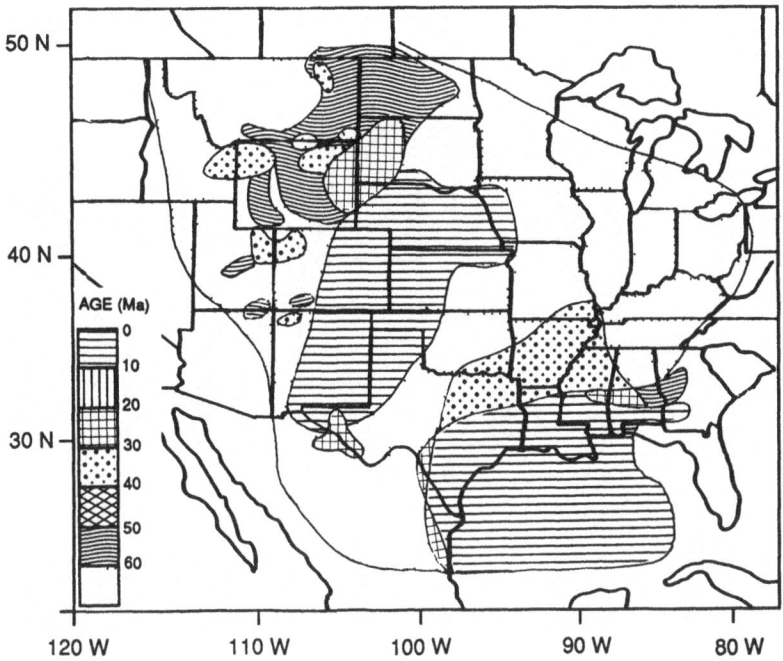

Figure 3. Numerical geologic map of northwestern Gulf of Mexico and its present drainage basin in terms of 10 my increments for past 60 my and single category for all older strata. Following convention for geologic maps of North America, Pleistocene glacial deposits of Midwest are not shown.

paleogeologic map of Levorsen and Berry (1967) except that the backstripping usually was done by removing all the material above an unconformity rather than that above a time horizon. The paleogeology revealed by backstripping forms part of our paleogeologic map, but the paleogeology of the source area must be added to complete the map.

In this volume (Wold and others, in press) we described how the paleotopographic base map for 10 Ma was generated by replacing sediment without considering the age or lithology of the material that had been eroded; in effect we constructed a generic paleotopographic map, reproduced here as Figure 5. The topography was generated by replacing the detrital sediment proportionally according to the ratio between the potential erosion volume of each grid square and the total potential erosion volume of all source squares. The elevations then were adjusted proportionally so that the mass of sediment eroded was equal to the mass

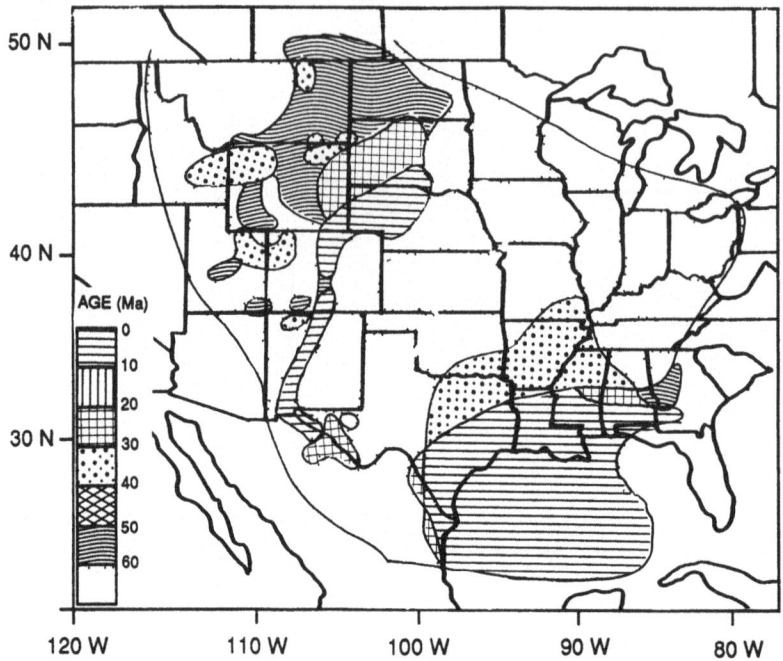

Figure 4. Numerical subcrop map showing ages of rocks below 10 Ma surface. Map is obtained by stripping away all strata having age of 10 Ma or less.

deposited. This process yielded a generic mass-balanced topography and the correct thicknesses of generic overburden. The thickness of generic overburden can be termed overburden space. The problem now is to determine the age of the sediment filling the overburden space at each site, that is, the age of the sedimentary materials before they were eroded. The key lies in knowing the amount of sediment of each age eroded to yield the younger sediment mass.

The detrital sediment is replaced on the source area sites by age following the inverse of the equation for detrital erosion. Replacement starts with the oldest sediment that was eroded. All areas that were not sites of deposition during the 0-10 Ma interval and that lie above the erosional baselevel of + 200 m are potential source areas. Next we select the subset of potential source areas covered by sediment equal in age to, or older than, those to be replaced. For example, Figure 3 shows that there is a mass of 0.179 x 1021 g sediment of age > 570 Ma (Precambrian) to be

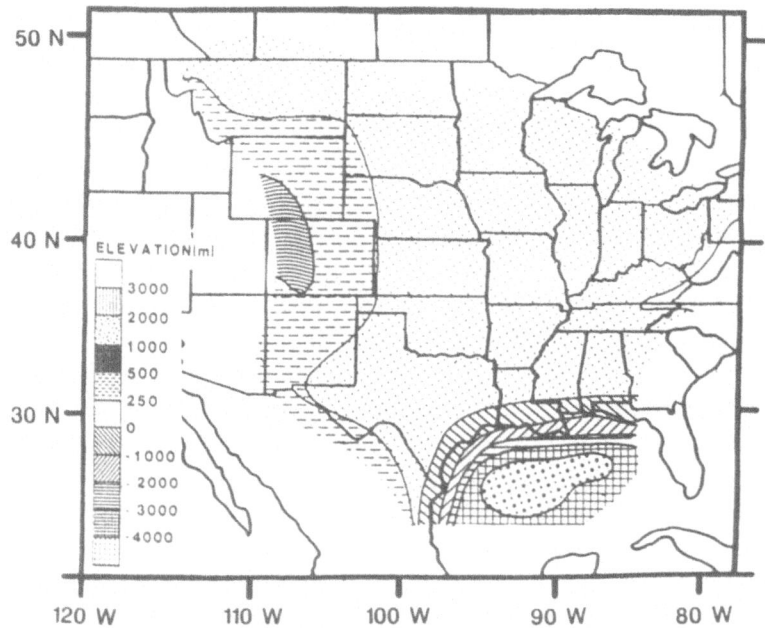

Figure 5. Mass-balanced paleogeographic reconstruction of northwestern Gulf of Mexico and its present drainage basin in interior of North America for 10 Ma (from Wold and others, in press).

spread over areas not covered by Phanerozoic sediments. However, for purposes of simplification, we lumped all sedimentary material older than 60 Ma into one category. There is 2.69×10^{21}g to be distributed over all of the grid squares not covered by sediment younger than 60 Ma and having an elevation > 200 m. This mass is distributed by multiplying the area of each potential source area grid square in this subset by its average elevation, giving an erosion potential volume. As noted in this volume (Wold and others, in press), the erosion potential volume is a direct measure of the relative mass of sediment eroded because we assume that erosion is a linear function of area and elevation. The erosion potential volumes are summed and the mass of detrital sediment to be replaced then is apportioned to each grid square in the ratio of its erosion potential volume to the sum of all erosion potential volumes. It turns out that the sediment mass of age > 60 Ma to be replaced is almost exactly the same as that required to fill the overburden space on the source squares covered by sediment or rock older than 60 Ma. If the sediment mass of age > 60

Ma had been smaller than that required to fill the overburden space, it would indicate that at 10 Ma some of the area where strata older than 60 Ma are exposed presently was covered by sediment having an age between 10 and 60 Ma. If the sediment mass of age > 60 Ma had been larger than the mass of overburden required, it would mean that at 10 Ma some of the area where strata older than 60 Ma are presently exposed was higher than our generic topographic reconstruction had indicated. To correct this, the elevation of the area having grid squares with sediments > 60 Ma must be raised by the amount required to achieve mass balance between the amounts eroded and deposited.

Next, the mass of 0.11×10^{21} g of 50-60 Ma sediment eroded into the 0-10 Ma sediment mass must be replaced onto those grid squares not covered by sediment younger than 50 Ma. Although this includes all the grid squares with sediments older than 60 Ma, the materials are replaced first onto those squares covered by sediment having an age of 50-60 Ma. This is done because these areas are the most obvious sources for sediment of this age. The sediment mass is apportioned according to the ratio of the potential erosion volume of each grid square to the total potential erosion volume of all grid squares. If the sediment mass of age 50-60 Ma is smaller than that required to fill the overburden space, it would indicate that at 10 Ma some of the area where rocks 50-60 Ma are exposed presently was covered by strata having an age between 10 and 50 Ma. If the sediment mass of age 50-60 Ma had been larger than the mass of overburden required, and if there had been a deficit of material of age > 60 Ma replaced earlier, so that there is an unfilled part of the overburden space, the excess 50-60 Ma material could be placed there. It would go on the highest grid squares with material > 60 Ma that are adjacent to grid squares covered by 50-60 Ma sedimentary material. If the sediment mass of age 50-60 Ma had been larger than the mass of overburden required, but there had not been a deficit of material of age > 60 Ma replaced earlier, it would indicate that at 10 Ma some of the area where strata older than 50 Ma are exposed presently was higher than the generic topographic reconstruction had indicated. To correct this, the elevation of the area having grid squares with sediments > 50 Ma must be raised by the amount required to achieve mass balance between erosion and deposition.

The reconstruction proceeds in increments until all of the overburden space is filled with sedimentary material by age and all necessary topographic adjustments have been made.

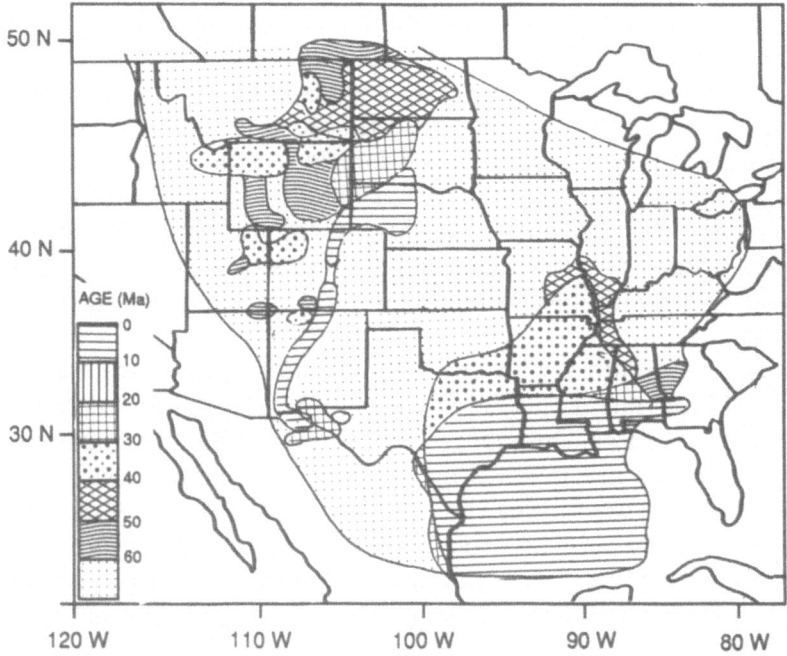

Figure 6. Reconstructed numerical paleogeologic map for 10 Ma. Major difference between this and numerical geologic map for present is in Montana, where sediments having age of 40-50 Ma are indicated to have existed. From overburden map (Fig. 12, Wold and others, in press) it can be seen that these eroded Eocene strata had thickness of about 200 m.

Combining that part of the numerical subcrop map for areas covered by sediment of age 0-10 Ma with the map showing the ages of overburden material replaced on the source areas produces a reconstructed numerical paleogeologic map for 10 Ma (Fig. 6).

REPLACEMENT BY AGE AND LITHOLOGY

More sophisticated replacements of overburden onto the source areas must include consideration of both the age and lithology of the material eroded. The major lithologies of sedimentary rock, sandstone, limestone, and shale, and igneous and metamorphic rocks respond differently to weathering and erosion. In advanced model calculation we have explored the effects of differential erosion rates. Considering the erodability of igneous rock to be 1, we assumed that limestones and sandstones are 3 times as easily eroded, and that shales are 9 times as easily eroded. In replacing the sedimentary material by lithology a significant new problem arises: which lithology(ies) should be replaced on a given site? To start with, lithologies can be replaced on similar lithologies exposed at the surface today, but this cannot be traced far back into the past without considering changing of the age and lithology of the overburden.

Many investigators who have worked on sedimentary cycling (eg. Garrels and Mackenzie, 1971) have concluded that different rock types cycle at different rates. On a large scale this cannot be true, because the sedimentary rocks are layered by age, and the older rocks cannot be eroded without first removing the overlying strata. Thus, once exposed a shale may weather and be eroded more rapidly than a sandstone or limestone. However, it cannot become exposed until the overlying sandstone or limestone has been removed. Lithologic changes act as a rate limiting agent in the system, making all sedimentary rocks recycle at similar rates. For this reason we believe that the method of reconstructing the masses of material of different age eroded to form a young sediment mass described here can be applied more broadly to sediments by major category. Some rocks may change their character as they recycle: arkose, which is texturally a sand may become a mudstone or shale upon erosion and redeposition. Other materials probably do not change significantly: quartz sandstones are recycled as quartz sand; limestones are recycled to form limestone, etc. We believe that by restricting the rock types to major categories it will be possible to reconstruct the overburden by both age and lithology.

Reconstructions of the nature of the overburden could be improved by incorporating more detailed knowledge of the stratigraphy. The overburden space should be filled from outcrop edges, restoring the strata to the thicknesses rocks of that age have in the adjacent subsurface. Such improvements would lead to greater accuracy in the estimates of paleogeopressures, heat flow, and a variety of other parameters.

CONCLUSIONS

Mass-balanced paleogeographic reconstructions remove sediments from their site of deposition and replace them onto their sites of origin.

The simplest version considers only the thickness of the replaced overburden, the overburden space. The mass/age and spatial distribution of the sediments contain the information required to determine the age and lithology of the rocks that filled the overburden space. By replacing the overburden onto the source areas in terms of age and lithology the geologic history of a region can be known in greater detail than previously suspected.

ACKNOWLEDGMENTS

This work has been supported by grants OCE 8409369 and OCE 8716408 from the National Science Foundation, and by gifts from Texaco, Inc.

REFERENCES

Garrels, R.M., and Mackenzie, F.T., 1971, Evolution of sedimentary rocks: Norton, New York, 397 p.

Hay, W.W., and Wold, C.N., 1990, Relation of selected mineral deposits to the mass/age distribution of Phanerozoic sediments: Geol. Rundschau, Bd. 79, Heft 2, p. 495-512.

Hay, W.W., C.N. Wold, C.N., and Shaw, C.A., 1989, Mass-balanced paleogeographic reconstructions: Geol. Rundschau, Bd. 78, Heft 2, p. 207-42.

Levorsen, A. I., and Berry, F.A.F., 1967, Geology of petroleum (2nd ed.): W. H. Freeman, San Francisco, 724 p.

Wold, C.N., C.A. Shaw, DeConto, R.M., and W.W. Hay, in press, Mass-balanced reconstruction of overburden: this volume.

Wold, C.N., and Hay, W.W., 1990, Estimating ancient sediment fluxes: Am. Jour. Sci., v. 290, no. 9, p. 1069-1089.

MASS-BALANCED RECONSTRUCTION OF OVERBURDEN

Christopher N. Wold, Christopher A. Shaw,
Robert M. DeConto, and William W. Hay
Cooperative Institute for Research in Environmental Sciences (CIRES)
University of Colorado, Boulder, CO

ABSTRACT

The mass-balanced paleogeographic reconstructions technique determines the thickness of the rock eroded from the source areas to supply sediment to the sedimentary basins. The thickness of this eroded overburden and the time of its erosion are valuable inputs for calculations of the thermal history of the rocks and the maturation of hydrocarbons.

INTRODUCTION

Paleogeographic maps classically have been 2-D reconstructions of the Earth's surface at a given time in the past. They show the distribution of geographic features, usually the distribution of land and sea, in a latitude-longitude framework. The more elaborate maps have differentiated between coastal plain areas, uplands and mountains, and shelf seas and ocean basins. Recognizing the importance of topography in influencing atmospheric circulation, the hydrologic cycle, and controlling climate, we set about to utilize the information inherent in the regional distribution of sediments to produce 3-D reconstructions of the surface of the Earth in the past, showing the distribution of a variety of properties in a three-dimensional latitude-longitude-elevation framework (Shaw, 1987; Shaw, 1989; Hay, Wold, and Shaw, 1989a, 1989b; Shaw and Hay, 1989). In the

Computerized Basin Analysis, Edited by J. Harff
and D.F. Merriam, Plenum Press, New York, 1993

course of producing paleogeographic reconstructions, a number of parameters unrelated to the description of the surficial geography are calculated; one of these is the thickness of the rock eroded from the source areas to supply sediment to the sedimentary basins. The thickness of this eroded overburden and the time of its erosion are valuable inputs for calculations of the thermal history of the rocks and the maturation of hydrocarbons.

MASS-BALANCED PALEOGEOGRAPHIC RECONSTRUCTIONS

Hay, Wold, and Shaw (1989a, 1989b) have defined a mass-balanced paleogeographic map to be a quantitative reconstruction of the Earth's surface at a moment in the past based on principles of mass-balance. Mass-balance requires conservation of mass; the erosional processes acting on a reconstructed paleogeographic surface through a given interval of time must remove a mass of detrital material equal to the mass of detrital sediment deposited in the adjacent basins. The distinction between detrital and chemical sediments is significant because the mechanical and dissolved transport of material are fundamentally different processes. The movement of detrital material is controlled mainly by gravity, and detritus usually is deposited downhill from its source area. Material carried in solution in water enters the world ocean and may be deposited anywhere. Detrital material is regionally confined, but the distribution of dissolved material is global. The paleogeographic reconstructions described here assume that the region under study is a closed system with respect to detrital sediment and that the erosion of detrital sediment is a function only of elevation. The global sediment mass is about 80% detrital material, so that, except for a few large carbonate platform areas, erosion and sedimentation on a regional scale are driven by the processes affecting detrital material. Hence, most paleogeographic reconstructions based on detrital mass-balance will need only relatively minor correction for the effects of chemical denudation.

Mass-balanced paleogeographic reconstructions cannot be made with a single leap back to an earlier time, but require successive reconstructions at time increments small enough to resolve changes in sediment sources and sinks. The existing distribution of sediments is a fragmentary geologic hologram; although the present distribution of sedimentary materials is only a fraction of its original extent, it contains information about the ancient source areas. In some regions the changing patterns of sediment accumulation recorded by the existing distribution of sedimentary materials contain all the information necessary to reconstruct the changing paleogeography.

DEFINING THE PROBLEM

Assuming that most of the sediment in the region under study is detrital, and that the region is essentially a closed system with respect to detrital sediment, determining the amount of overburden that has been removed during some time interval from a given site within the source area involves answering two questions: (1) what is the total amount of sediment that has been removed regionally? and (2) how much of this total has been removed from a particular site?

DISCUSSION

The answer to the first question, determining the total removed on a regional scale during a given interval, involves compilation of the mass of detrital sediment representing the interval present today in the associated depositional basins and from this to deduce the mass originally deposited. The mass existing today is not the same as the mass originally deposited because some the material has been reeroded and redeposited. The mass/age distribution of the sediments is an abstract of the history of the erosional and depositional history of the region. It contains the information required to reconstruct rates of sedimentary cycling and the ancient sediment fluxes. The answer to the second question, determining how much of the total sediment mass was removed from a particular site, depends on factors controlling differential erosion within a source area. Recent investigations exploring the relation between the suspended load of rivers and elevation, climate and other environmental factors have concluded that the most important control on the suspended load is the elevation of the source area (Hay and others, 1987; Pinet and Souriau, 1988). Accordingly, in the example given here we consider elevation to be the sole criterion for replacement of detrital sediment onto the source area. More sophisticated replacements of overburden onto the source areas can include consideration of the lithology and age of the material eroded, as will be discussed in the next paper (Hay and Wold, in press) in this volume.

SEDIMENT CYCLING

Global Analysis

The global mass/age distribution of sediments shows an exponential decay with age. Gilluly (1969) recognized that the distribution of sedimentary rocks as a function of age is a reflection of the cannibalistic nature of erosion and sedimentation processes. Sedimentary rocks erode more readily than igneous and metamorphic rocks and cover a larger area of the land surface; hence most sediments are derived from the erosion of older sedimentary rocks. The relation between area, volume, or mass of sedimentary rock and age is that of an exponential decay having the general form

$$y = Ae^{bt} \tag{1}$$

where y is the area, volume, or mass of sedimentary rock of age t presently existing, A is the 0-age intercept of the decay curve (in effect, the 0-age flux), and b is the proportion of the total area, volume, or mass recycled during each age increment.

Veizer (1988) suggested that the sedimentary system is 90% ± 5% cannibalistic. Wold and Hay (1990), from analysis of the data of Ronov (1980) and Budyko, Rovov, and Yanshin(1987), suggested that averaged for the Phanerozoic the sedimentary system has been about 90% cannibalistic, with the remaining 10% being mainly a gain of material from the weathering and erosion of volcanics that is balanced almost exactly by the loss of material to subduction. Although these estimates are based on the entire global sediment mass, analysis of only the detrital portion of the global sediment mass yields similar results.

Hay, Wold, and Shaw (1989a), Hay and Wold (1990), and Wold and Hay (1990) discuss reconstructing the masses of sediments that existed in the past and estimating ancient sediment fluxes. The method involves the assumption that the mass/age distribution of sedimentary materials represents a closed system. It is summarized here using their terminology.

An exponential decay curve is fit to the mass/age distribution for existing sediment. The curve is

$$M_{ij}^{\#} = M_p^{\#} \, e^{b(i+j)/2} \tag{2}$$

where M^*_{ij} is the mass of sediment deposited during the interval from age i to age j that would be in existence if there were no fluctuations in the rate of sediment cycling. M^*_p is the average long-term sediment mass flux [the intercept of the exponential decay curve with the 0-age axis, or A in Equation (1)]; b is the proportion of the sediment mass recycled per unit time; and $(i + j)/2$ is the age of the mid-point of the time interval [t in Equation (1)]. The actual mass of sediment existing today, M_{ij}, usually differs from M^*_{ij} because the rate of sediment cycling has changed continuously, fluctuating about a mean value represented by M^*_p and b. The estimate of the amount of sediment deposited during the interval i to j that existed at the end of the interval, M^*_{ij}, is

$$M^{\#}_{ij} = M^{\#}_p \frac{M_{ij}}{M^{\#}_{ij}} \qquad (3)$$

Although this method provides an estimate closely approximating the ancient fluxes of material and can be used as a first approximation in estimating past overburden, it is not exact and is least precise when there are large changes in flux rates.

Regional Analyses

Regional mass/age distributions may be more irregular than the global distribution, because they are responding to local tectonics. If the mass/age distribution is for an area that is a closed system, it usually will have the form of an exponential decay (b is negative). If, however, the area for which the mass/age distribution is compiled does not approximate a closed system, the mass/age distribution may seem to be random or may show no decay with increasing age. If the compilation is solely for a basin it may even have the appearance of an exponential growth curve (b is positive). In the situation of passive margins, their early history of rapid subsidence provides large volumes of accommodation space. The high local relief of the margins of the young ocean provides a ready sediment supply. As the passive margin ages, subsidence rates, volumes of accommodation space, and erosion of the marginal source areas all decline. If the mass/age distribution is compiled for only the former rift basin, it will show preserved sediment masses that increase with age up to the rifting event. Experience in compiling regional mass/age distributions for parts of North America and Europe suggests that source and sink areas change markedly through time. The assumption that a region selected from present geography is a closed system needs to be reexamined as the paleogeographic reconstructions proceed.

Tectonic events, such as mountain building or regional uplift, accelerate the sediment cycling rate. This involves increases in both the amounts of sediment eroded (governed by b), and deposited (M*). As b increases, it implies preferential erosion of young sediment; in effect, an event that results in increased erosion rates selectively destroys sediments that were young at the time (Wold and Hay, 1990). The effect on the mass/age distribution is to produce a large mass of young sediments by selectively reducing the mass of slightly older sediments.

Figure 1 shows the mass/age distribution for the northwestern Gulf of Mexico and its present drainage basin. It also shows an exponential decay curve fit to the data and an estimate of Phanerozoic sediment fluxes, reconstructed using the method of Wold and Hay (1990). The Laramide, Sevier, Ouachita, and older orogenies appear as large pulses of sediment deposition.

Through diversion of drainage, source areas alternately may feed sediment to two or more basins. These effects can be large. Figure 2 shows the mass/age distribution of sediments along the Atlantic margin of the United States and in the offshore abyssal plains. The Appalachian region generally has been regarded as the major source for the sediments that accumulated on the Atlantic margin of the U.S. during the Mesozoic and Cenozoic. However, DeConto, Hay, and Shaw (1989) and Hay, DeConto, and Wold (1990) have suggested that the large mass of sediment that accumulated along the U.S. Atlantic margin during the Cenozoic exceeds the amount that expected from erosion of the Appalachians. The only potential source area for those sediments that has the required area and elevation is the continental interior. The implication is that during much of the Cenozoic the St. Lawrence River drained the interior of North America. Capture of this large sediment source area by the Mississippi River did not occur until the Quaternary. Thus neither the mass/age compilation for the Atlantic margin (Fig. 2) nor the regional mass/age compilation for the northwestern Gulf of Mexico and the continental interior (Fig. 1) represents a closed system during the entire Cenozoic.

An exponential decay curve fit to the regional mass/age distribution can be used to calculate the masses of sediment that existed in the past but have been subsequently eroded to form younger sediment. If the data are so irregular that the fit of a decay curve is suspect, the global decay rate might be assumed for want of any better indication. If the mass/age

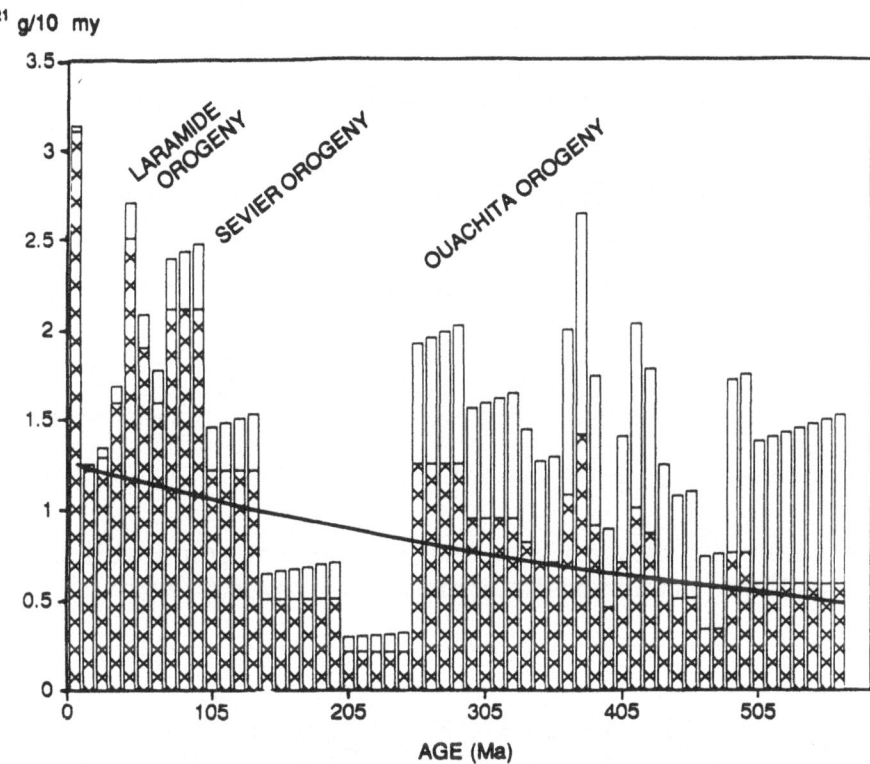

Figure 1. Mass/age distribution of sediments and sedimentary rock in northwestern Gulf of Mexico and its present drainage basin in interior of North America, shown as bars with cross hatching; ages have been normalized to 10 my increments. Exponential decay curve has been fit to data by least-squares method; M^*_p = 1.261 X 1021 g, b = 0.00168. Open bars are estimate of amounts of sediment of that age that once existed but have been destroyed subsequently by erosion, reconstructed using method of Wold and Hay (1990). Laramide, Sevier, Ouachita, and older orogenies appear as large pulses of sediment deposition.

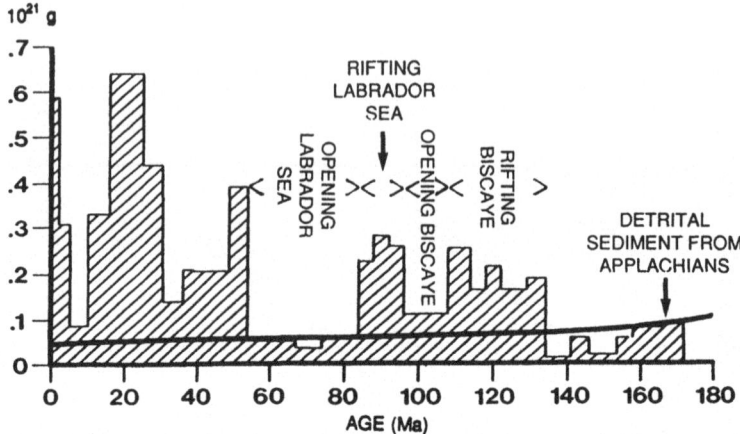

Figure 2. Mass/age distribution of sediments and sedimentary rocks along Atlantic margin of United States and in offshore continental slope, rise and abyssal plains. Line represents amount of sediment expected to be derived from simple topographic decay of Appalachians, assuming that present-day detrital erosion/elevation relation can be applied throughout Cenozoic and Mesozoic.

distribution has the form of an exponential that grows with age, it is obvious that it is an impossibility in the real world.

THE RATE OF EROSION OF DETRITAL SEDIMENT AS A FUNCTION OF ELEVATION

Assuming that the most important factor controlling the suspended load is the elevation of the source area (Hay and others, 1987; Pinet and Souriau, 1988). Hay, Wold, and Shaw (1989a) suggested a simple formula for the relation between the rate of denudation and elevation:

$$\frac{\delta Tde}{\delta t} = (H - Ed_b)\, Ktd_e \qquad (4)$$

where $\delta Tde/\delta t$ is the rate of mechanical denudation expressed in terms of the thickness of detrital sediment eroded per meter of elevation above the detrital erosional base level; per year. H is the elevation of the site; Ed_b

is the elevation of the erosional base level for detritus above sealevel; and Ktd_e is an erosion/elevation constant expressing the rate of denudation in terms of thickness of the solid phase of the detritus eroded per incremental meter of elevation per year. Hay, Wold, and Shaw (1989a) indicated that the global average for Ed_b is 200 m and that for Ktd_e is 0.113×10^{-6} m m$^{-1}_{elev}$ yr^{-1}.

Topographic Decay

If there are no sedimentary materials younger than the age of the youngest desired reconstruction to provide information on the internal history of the source area, no details can be known. However, it may be useful and instructive to explore what the past elevation and sediment fluxes from the source area would have been assuming that its history has been one of simple erosion and isostatic uplift in response to unloading. Further, it is possible to estimate the past elevations, sediment flux rates, and the total amounts of sediment removed from knowledge of the modern detrital sediment erosion rates and assuming that the regional elevation/denudation rate has held constant in the past. From this it is possible to calculate elevations in the past, taking into account isostasy and sealevel changes. The term "simple topographic decay" describes the process by which a region continuously looses elevation to erosion, with the elevation loss partially compensated by isostatic uplift in response to the loss of mass. Figure 3 shows the loss of elevation and the thickness of overburden removed from a region undergoing topographic decay, based on the global average Ktd_e of 0.113×10^{-6} m m$^{-1}_{elev}$ yr^{-1} and Ed_b of 200 m, and mantle and sediment densities of 3300 kg m^{-3} and 2200 kg m^{-3} respectively. Mountains of 10 km elevation are reduced to 5 km in only 9 my and to 1 km in 32 my. With a ratio of the densities of sediment to mantle of 2:3, the thickness of overburden removed is 3 times greater than the isostatically compensated loss of elevation. Analysis in terms of simple topographic decay gives an estimate of the mass of sediment eroded; this can be compared with the amount of sediment observed and calculated from sediment cycling to have been deposited.

One area that has been amenable to analysis in terms of simple topographic decay is the Appalachian region. The Appalachian region contains no sediments younger than Paleozoic, hence there is no direct information about its Mesozoic-Cenozoic history. According to Curtis, Culbertson, and Chase (1973) the present annual suspended sediment load of rivers draining the Appalachian region to the Atlantic is about 12 x 10^{12} g. The area supplying this sediment is 0.637 x 10^6 km²; the present rate of denudation of the Appalachian region is therefore 18.8 g m⁻² yr⁻¹. This is about an order of magnitude less than the present global average denudation rate of 162 g m⁻² yr⁻¹ used by Hay, Wold, and Shaw (1989a), but this is to be expected because the rocks of the Appalachians are well indurated and erosion resistant. For the Appalachians Ktd_e is about 0.024 x 10^{-6} m m⁻¹$_{elev}$ yr⁻¹.

Figure 4 shows the elevation of the Appalachians in the past, and the amount of overburden removed since they formed. The model assumes that the present erosion/elevation relation has existed throughout the later Phanerozoic and that for the isostatic calculation the density of the mantle is 3300 kg m⁻³ and the density of sediments is 2200 kg m⁻³.

Figure 2, the mass/age distribution for sediments on the U.S. Atlantic margin and in the basins of the western North Atlantic, also shows the amount of sediment expected from erosion of the Appala-

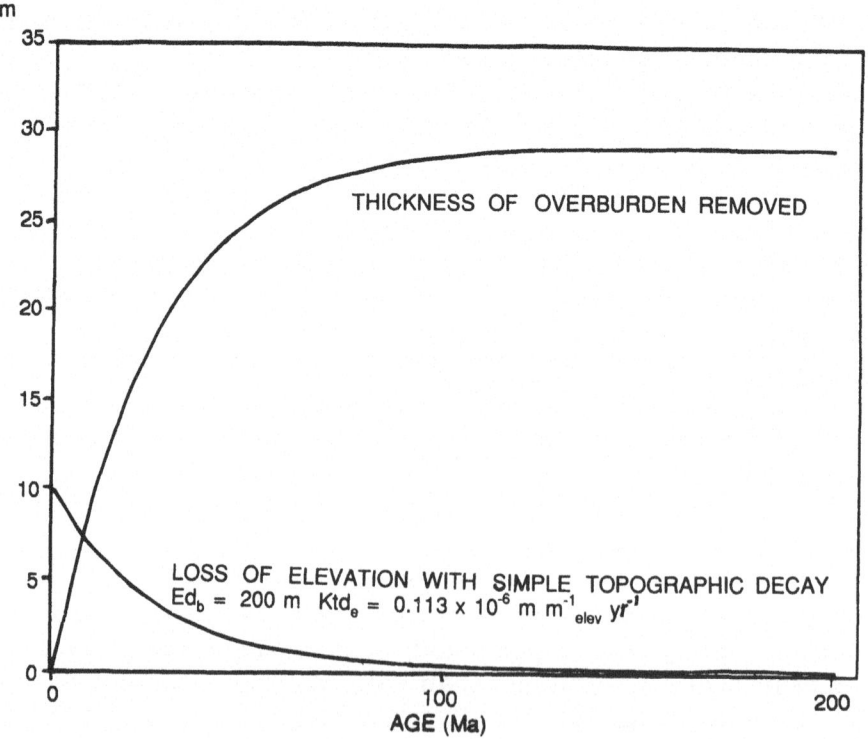

Figure 3. Loss of elevation and thickness of overburden removed from region undergoing topographic decay, based on global average Ktde of 0.113 x 10-6 m m-1elev yr-1 and Edb of + 200 m; density of mantle is assumed to be 3300 kg m-3, that of sedimentary rock to be 2200 kg m-3.

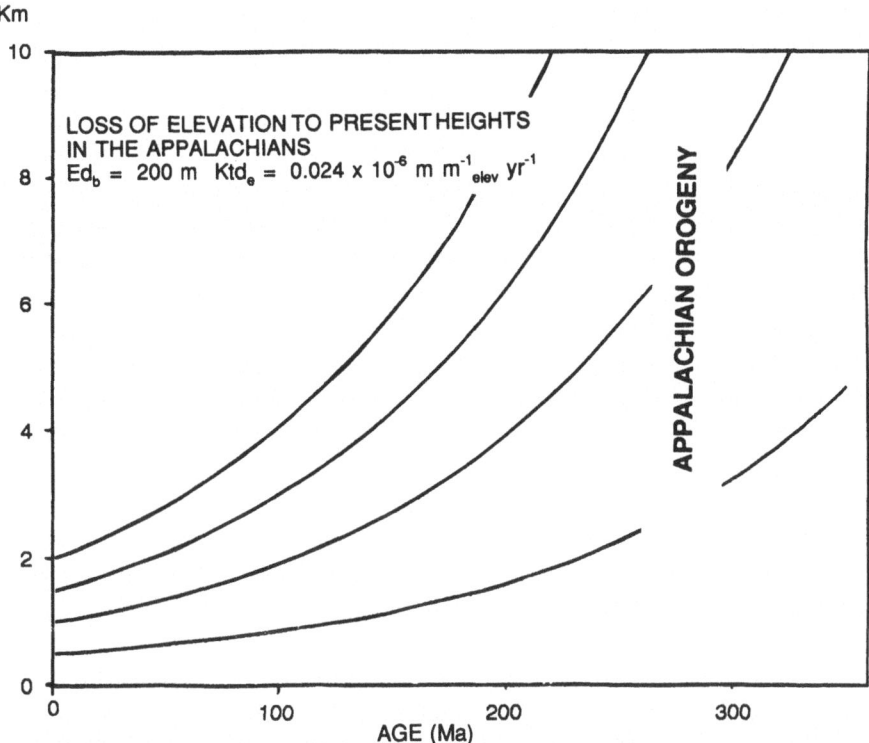

Figure 4. Elevation of Appalachians in past, and amount of overburden removed since they formed. Ktde is 0.024×10^{-6} m m$^{-1}_{elev}$ yr^{-1}, Ed$_b$ is + 200 m; density of mantle is assumed to be 3300 kg m-3, that of sedimentary rock to be 2200 kg m-3. Several different lines are shown, all based on this equation, but working backwards from different present-day elevations.

chians. DcConto, Hay, and Shaw (1989,1990), and Hay, DeConto, and Shaw (1990) concluded that the masses of detrital sediment that accumulated in the Atlantic off the U.S. during much of the Cenozoic and Cretaceous are too large to have been derived solely from the Appalachians. The excess sediment must have been supplied from other source areas, probably the interior of North America during the Cenozoic, and from sources north of the young Atlantic Ocean during the Cretaceous.

Replacement of Sediment onto Source Areas

The distribution of sedimentary materials in the source region provides information on its history. Hay, Wold, and Shaw (1989a) and Shaw (1989) present detailed accounts of the method of replacing overburden onto the source area. The total mass of detrital sediment that accumulated during each interval of time is determined from analysis of the mass/age distribution of existing sediment. The masses of detrital sediment then are redistributed on the potential source area in discrete time steps, starting from the present and working backwards. Potential source area sites are those sites that were not sites of deposition at the time and lie above the erosional baselevel. The sediment is redistributed onto these sites using the inverse of the equation relating denudation to elevation. In essence, all grid squares not occupied by sediment of the age to be replaced form a set of potential source areas. Of this set, all grid squares having an average elevation below the erosional baselevel are excluded. For the remaining subset, the area of each square is multiplied by its average elevation, this gives a volume that can be termed the erosion potential volume. The erosion potential volume is a direct measure of the relative mass of sediment that will be eroded if erosion is a linear function of both area and elevation. The erosion potential volumes are summed. The mass of detrital sediment to be replaced then is apportioned to each grid square in the ratio of its erosion potential volume to the sum of all erosion potential volumes.

The initial surface on which sediment is replaced is the present land surface; each successively older interval replaces sediment eroded during that interval onto the surface reconstructed for the end of the interval. After each replacement the resulting topographic surface then is subjected to erosion, using the same equation relating denudation and elevation to determine whether the sediment yield is greater than, equal to, or less than the amount actually delivered. If the yield is too large or too small the elevation at all sites is proportionally reduced or increased so that the erosion of the surface will yield exactly the amount of sediment actually derived from it. This adjusted surface serves as the base for the next replacement. The adjustment of the topographic surface is considered tectonic or epeirogenic uplift or subsidence.

Figure 5 shows the present average elevation of the source area for detrital sediments accumulating in the northwestern margin and most of the deep basin of the Gulf of Mexico. Figure 6 shows a mass-balanced paleogeographic reconstruction of the same area for 10 Ma. Figure 7 shows the thickness of overburden calculated to have been removed from this region during the past 10 m.y. Although the coarse 1° x 1° resolution of the model does not show details, the thicknesses of overburden shown to have been removed from central Colorado correspond to the difference in elevation between the Great Plains and the erosion surfaces of the Front and Rampart Ranges.

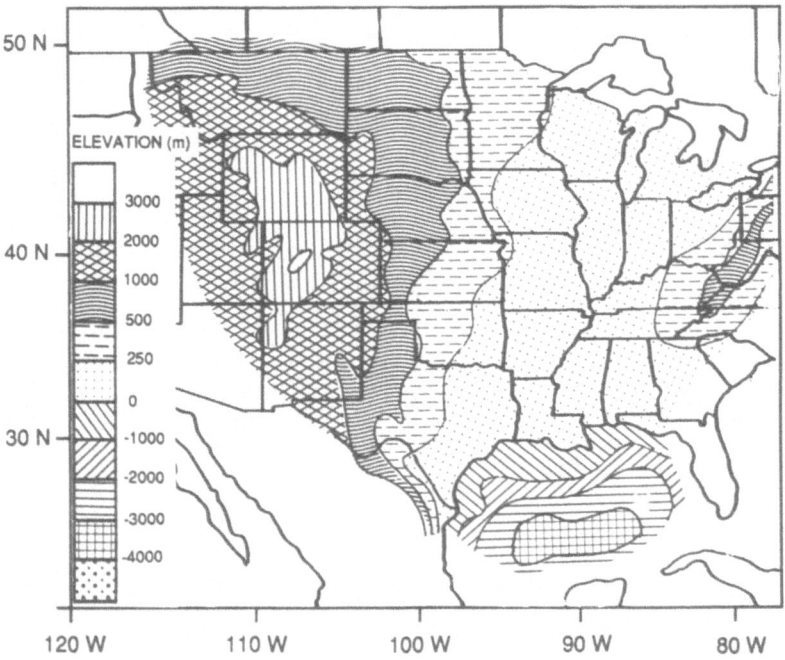

Figure 5. Present average elevations of northwestern Gulf of Mexico and its present drainage basin in interior of North America. North American interior was source area for detrital sediments that accumulated on northwestern margin and in most of deep basin of Gulf of Mexico during later Cenozoic.

CONCLUSIONS

The process of making mass-balanced paleogeographic reconstructions involves replacement of eroded material to its site of origin. In the simplest version only the thickness of the replaced overburden is known. Although neither its age nor lithology is known the information on thickness may be useful in calculating thermal maturation in the underlying rocks. The replacement of overburden onto source areas using mass-balanced paleogeographic techniques gives reasonable results. If the ages and lithologies of the material being replaced can be determined, as suggested in the succeeding paper (Hay and Wold, this volume) the geologic history of a region can be known in greater detail than previously suspected.

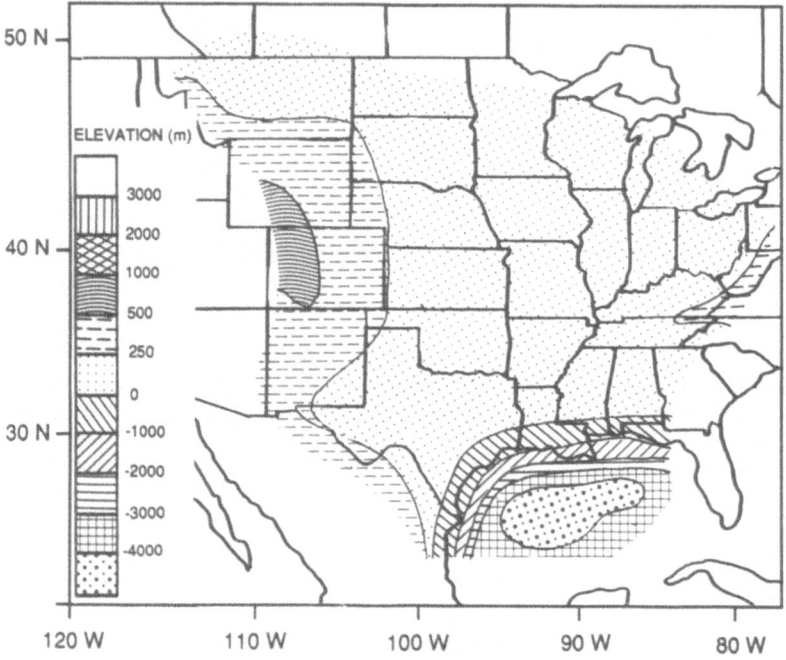

Figure 6. Mass-balanced paleogeographic reconstruction of same area for 10 Ma, showing much lower relief hindcast from much lower sediment supply at that time. Comparison with Figure 2 shows that sediment was not being delivered to Atlantic margin at this time.

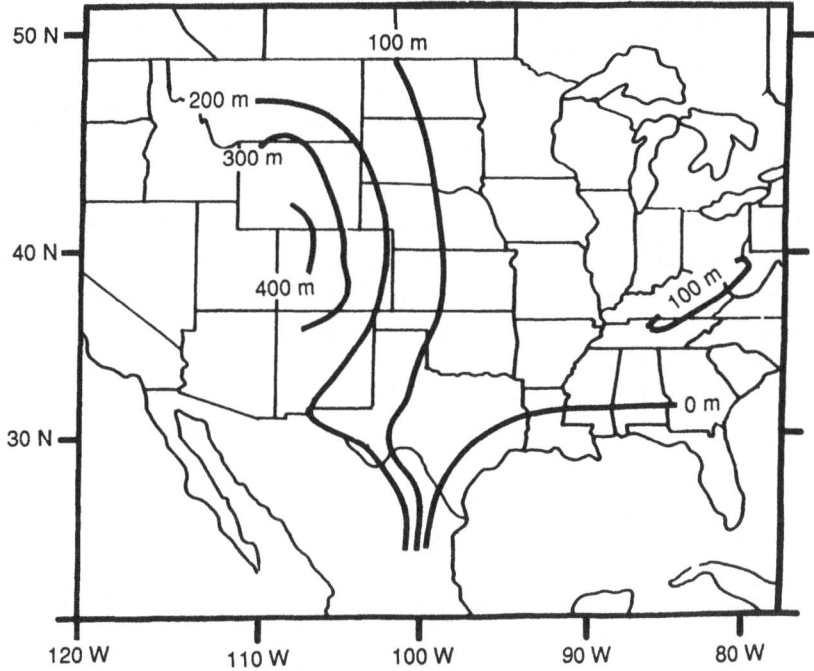

Figure 7. Thickness of overburden removed from interior of North America during past 10 m.y.

ACKNOWLEDGMENTS

This work has been supported by grants OCE 8409369 and OCE 8716408 from the National Science Foundation, and by gifts from Texaco, Inc.

REFERENCES

Budyko, M.I., Ronov, A.B., and Yanshin, A.L., 1987, History of the Earth's atmosphere: Springer-Verlag, New York, 139 p.

Curtis, W.F., Culbertson, J.K., and Chase, E.B., 1973, Fluvial sediment discharge to the oceans from the conterminous United States: U.S. Geol. Survey Circ. 670, 17 p.

DeConto, R.M., Hay, W.W., and Shaw, C.A., 1989, Detrital sediment fluxes in eastern and central North America and the adjacent Atlantic and Gulf of Mexico during the Cenozoic and Cretaceous: Geol. Soc. America, Abst. with Programs, v. 21, no. 6, p. A140.

Gilluly, J., 1969, Geological perspectives and the completeness of the geologic record: Geol. Soc. America Bull., v. 80, no. 11, p. 2303-2312.

Hay, W.W., DeConto, R.M., and Wold, C.N., 1990, Source of detrital sediments during the early opening of ocean basins: Geol. Soc. America, Abst. with Programs, v. 22, no. 7, p. A299.

Hay, W.W., Rosol, M.J., Jory, D.E., and Sloan, II, J.L., 1987, Tectonic control of global patterns of detrital and carbonate sedimentation, *in* Doyle, L.J., and Roberts, H.H., eds., Carbonate clastic transitions: Developments in Sedimentology, v. 42, Elsevier, Amsterdam, p. 1-34.

Hay, W.W., and Wold, C.N., 1990, Relation of selected mineral deposits to the mass/age distribution of Phanerozoic sediments: Geol. Rundschau, Bd. 79 , Heft 2, p. 495-512.

Hay, W.W., Wold, C.N., and Shaw, C.A., 1989a, Mass-balanced paleogeographic maps: background and input requirements, *in* Cross, T., ed., Quantitative dynamic stratigraphy: Plenum Press, New York, p. 261-275.

Hay, W.W., Shaw, C.A., and Wold, C.N., 1989b, Mass-balanced paleogeographic reconstructions: Geol. Rundschau, Bd. 78, Heft 1, p. 207-242.

Pinet, P., and Souriau, M., 1988, Continental erosion and large-scale relief: Tectonics, v. 7, no. 3, p. 563-582.

Ronov, A.B., 1980, The earth's sedimentary shell (quantitative patterns of its structure, compositions, and evolution)-The 20th V.I. Vernadski Lecture (in Russian), *in* Yaroshevskii, A.A., ed., The Earth's sedimentary shell (quantitative patterns of its structure, compositions, and evolution): Nauka, Moscow, v. 80 (English translation in Intern. Geol. Rev., v. 24, p. 1313-1388, 1982; and Am. Geol. Inst. Reprint Series, v. 5, 73 p., 1983).

Shaw, C.A., 1987, Balanced paleogeographic reconstructions of the northwestern Gulf of Mexico margin and its western-central North American source area since 65 ma: unpubl. masters thesis, Univ. Colorado, 285 p.

Shaw, C.A., 1989, Mass balanced paleogeographic modeling: examples from the western North Atlantic Ocean and Gulf of Mexico: unpubl. doctoral dissertation, Univ. Colorado, 381 p.

Shaw, C.A., and Hay, W.W., 1989, Mass-balanced paleogeographic maps: modeling program and results. *in* Cross, T., ed., Quantitative dynamic stratigraphy, Prentice-Hall, Inc., Englewood Cliffs, p. 277-291.

Veizer, J., 1988, The evolving exogenic cycle, *in* Gregor,.C.B., Garrels, R.M., Mackenzie, F.T., and Maynard, J.B., eds., Chemical cycles in the evolution of the Earth: Wiley-Interscience, New York, p. 175-220.

Wold, C.N., and Hay, W.W., 1990, Estimating ancient sediment fluxes: Am. Jour. Sci. v. 290, no. 9, p. 1069-1089.

THE SIMULATION OF LARGE-SCALE SEDIMENTARY STRUCTURES

P.A. Dowd
University of Leeds,
Leeds, England, U.K.

ABSTRACT

The paper discusses a number of methods for working with specific aspects of the simulation of multiple layered formations in sedimentary basins.

Simulations can be divided into three parts: basin morphology, morphology of formations within the basin, and variables within the formations. The emphasis in this paper is on the simulation of the morphology of the formations and their constituent layers or units. The layers are defined by their thicknesses at specified locations. Methods are discussed for ensuring that specific relationships between layers are reproduced in simulations.

The simulations may be used as exploration models or as a basis for production planning models.

INTRODUCTION

In the study and evaluation of many large sedimentary structures models must be constructed from sparse data. In the North Sea, for cxamplc, a typical data dcnsity is of the order of 1 well per 100 km² with additional geophysical data available on a smaller, but nevertheless sparse, grid. There are two approaches to constructing models: estimation from the available data or simulation.

Computerized Basin Analysis, Edited by J. Harff
and D.F. Merriam, Plenum Press, New York, 1993

Estimation usually involves some form of averaging which, when combined with sparse data, will almost always yield extremely smoothed estimates and hence a model which has little resemblance to reality. The aim of simulation, on the other hand, is to reproduce the variability observed in the data values together with any other characteristics (eg. histograms) deemed important; *conditional* simulation imposes the additional property that the simulated model must take on the actual data values at the data locations.

In the applications discussed here simulation refers to *stochastic* simulation in space and the objective is to provide a probabilistic model consisting of values at given locations of one or more variables (eg, layers in a formation together with physical properties of those layers) rather than simulating a geological process through time and space. The simulation method used here is the geostatistical method known as *conditional simulation* which is based on the application of the *turning bands* method and *kriging*.

Modeling Sedimentary Structures

Although the geostatistical simulation technique known as the turning bands method now is well known and widely used, problems are encountered when applying the method to large simulations with implicit or explicit constraints on some or all of the variables.

In the methods discussed here the morphology of the formation is modeled by the thicknesses of component units at given locations; however, there are other approaches, see for example Soares (1985, 1988) for a method based on indicators. Let $Z_i(x)$ be the thickness of unit i at location x and let $Z(x)$ be the thickness of the total formation at the same location, then :

$$Z(x) = \sum_i Z_i(x)$$

The thicknesses of the individual units and of the total formation can be simulated by the turning bands method and the simulations then can be conditioned to data values by the kriging technique described, for example, in Journel and Huijbregts (1978). If the individual unit thicknesses are correlated and if the coregionalization can be represented by a linear coregionalization model:

$$\sigma_{ij}(h) = \Sigma_k\, b_{ijk}\, C_k(h)$$

then the simulation of the coregionalization can be achieved by taking weighted sums of independently simulated random variables. Modeling is even simpler in the situation of *intrinsic coregionalization* :

$$\sigma_{ij}(h) = b_{ij}\, C_k(h)$$

The particular problems addressed here can be summarized as follows :

(1) Relationships between component units of sedimentary formations may not be quantifiable in the sense of a significant correlation coefficient or cross-covariances. For example, the thickness of a unit in a particular region may indicate that it has been pinched out on the limb of a fold; obviously the remaining units in the formation must conform to the same structure.

(2) The number of component units of a formation may be so large as to make a complete conditional simulation, based on cokriging, prohibitive in terms of modeling the coregionalization and in terms of the computing time required for conditioning

(3) There are many compatibility requirements between individual units and sums of units in a formation as well as sums of formations. For example, groups of units may constitute important subformations :

$$Z'(x) = \sum_{j=n_1}^{n_2} Z_j(x)$$

where $Z'(x)$ is the total subformation thickness at any given location. It is important to ensure that at all locations the sum of the conditionally simulated unit thicknesses is the same as the conditionally simulated total subformation thickness. The usual situation is the one in which $Z'(x)$ is the thickness of the total formation.

SIMULATION

The simulation can be divided into three parts :

(1) simulation of basin morphology
(2) simulation of formations and component units within the basin
(3) simulation of variables (eg, porosity, permeability) within the units of the formation

Basin Morphology

Basin morphology is simulated as a surface, usually subcircular in plan, and the formation units are simulated as thicknesses at specified locations within the simulated basin as illustrated in Figure 1. Variables within units are defined as values measured on constant, specified volumes (eg, drill-core segments).

Depending on the complexity of the basin morphology it may be simulated either geostatistically (for example, specifying depth as a regionalized variable) or simply by a mathematical function (for example, a noncylindroidal folded surface). In general, the formations within the basin are the more important, and the more difficult part of the simulation and this is achieved by geostatistical simulation of the thicknesses of the component units of the formations.

There may be a strong relationship between the basin morphology and the thicknesses of the formations. In such situations this relationship must be reproduced in the simulation. However, truncation, folding, or

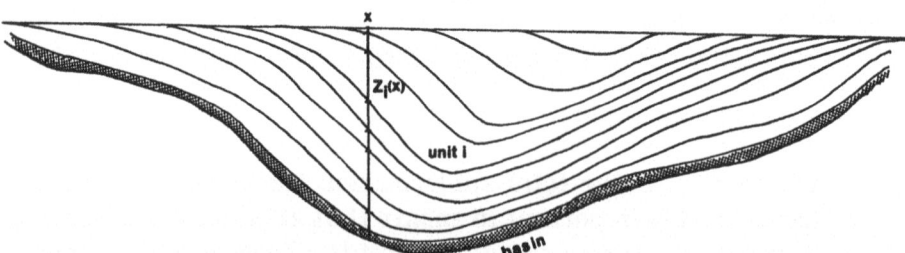

Figure 1. Cross-sectional view of sedimentary basin with formations (i) defined by thicknesses, $Z_i(x)$, at specified locations, x.

other deformation of the formations may have severely reduced, or eliminated, any relationship with the basin morphology.

Formations

Whatever the relationship with the basin morphology, the formation thicknesses almost always will be related. This relationship may be quantifiable in the form of significant correlation coefficients and cross-variograms or cross-covariances describing the coregionalization of the thicknesses of the formations. On the other hand, the relationship may be present only in the form of *inherent* constraints. Even when an explicit coregionalization is present the number of units in the formation may make modeling prohibitive.

The total thickness of a formation may be equal in significance to that of the individual component units. Anomalies may occur if formation thickness is simulated independently of the thicknesses of the component units.

Further problems arise when conditioning the simulations of a large number of units to data values. A complete cokriging of the thickness of each constituent unit will ensure that estimated (conditioned) thickness of the total formation will be the same as the sum of the estimated (conditioned) thicknesses of the individual units. However, when there is a large number of units conditioning by cokriging may be prohibitive in terms of modeling and computing time. Conditioning the simulated thicknesses of the units and the total formation by kriging each individually will not ensure that the sum of the estimates is equal to the estimate of the sum with the result that the conditionally simulated model of the formation will display serious anomalies.

SIMULATION OF INHERENT CORRELATIONS

In an earlier paper, Dowd (1984) presented two methods for working with inherent correlations as outlined. Both methods simulated auxiliary variables based on proportions and the more successful of the two methods defined the variables as follows.

If $Z_i(x)$ is the thickness of unit i at location x and $Z(x)$ is the thickness of the total formation at location x $\{Z(x) = \Sigma_i Z_i(x)\}$ then the proportion of unit i in the formation is defined by :

$$P_i(x) = \frac{Z_i(x)}{Z(x)}$$

and $\Sigma_i P_i(x) = 1$.

Now define a new set of regionalized variables :

$$R_1(x) = P_1(x), \qquad R_2(x) = \frac{P_2(x)}{1 - P_1(x)}, \qquad R_i(x) = \frac{P_i(x)}{1 - \displaystyle\sum_{j=1}^{i-1} P_j(x)}$$

and $0 \le R_i(x) \le 1$

$$1 - \sum_{j=1}^{k} P_j(x) \ne 0 \qquad \text{for } k = 1 \text{ to } n-2$$

To ensure that this second constraint is always satisfied it is sufficient that the index n should be assigned to a thickness which always is present, that is $Z_n(x)$ is strictly greater than zero :

$$Z_n(x) > 0 \Rightarrow P_n(x) = 1 - \sum_{j=1}^{n-1} P_j(x) > 0 \Rightarrow 1 - \sum_{j=1}^{k} P_j(x) \ne 0 \ \forall k = 1 \text{ to } n-2$$

The simulation of the thicknesses then is done by simulating $Z(x)$, $R_1(x)$, $R_2(x)$,..., $R_{n-1}(x)$ and then deducing $Z_1(x)$, $Z_2(x)$, ..., $Z_{n-1}(x)$ and

$$Z_n(x) = Z(x) - \sum_{i=1}^{n-1} Z_i(x)$$

The steps involved are :

(i)   transform the variables $Z(x)$, $(R_i(x), i = 1 ,..., n-1)$ into standard gaussian variables

(ii) calculate variograms and model the coregionalization of these standard gaussian variables
(iii) conditionally simulate values of the variables at the specified locations (the conditioning data are the transformed data values $Z(x_j), R_i(x_j)$)
(iv) take the inverse transforms
(v) calculate the values of $\{Z_i(x), \ i = 1, \ ...., n\}$

If the ratio variables are coregionalized the coregionalizations must be modeled. However, if there is a large number of units there may be be a prohibitive number of cross-variances to model. Note that this method satisfies the compatibility constraint, that is, the conditionally simulated total formation thickness is equal to the sum of the conditionally simulated component unit thicknesses. An example of an application of this method is given in Dowd (1984).

COREGIONALIZATIONS OF MULTIPLE LAYERS

This second method is a practical application of a simplification of cokriging developed by Matheron (1979) and may be applied in situations where there is an explicitly quantifiable, nonintrinsic coregionalization of the thicknesses of the units which make up the formation(s).

Matheron's intention was to develop a simplified form of cokriging for the situation in which the cross-covariances ($\sigma_{ij}(h)$ for $i \neq j$) are unknown or are too numerous to model.

Let $\sigma(h)$ be the covariance function of $Z(x)$ and let $\sigma_{ij}(h)$ be the cross-covariance function of the component variables $Z_i(x)$ and $Z_j(x)$ (with $\sigma_{ii}(h)$ the covariance function of $Z_i(x)$) :

$$\sigma(h) = \Sigma_i \ \Sigma_j \ \sigma_{ij}(h)$$

Matheron postulates a model of the form :

$$Z_i(x) = Y_i(x) + \omega_i Z(x)$$

in which the random function $Z_i(x)$ is split into two orthogonal random functions and the $Y_i(x)$ are in intrinsic coregionalization. Intrinsic coregionalization of the $Y_i(x)$ is an assumption; orthogonality is ensured if :

$$\sum_j \sigma_{ij}(h) = \omega_i \sigma(h)$$

as can be demonstrated by :

$$E[Z(x)Y_i(x)] \quad = \quad E[Z(x) \cdot \{Z_i(x) - \omega_i Z(x)\}]$$

$$= \quad E[\sum_j Z_j(x)Z_i(x)] - \omega_i E[Z(x)]^2$$

$$= \quad \sum_j s_{ij}(0) - \omega_i \sigma(0)$$

$$= \quad \sum_j \sigma_{ij}(0) - \sum_j \sigma_{ij}(0)$$

$$= \quad 0$$

The orthogonality condition also ensures autokrigability, that is, the cokriging of $Z(x)$ from values of its constituent units at data locations $\{x_k\}$ is the same as kriging it from values of $Z(x_k)$ at the same data locations. Autokrigability ensures the compatibility condition, that is, that the conditionally simulated total formation thickness is equal to the sum of the conditionally simulated component unit thicknesses. The compatibility condition is satisfied automatically if the Z_i are in intrinsic coregionalization.

The cross-covariance function of the $Y_i(x)$ can be determined by putting $C_{ij}(h) = cov(Y_i(x)Y_j(x+h))$ and then :

$$C_{ij}(h) = cov [(Z_i(x) - \omega_i Z(x)).(Z_j(x+h) - \omega_j Z(x+h))]$$

$$= cov(Z_i(x)Z_j(x+h)) - \omega_i \omega_j cov(Z(x)Z(x+h)) - \omega_i cov(Z_j(x+h)Z(x))$$
$$- \omega_j cov(Z_i(x)Z(x+h))$$

Imposing the orthogonality/autokrigability condition gives :

$$C_{ij}(h) = \sigma_{ij}(h) - \omega_i \omega_j \sigma(h)$$

an expression for ω_i can be determined by setting $h = 0$ in the orthogonality/autokrigability condition:

$$\omega_i = \frac{\sum_j \sigma_{ij}(0)}{\sigma(0)}$$

where $\sigma_{ij}(0)$ is the cross-covariance between Z_i and Z_j and $\sigma(0)$ is the variance of Z. Note that $\sum_i \omega_i = 1$

The Simulation

Define a new variable for each unit of the formation :

$$Y_i(x) = Z_i(x) - \omega_i Z(x)$$

with ω_i defined so that $\sum_i \omega_i = 1$ and $\sum_i Y_i(x) = 0$.

Z(x) and $Y_i(x)$ are orthogonal and therefore can be simulated independently; the coregionalizations of the $Y_i(x)$ must be simulated but it may be possible to use simplified forms of the cross-covariances as shown next. Let Zs(x) and Ysi(x) be the simulated realizations of Z(x) and $Y_i(x)$.

Whether the simulation is conditioned to data values, the compatibility condition :

$$\sum_i Zs_i(x) = Zs(x)$$

is satisfied.

Conditioning

The conditioning is done individually on each unit in the usual manner :

$$Zcs_i(x) = Z_i^*(x) + [Zs_i(x) - Zs_i^*(x)]$$

where $Zcs_i(x)$ is the conditionally simulated random variable at location x, $Z_i^*(x)$ is the random variable kriged from the random variables at the data locations, $Zs_i(x)$ is the simulated random variable at location x and $Zs_i^*(x)$ is the random variable kriged using the simulated random variables at the data locations.

The additional requirements of the conditioning are the cokriged estimates of :

$$Zs_i(x) = Ys_i(x) + \omega_i Zs(x)$$

and

$$Z_i(x) = Y_i(x) + \omega_i Z(x)$$

As $Ys_i(x)$ and $Zs(x)$ are orthogonal, the cokriging of $Zs_i(x)$ is the sum of the cokriging of $\omega_i Zs(x)$ and the cokriging of $Ys_i(x)$. Similarly, the cokriging of $Z_i(x)$ is the sum of the cokriging of $\omega_i Z(x)$ and the cokriging of $Y_i(x)$. Because of the orthogonality, the cokriging of $Zs(x)$ and of $Z(x)$ are identical to their kriging. Let these estimates be :

$$Zs^*(x) = \sum_k \lambda_k Zs(x_k) \quad \text{and} \quad Z^*(x) = \sum_k \lambda_k Z(x_k)$$

The coefficients are identical for each estimator and are determined by solving the usual kriging equations using the covariance function, $\sigma(h)$, for the total thickness of the formation :

$$\sum_k \lambda_k \sigma(x_k, x_l) - \mu = \sigma(x_l, x)$$

$$\sum_k \lambda_k = 1$$

As the $Ys_i(x)$ and the $Y_i(x)$ are coregionalized intrinsically, the cokriging of $Ys_i(x)$ and of $Y_i(x)$ are identical to kriging :

$$Ys_i^*(x) = \sum_k \beta_k Ys_i(x_k) \quad \text{and} \quad Y_i^*(x) = \sum_k \beta_k Y_i(x_k)$$

As $C_{ij}(h)$ is a cross-covariance model of an intrinsic coregionalization it can be expressed as :

$$C_{ij}(h) = b_{ij} C(h)$$

and it is only the $C(h)$ component which has any influence on kriging weights. The coefficients β_k are identical for both estimators and are determined by solving the usual kriging equations using the intrinsic component of the cross-covariance function :

$$\Sigma_k \beta_k C(x_k, x_l) - \mu' = C(x_l, x)$$

$$\Sigma_k \beta_k = 1$$

Matheron (1979) shows that, by minimizing the weighted error variance $\Sigma_i W_i var[Zs_i(x) - Zs_i^*(x)]$, $C(h)$ can be replaced in the kriging equations by :

$$K(h) = \Sigma_i \frac{1}{\omega_i} \sigma_{ii}(h) - \sigma(h)$$

and that this applies whether the autokrigability/orthogonality condition is imposed provided that in the latter situations the weights are taken as ω_i. To ensure that $K(h)$ is a positive definite function the weights ω_i must be nonnegative, that is, $\omega_i \geq 0$.

The total cokriging of $Zs_i(x)$ is thus :

$$Zs_i^*(x) = \omega_i Zs^*(x) + Ys_i^*(x)$$

$$= \omega_i \Sigma_k (\lambda_k - \beta_k) Zs(x_k) + \Sigma_k \beta_k Zs_i(x_k)$$

Similarly, the total cokriging of $Z_i(x)$ is :

$$Z_i^*(x) = \omega_i \Sigma_k (\lambda_k - \beta_k) Z(x_k) + \Sigma_k \beta_k Z_i(x_k)$$

Note that the kriging weights (λ_j and β_j) are the same for each unit i of the formation.

The steps involved are :

(i) From the data values at locations $\{x_k\}$ calculate the values of $Y_i(x_k)$

(ii) Estimate the covariance (or variogram) of $Z(x)$ and calculate ω_i

(iii) Estimate the covariances (or variograms) of $\{Y_i(x_k), i = 1, ..., n\}$. These covariances can be compared with the theoretical model of the intrinsic component :

$$K(h) = \Sigma_i \frac{1}{\omega_i} \sigma_{ii}(h) - \sigma(h)$$

The procedure seems robust enough for even approximate agreement here to allow use of the intrinsic model in practice.

(iv) Simulate the values of Y_i and Z on the basis of the covariance models fitted in (iii) and (ii) respectively.

(v) Calculate the values of $Zs_i(x)$

(vi) Condition the simulated Zs_i and Zs values to their respective data values

In addition, gaussian transforms may be required prior to the simulation together with corresponding inverse transforms of the simulated values.

SIMULATION OF VARIABLES WITHIN FORMATIONS

Once the morphology of the model has been simulated the next step is to provide a simulation of variables (such as porosity, permeability, solids, water, and oil) within the formations. These variables usually are correlated highly but may have complicated relationships with the morphology, for example the coregionalizations may change with depth as is the situation in the case study of the Athabasca Tar Sands summarized in Figure 2 (cf. Dowd and Royle, 1977).

Some of these variables show little or no relationship with the layers of the formations as for example when productive intervals of hydrocarbons extend across a sequence of formations. Variables may have been measured within layers or independently of layer boundaries. Unless there are strong reasons for not doing so, measurements should be composited into equal lengths respecting layer boundaries; where variability between the boundaries of a layer is not important compositing is done over the complete layer thickness to give an accumulation.

Variables such as permeability may be simulated by auxiliary relationships with other variables, for example using a log-log or semilog relationship between permeability and porosity.

The simulation of these variables may be two-dimensional (accumulations) or three-dimensional (values defined on fixed supports) depending on whether there is any vertical differentiation of sample values and whether such differentiation is relevant. In general, for large basins the

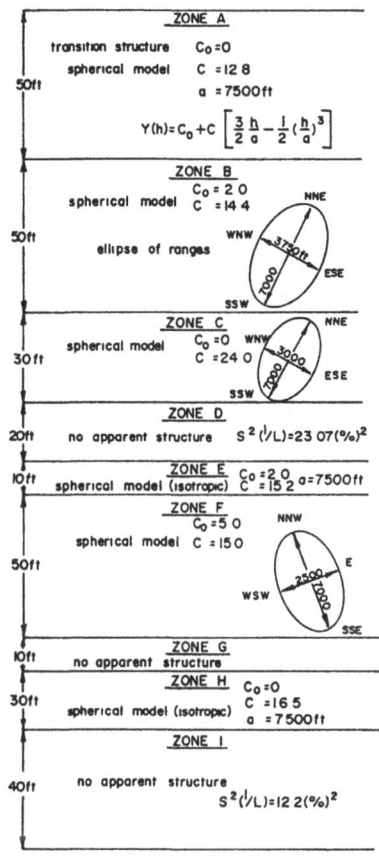

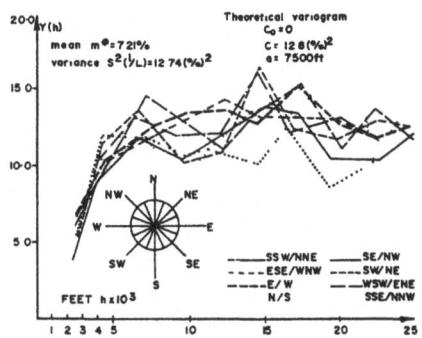

horizontal variograms for oil Zone A

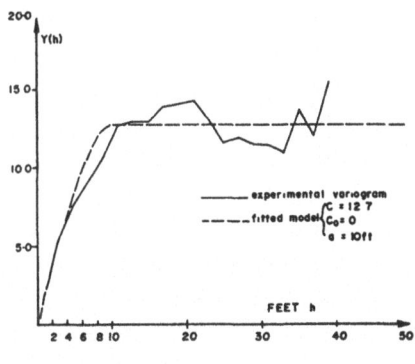

vertical variogram for oil Zone A

Figure 2. Summary of zonal structures revealed by variogram analysis of hydrocarbon content in lease of Athabasca Tar Sands together with experimental variograms for Zone A.

importance of lateral variability within layers far outweighs that of vertical variability between the layer boundaries and most simulations of these variables are two dimensional.

CONCLUSIONS

Relatively simple techniques can be used in conjunction with the standard simulation methods to provide simulations which conform more closely to reality. Inherent correlations and compatibility conditions are important constraints which must be included in simulations.

Simulated models generated by these methods can be used as a basis for production planning models or generating values of variables in a block model. The simulations also may be used as exploration planning models.

REFERENCES

Dowd, P.A., 1984, Conditional simulation of inter related beds in an oil deposit, *in* Verly, G., David, M., Journel, A.G., and Marechal, A., eds., Geostatistics for natural resource characterization NATO: ASI Series C : Mathematical and Physical Sciences, v. 122, pt. 2: D. Reidel Publ. Co., Dordrecht, Netherlands, p. 1031 - 1043.

Dowd, P.A., and Royle, A.G.,1977, Geostatistical applications in the Athabasca Tar Sands: Proc. 15th APCOM Symposium, Australasian Inst. Mining and Metallurgy Publ., Melbourne, Australia, p. 235 - 242.

Journel, A., and Huijbregts, C., 1978, Mining geostatistics: Academic Press, London. 600 p.

Matheron, G., 1979, Récherche de simplification dans un problème de cokrigeage: Centre de Géostatistique et Morphologie Mathématique, Fontainebleau, France, Note N-628, 19 p.

Soares, A., 1985, Simulação morfologica de jazigos minerais : aplicação a um jazigo lignitico: unpubl. masters thesis, Instituto Superior Tecnico, Lisbon, Portugal, 76 p.

Soares, A., 1988, Conditional simulation of indicator data : a case study of a multiseam coal deposit, *in* Chung, C.F., Fabbri, A.G., and Sinding-Larsen, R., eds., Quantitative Analysis of Mineral and Energy Resources: D. Reidel Publ. Co., Dordrecht, Netherlands, p. 375-384.

A QUANTITATIVE BASIN ANALYSIS SYSTEM FOR PETROLEUM EXPLORATION

S. Cao, S. Bachu, and A. Lytviak
Alberta Research Council,
Edmonton, Alberta, Canada

ABSTRACT

A computer-based Quantitative Basin Analysis System (QBAS) has been developed for the study of sedimentary basins. The QBAS consists of three major components: database, data processing, and modeling. Usually available data in a sedimentary basin, such as stratigraphic picks, core analyses, drillstem tests, analyses of organic matter, bottom-hole temperatures, and analyses of formation waters, are stored and manipulated in a relational well database. The data are processed and synthesized to higher levels using different graphical and mathematical techniques. The data processing also provides the regional and local scale information of geological, geochemical, geophysical, geothermal, and hydrogeological parameters which are prerequisites for basin modeling. In the modeling module, both one- and two-dimensional models are used for the simulation of sediment compaction and fluid flow, and for the reconstruction of burial history; a thermal inversion method is used for the reconstruction of thermal history; and a kinetic approach is used for the simulation of hydrocarbon generation from organic matter. Hubbert's method is used for the evaluation of hydrodynamic entrapment of hydrocarbons. The QBAS provides the explorationists with the past and present-day conditions in a sedimentary basin in terms of geology, geothermics, hydrogeology, and hydrocarbon generation, migration, and accumulation.

Computerized Basin Analysis, Edited by J. Harff
and D.F. Merriam, Plenum Press, New York, 1993

INTRODUCTION

Quantitative basin analysis for petroleum exploration is a systematic approach to understand and reconstruct those processes related to the generation, migration, and accumulation of hydrocarbons in a sedimentary basin, using various mathematical and computer models based on the principles of geology, geophysics, geochemistry, hydrodynamics, and thermodynamics. In the past decade, many comprehensive models have been developed for quantitative basin analysis through modeling, and application of these models have demonstrated that basin modeling is a useful tool in petroleum exploration (Welte and Yukler, 1981; Ungerer and others,1984;Bethke,1985;Cao and Lerche,1987;Nakayama and Lerche, 1987).

Fundamental to basin analysis is the acquisition, interpretation, and evaluation of a diverse suite of point data into synthesized form so that it can be used in various studies. In the Western Canada Sedimentary Basin, there are hundreds of thousands of wells drilled with immense amounts of geological, geophysical, geochemical, and hydrogeological data available for comprehensive basin studies. A Quantitative Basin Analysis System (QBAS) has been developed at the Alberta Geological Survey (AGS), Alberta Research Council, Canada, in order to provide a quantitative description of basin evolution and evaluate the potential for hydrocarbon resources. The QBAS is a computer-based system to capture, store, and process large amounts of data, and to simulate those processes related to hydrocarbon generation, migration, and accumulation in a sedimentary basin.

The QBAS consists of three major components: database, data processing, and modeling (Fig. 1). Although there is a general flow of information from point data in the database to the highly synthesized forms used as input for or results from modeling, there is a permanent feedback at various levels between the three modules. The whole process has an interactive aspect because of the permanent need to update, reconsider, and reevaluate existing information in view of new data, concepts, methods, and results. Data such as stratigraphic picks, core analyses, drillstem tests, bottom-hole temperatures, and analyses of organic matter and formation waters, are stored and manipulated in a relational database (Lytviak, 1987). The data are processed and synthesized using graphic and mathematical techniques (Bachu and others, 1987). The data processing also provides the regional and local scale geological, geophysical, geochemical, geothermal, and hydrogeological parameters which are prerequisites for basin modeling. In the modeling

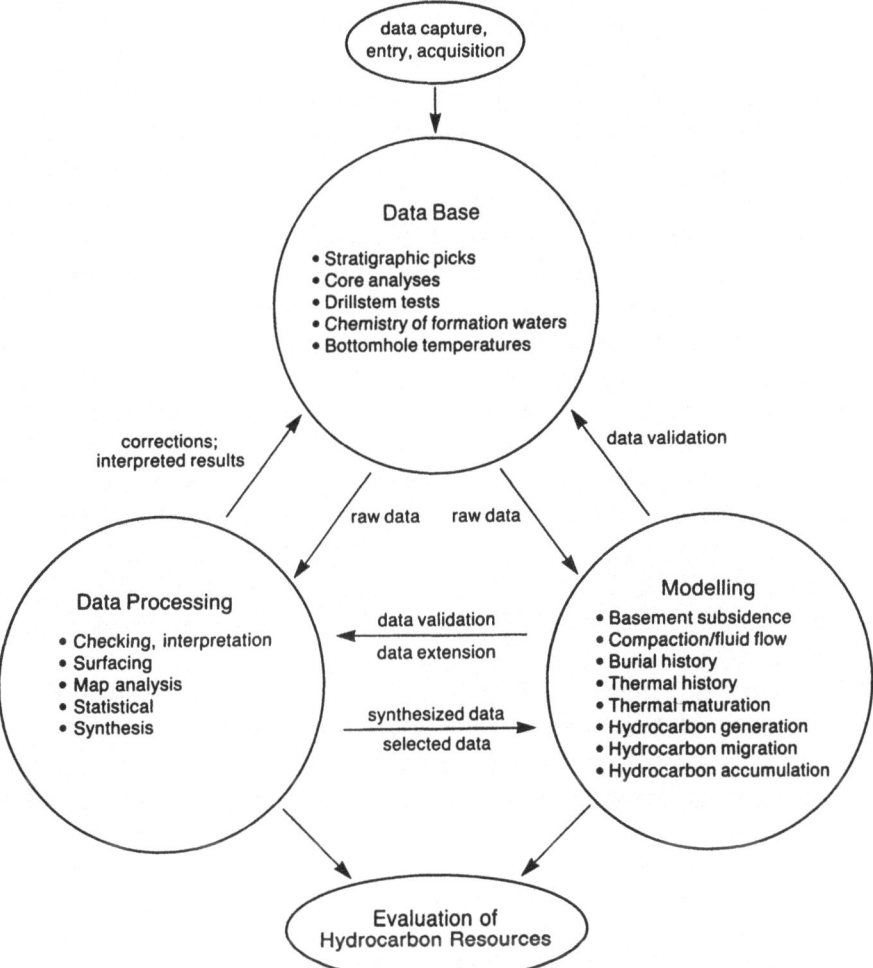

Figure 1. Diagram of Quantitative Basin Analysis System for petroleum exploration.

module, both one- and two-dimensional models are used for the simulation of sediment compaction and fluid flow, from which the burial history
is reconstructed. A thermal inversion method is used for the reconstruction of thermal history, and a kinetic approach is used for the simulation
of hydrocarbon generation from organic matters. A hydrodynamic method
is used for the evaluation of hydrodynamic entrapment of hydrocarbons.
This paper introduces the QBAS with some application examples in the
Peace River Arch area, western Canada.

DATABASE

The QBAS uses the existing Alberta Geological Survey Well Database (AGSWDB) as its central database to provide a set of basic tools for adding, deleting, and retrieving data. The AGSWDB used the INGRES commercial database management system and contains on-line data for more than 200,000 wells. The INGRES implements a relational data model, which has the attribute that new relations can be easily created. This is coupled with the widely known and relatively standard SQL language for data definition and retrieval, thus allowing much flexibility in supporting nonstandard ad-hoc data retrievals. This ability is valuable in an environment oriented toward research, development, and exploratory work. A central capability of the AGSWDB is to support simultaneously standard organization-wide data sets, project specific data sets, and private subsets. Thus, many data are shared whereas data manipulation of a private data subset does not affect other users. Furthermore, the separation of data description and storage from applications allows major additions or restructuring of the database without necessitating modification of the application software.

The QBAS database holdings consist of two groups: those implemented as part of the central AGSWDB and those implemented as straight VMS files. In general, those data seen to be of use for a wider range of programs during a longer period of time, such as stratigraphic picks, core analyses, bottom-hole temperatures, etc., are implemented in the AGSWDB. Those which are specific to single projects or are relevant for only a short time are kept in form of VMS files. In many situations, the project data in the VMS files were extracted from the AGSWDB and are kept separate in order to keep them static throughout subsequent AGSWDB updates. These extractions incorporate project-specific processing or are augmented by determinations or assumptions which are not necessarily applicable beyond the immediate project.

The Peace River Arch area (Fig. 2) with an area of about 220,000 km², is one of the most active petroleum exploration areas in the Western Canada Sedimentary Basin. Hydrocarbons have been located in units ranging from

Devonian to Cretaceous with a total hydrocarbon accumulation of about 275 billion cubic meters (Hitchon, 1984a). Some 27,000 wells have been drilled in the Peace River Arch area, providing a good control for the regional-scale geological, hydrogeological, and geothermal framework. Many types of data are collected at each well, being broadly divided into four categories: (1) geological (e.g. stratigraphic picks, lithology);

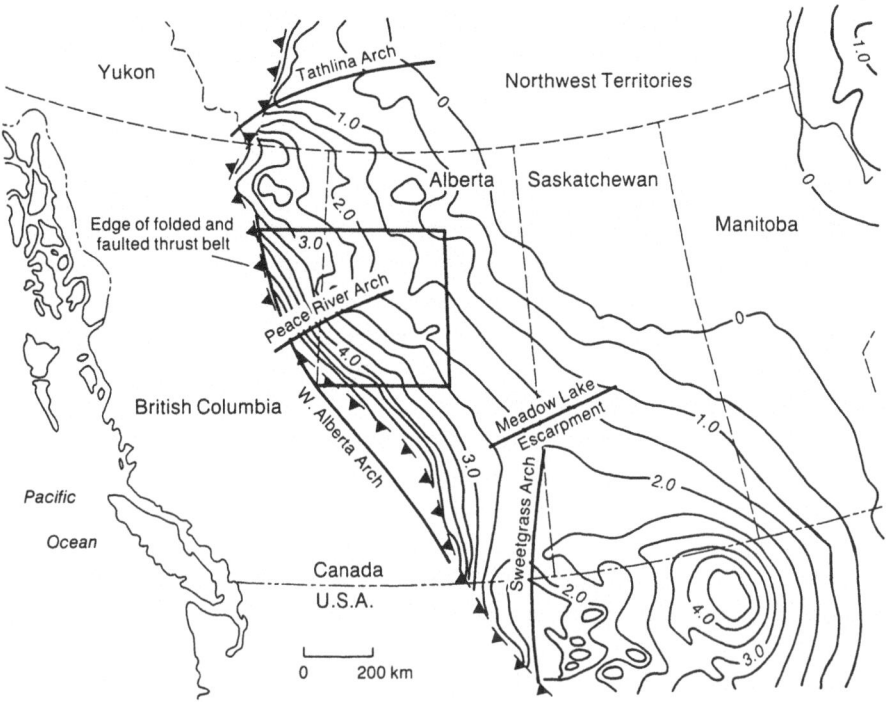

Figure 2. Location of Peace River Arch area, western Canada. Isopach (C.I. = 0.5 km) represent total preserved thickness of Phanerozoic strata east of Cordillera (based on Porter , Price, and McCrossan, 1982).

(2) hydrogeological (e.g. rock permeability and porosity, formation pressure); (3) geothermal (e.g. bottom hole temperature, heat flow); and (4) geochemical (e.g. analyses of formation fluid, vitrinite reflectance). Stratigraphic data from about 22,000 wells, 480,000 core analyses, 42,000 bottom-hole temperatures, 41,000 drillstem tests, and 30,000 formation water analyses from the area were collected and stored in the QBAS database. In addition, approximately 400 samples of vitrinite reflectance measurements and 1,500 samples of Rock-Eval data were analyzed in order to characterize the source rocks and evaluate the source rock potential for hydrocarbons.

DATA PROCESSING

The information contained in the QBAS database is actually a mass of raw point data derived from drilled wells, each measured or observed value being associated with the corresponding coordinates that define its position in space and time. In order to process and synthesize these large amounts of point data into higher level of information characterizing the geology, hydrogeology, and geothermics of the basin, a set of automatic data processing techniques and methods were used (Bachu and others, 1987). To guarantee the reliability and accuracy of data processing, one must have a clear understanding and definition of the criteria and algorithms used for data selection and processing, or error detection. In fact, processing the data from point into synthesized form is an interactive hybrid loop of human and computer interaction because the data have to be examined and the process has to be repeated until the respective data set is free of errors.

In general, variables with a continuous variation, such as formation elevation, fluid pressure, and temperature, are analyzed for regional trends and represented as maps. Parameters which depend on rock properties, such as permeability and porosity, are scaled-up from core or well scale to the formation and basin scale using statistical methods (Dagan, 1989). The stratigraphic and lithological data are processed first to define the framework of geology, stratigraphy, and geometry of the sedimentary rocks. The concepts of stratigraphic relationships (on-lap, off-lap or truncation phenomena) are well defined and applied in automatic data processing. The entire process has basically three steps: (1) interpreting and defining major geological events and stratigraphic considerations; (2) data processing; and (3) checking the resulting computer grids and maps for internal consistency. The lithological, hydrogeological, hydrochemical, and geothermal data then are processed by hydrostratigraphic units, in order to define the hydrogeological and geothermal regimes in the sedimentary succession. Drillstem test data are interpreted in a consistent and fast manner using a specially designed interactive and graphic software package to obtain formation pressures and permeabilities. Permeability and porosity data from core analyses are processed to obtain representative values of the cored intervals. Various culling methods are used in processing the hydrochemical data (i.e. formation water analyses) because of their generally questionable quality (Hitchon, 1984b).

A special package of cartographic software transforms the data location from original Dominion Land Survey (DLS), National Topographic Series (NTS), or Latitude-Longitude coordinates into Cartesian coordinates, allowing for posting, mapping, and transformation of the irregular distribution of data points into a regular grid of values. Thus, the processed data can be displayed as various distribution and contour

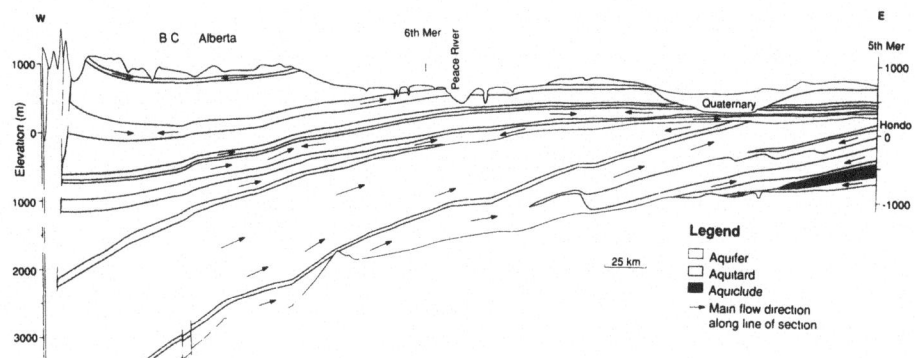

Figure 3. Dip cross section showing hydrostratigraphic geometry and fluid flow pattern in the Peace River Arch area, western Canada (from Hitchon, Bachu, and Underschultz, 1990).

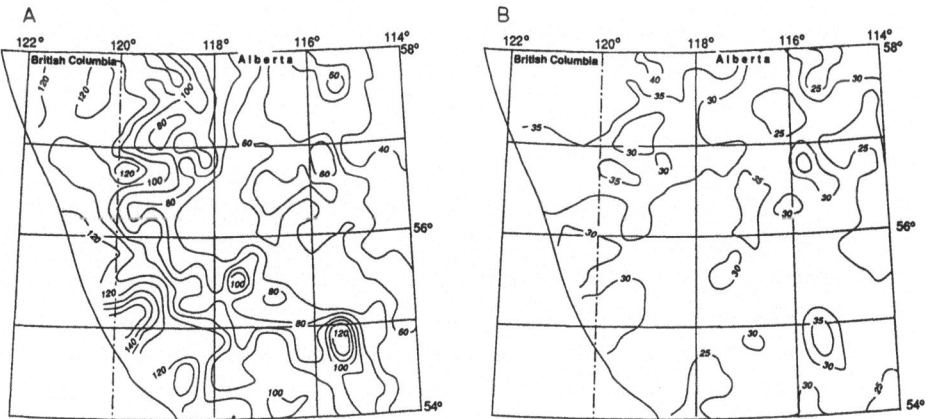

Figure 4. Geothermal regime in Peace River Arch area, western Canada: (A) distribution of temperature at Precambrian basement, and (B) distribution of geothermal gradient of entire sedimentary cover.

maps, cross sections, and a variety of statistical distribution plots and mathematical relations which differ with the parameters and variables being studied.

The data collected from 27,000 wells drilled in the Peace River Arch area were processed using the previously described methods and software. A regional-scale geological framework comprised of 60 stratigraphic units was defined, in which 32 aquifers, 23 aquitards, and 5 aquicludes were identified based on lithology and the hydrogeological and geochemical properties of formation fluids and rocks (Hitchon, Bachu, and Underschultz, 1990). Figure 3 is the dip cross-section showing hydrostratigraphic geometry and flow pattern. Figure 4 shows the distributions of temperature at the Precambrian basement and the geothermal gradient of the entire sedimentary cover. The present-day geological framework and hydrogeological and geothermal regimes were used as a basis for basin modeling.

MODELING

A comprehensive quantitative basin analysis should include the following four aspects:

(1) reconstruction of the geohistory, which includes burial history, basement subsidence, sediment deposition and compaction, fluid flow, and changes in porosity, permeability, and pressure in time and space;

(2) reconstruction of the thermal history, which includes the variation of the heat flux and temperature (thermal gradient) in time and space, and the thermal maturation history;

(3) reconstruction of the hydrocarbon generation history, through modeling the volume of hydrocarbons generated during geological time and determining the "oil and gas windows"; and

(4) reconstruction of hydrocarbon migration and accumulation history, which includes the time, direction, and volumes of hydrocarbon migration, and the possible traps for hydrocarbon accumulation.

Because of differences in the level of understanding of the geological, geochemical, and geophysical processes in sedimentary basins, and in the mathematical methods and numerical implementations, many models

with different features have been developed to simulate processes related
to basin evolution, and hydrocarbon generation, migration, and accumulation (Welte and Yukler, 1981; Goff, 1983; Ungerer and others, 1984;
Bethke, 1985; Cao and Lerche, 1987; Nakayama and Lerche, 1987). The
major modeling capabilities of the QBAS are presented briefly here,
whereas detailed descriptions are given in the corresponding references.

. Geohistory Model

In this model, the simulation of sediment compaction is crucial
because sediment compaction either dominates or influences all other
aspects of basin evolution. Although the effects of diagenesis on compaction,
porosity, and permeability are undoubtedly significant, they are subor-
dinate to the effect of overburden weight causing mechanical compaction.
Many processes such as burial history, thermal history, hydrocarbon
generation, migration, and accumulation, etc., are either directly or
indirectly dependent upon sediment compaction. The fluid movement
caused by increasing overburden can be described by a time dependent
partial differential equation (Lerche, 1990) which combines Darcy's law
and the equation for the conservation of fluid mass with a transformation
of coordinates from the physical depth to fully compacted depth. The
changes in sediment thickness, porosity, permeability, pressure, and
fluid flow rate with time and space are obtained by solving this time-
dependent compaction and fluid flow partial differential equation in both
one dimension (Cao and Lerche, 1989) and two dimensions (Nakayama
and Lerche, 1987). Figure 5 shows the flowchart for simulation of
sediment compaction and fluid flow. The basement subsidence caused by
tectonism only can be calculated using models based on isostatic compensation (Steckler and Watts, 1978; Goff, 1983). These models assume that
the total basement subsidence is caused by tectonism, the weight of the
sediments, and the weight of the water above the sediment surface. Also,
the cementation, dissolution, and fracturing caused by abnormally high
pore pressure are simulated by changing the formation permeability
(Nakayama and Lerche, 1987; Cao and Lerche, 1989). The data required
to run the geohistory model are depth and age of each formation base, and
lithology, and paleowater depth of each formation.

In the Peace River Arch area, twelve wells were selected for one-
dimensional modeling and six west-east cross-sections based on 75 wells
were constructed for two-dimensional modeling. Figure 6A shows the
burial curves reconstructed for the well 10-35-78-13W6, located on the

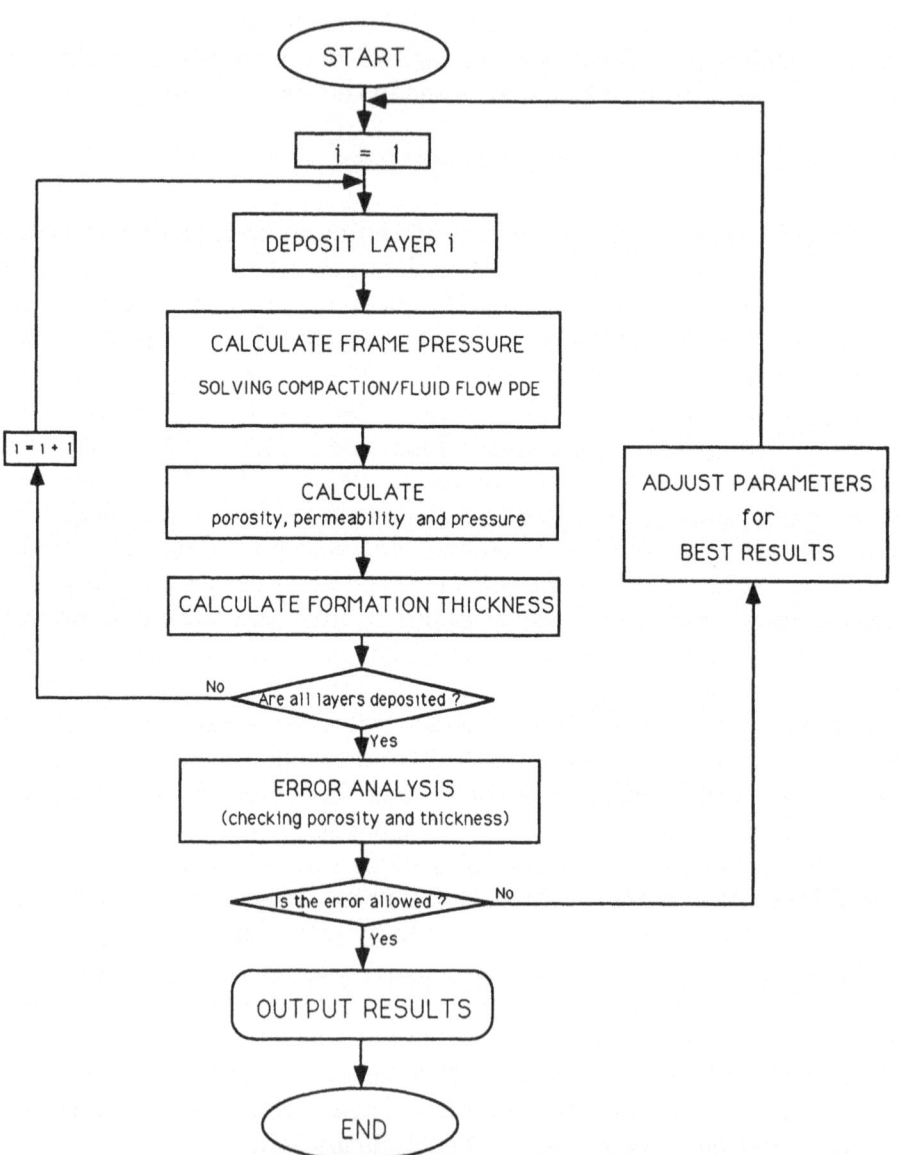

Figure 5. Flowchart for simulation of sediment compaction and fluid flow.

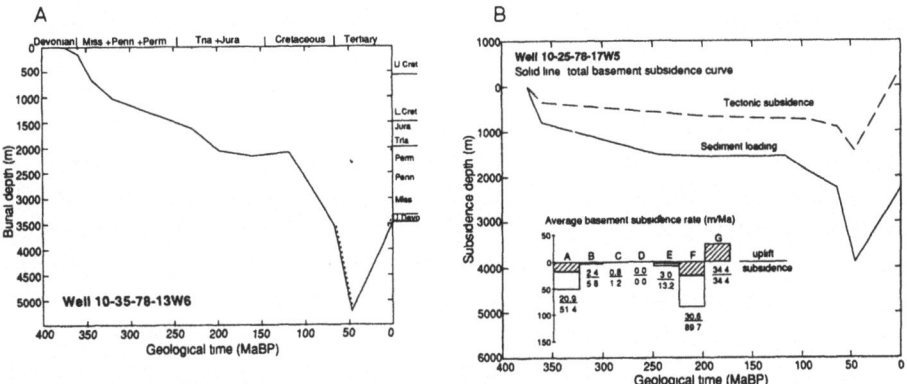

Figure 6. Reconstruction of geohistory for well 10-35-78-13W6 in Peace River Arch area, western Canada: A, burial curves; B, basement subsidence rate.

main structural axis of the Peace River Arch. During the passive-margin stage, continuous subsidence of the cratonic platform took place from Devonian to Mississippian time, with dominating transgressive marine carbonate deposition, and several subsidence-uplift cycles took place from Pennsylvanian to Early Jurassic time, with transgressive and regressive sediment deposition. The Peace River Arch area transformed during Middle Jurassic time from a passive margin into a foreland basin when foreign terranes began to accrete along the western margin of the Canadian craton (Porter, Price, and McCrossan, 1982). As a result, the area subsided from tectonic load to easterly directed thrusting and sediments were deposited in the foreland basin. This event is shown in the burial curves by fast subsidence and high deposition rates with huge marine and nonmarine clastic sediments deposition during Cretaceous and Paleocene time (Fig. 6B). From the end of the Laramide orogeny (Early Eocene) to present, regional uplift affected the whole area and most of the Paleocene sediments were eroded.

Thermal History Model

Based on the burial history obtained from the geohistory model, the thermal history model reconstructs the thermal history (paleoheat flux and paleotemperature) by comparing predicted thermal indicator values (such as vitrinite reflectance) to corresponding measurements down a borehole and adjusting accordingly the paleoheat flux to minimize the differences (Lerche, 1990). The variation of the heat flow in time is assumed to be of the form $Q(t) = Qo(1 + BETA)$, where $Q(t)$ is the heat flow at time t in the past, Q_0 is the present heat flow, t is geological time in million years, and BETA is a parameter to be determined. Because the heat flow function is linear, a parameter Qm is used in the thermal inversion model to set a limit for $Q(t)$ if $Q(t)$ goes beyond certain extreme values. Figure 7 shows the thermal inversion procedure used in the QBAS. The output are: (1) heat-flow variation in time, (2) temperature variation with time and depth, and (3) the variation of a thermal indicator such as vitrinite reflectance, Ro, or Time Temperature Index (TTI), with time and depth. The input data required to run the thermal history model are temperature at the sediment surface, bottom-hole temperature, and some thermal indicator measurements with depth.

Figure 8A shows the temperature history curves on the Peace River Arch at the well 10-35-78-13W6. The regional distribution of paleotemperature for a given formation can be estimated through contouring the calculated paleotemperatures at many locations. Generally, the geothermal gradients seem to have been lower in the past than at present. The thermal maturation in the past can be estimated by superimposing calculated paleovitrinite reflectance on the reconstructed burial curves (Fig. 8B). The paleothermal maturation study shows that the Paleozoic source rocks in most of the region entered the "oil window" during Late Cretaceous and Early Tertiary time, and that Cretaceous source rocks matured during the regional uplift since Eocene time.

Hydrocarbon Generation Model

The hydrocarbon generation model is based on the kinetics of kerogen degradation whose general scheme of evolution is sketched in Figure 9. Two alternate approaches are used in the model to simulate hydrocarbon generation from kerogen. The first simulates the formation of oil from kerogen in six parallel reactions (first stage), and the formation of gas cracked from oil in a single reaction (second stage) (Tissot and Welte, 1984, Tissot, Pelet, and Ungerer, 1987). The second is a modified approach (Cao and Lerche, 1989), which adds the major gaseous products from both kerogen degradation and oil cracking. Based on the burial history and thermal history reconstructed in the previous models, the hydrocarbon generation model calculates the amount of hydrocarbon generated with time and depth. The data required to run the hydrocarbon generation model are the content of different kerogen types for each formation.

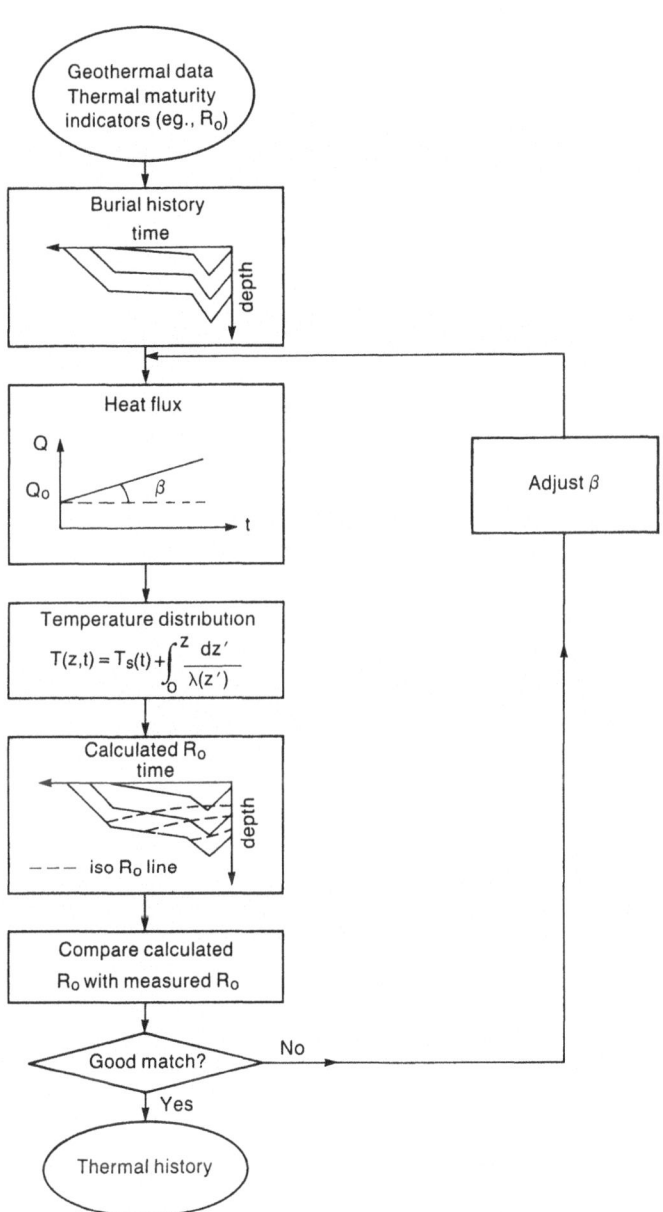

Figure 7. Diagram of thermal inversion procedure for thermal history reconstruction.

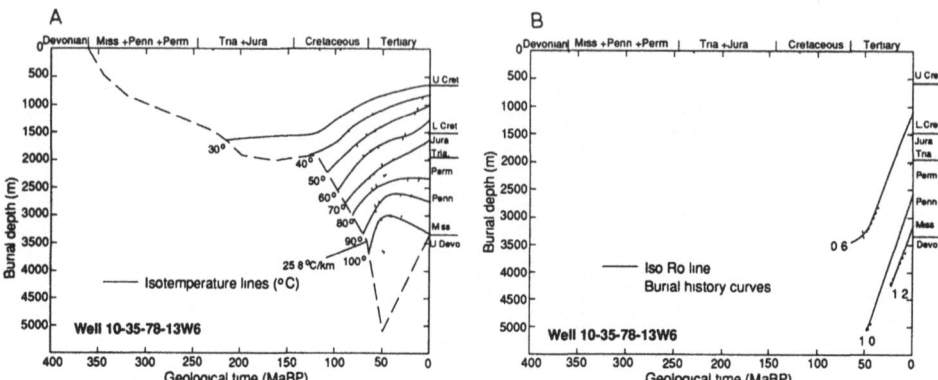

Figure 8. Variation of: A, temperature; B, vitrinite reflectance in time and space for well 10-35-78-13W6 in Peace River Arch area, western Canada.

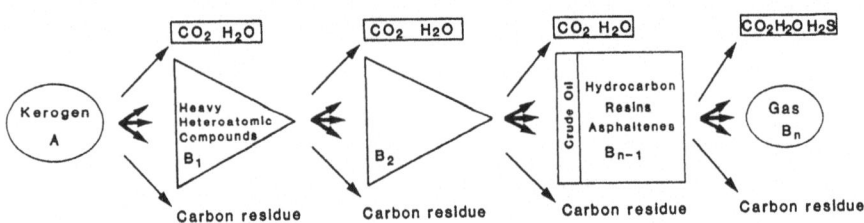

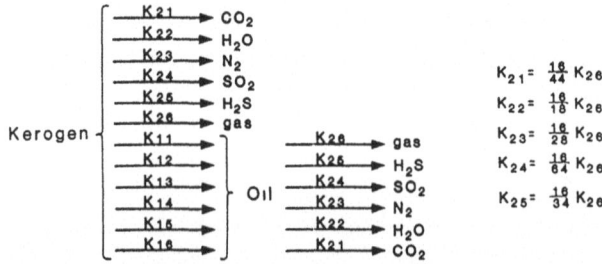

Figure 9. Hydrocarbon generation model: A, general scheme of kerogen degradation; B, detailed kinetic framework of hydrocarbon generation (modified from Tissot and Welte, 1984).

The modeling results of kinetic model for hydrocarbon generation area sensitive to the kinetic parameters used in modeling (Tissot, Pelet, and Ungerer, 1987). The kinetic parameters should be calibrated based on geochemical data available in the study are. Because the kinetic parameters for the source rocks in the Peace River Arch area have not been well defined, no application examples of the hydrocarbon generation model will be given here.

Hydrodynamic Model

Hydrocarbon migration and accumulation are controlled mainly by three factors: buoyancy, capillary pressure, and hydrodynamics. The hydrodynamic or U-V-Z method (Hubbert, 1953; Dahlberg, 1982) is a practical approach to evaluate the hydrodynamic effect of groundwater flow on hydrocarbon migration and accumulation and is used in the modeling module of the QBAS. In the U-V-Z method, the calculated U-surfaces are oil or gas equipotential surfaces, the calculated V-surfaces are water equipotential surfaces, and the Z-surfaces are formation top surfaces. The U-V-Z surfaces are used to evaluate the hydrodynamic entrapment of hydrocarbons for given densities of hydrocarbons and water in a formation. Paleohydrodynamic effects on hydrocarbon migration and accumulation are evaluated using paleo U-V-Z surfaces calculated from reconstructed paleostructure and paleopressures. Figures 10A and 10B show the contours of the present-day U- and V- surfaces at the base of Cretaceous, indicating the regional flow direction of oil and water. Figure 10C shows the schematic depiction of oil migration and accumulation in the area. Figure 10 shows that regional groundwater flow is in the same direction as regional oil flow, and major hydrocarbon occurrences are in the pathway of the regional flow of oil and water.

SUMMARY AND CONCLUSIONS

The Quantitative Basin Analysis System integrates conventional basin analysis with modern computer techniques in database management, data processing, and basin modeling. It provides the petroleum explorationists with a useful tool to synthesize and evaluate the present and past geological framework, hydrogeological, and geothermal regimes, and hydrocarbon generation, migration, and accumulation in sedimentary basins. For those sedimentary basins with significant hydrocarbon reserves and extensive drilling such as the Western Canada

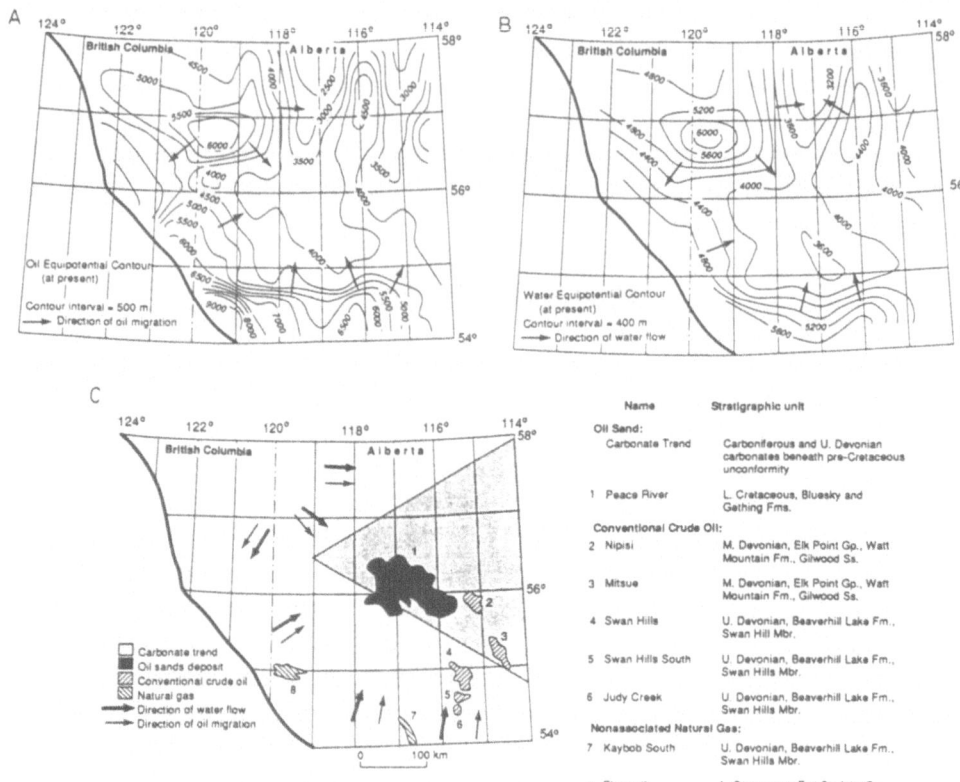

Figure 10. Application of Hubbert's method for hydrodynamic entrapment of hydrocarbons in Peace River Arch area, western Canada: A, U-surface; B, V-surface contours at base of Cretaceous; C, schematic depiction of oil migration and accumulation.

Sedimentary Basin, the QBAS is useful even more because of the huge amounts of available data which cannot be processed easily and rapidly by classical methods and therefore require extensive use of automatic databases and data-processing techniques. Nevertheless, the Quantitative Basin Analysis System is useful for basins where the data are scarce as well. A good understanding and knowledge of the present-day geological framework and hydrogeological and geothermal regimes in a sedimentary basin also increase the credibility and reliability of basin modeling by providing high-quality information for model input and calibration. Thus, the whole picture of the basin evolution, especially for hydrocarbon generation, migration, and accumulation, can be understood better by knowing the present and past situation of the basin.

REFERENCES

Bachu, S., Sauveplane, C.M., Lytviak, A.T., and Hitchon, B., 1987, Analysis of fluid and heat regimes in sedimentary basins: techniques for use with large databases: Am. Assoc. Petroleum Geologists Bull., v. 71, no. 7, p. 822-843.

Bethke, C.M., 1985, A numerical model of compaction-driven ground water flow and heat transfer and its application to the paleohydrology of intracratonic sedimentary basins: Jour. Geophys. Res., v. 90, no. 8, p. 6817-6828.

Cao, S., and Lerche, I., 1987, Geohistory, thermal history and hydrocarbon generation history of the northern North Sea Basin: Energy Exploration and Exploitation, v. 5, no. 4, p. 315-355.

Cao, S., and Lerche, I., 1989, Geohistory, thermal history and hydrocarbon generation history of Navarin Basin COST No. 1 Well, Bering Sea, Alaska: Jour. Petroleum Geology, v. 12, no. 3, p. 325-352.

Dagan, G., 1989, Flow and transport in porous formation: Springer-Verlag, Berlin, 465 p.

Dahlberg, E.C., 1982, Applied hydrodynamics in petroleum exploration: Springer-Verlag, New York, 171 p.

Goff, J.C., 1983, Hydrocarbon generation and migration from Jurassic source rocks in the east Shetland Basin and Viking graben of the northern North Sea: Jour. Geol. Soc. London, v. 140, pt. 3, p. 445-474.

Hitchon, B., 1984a, Geothermal gradients, hydrodynamics, and hydrocarbon occurrences, Alberta, Canada: Am. Assoc. Petroleum Geologists Bull., v. 68, no. 6, p. 713-743.

Hitchon, B., 1984b, Graphical and statistical treatment of standard formation water analysis, in Hitchon, B., and Wallick, E.I., eds., Proc. First Canadian/American Conference on Hydrogeology: Practical Applications of Ground Water Geochemistry: National Water Well Associations, Dublin, Ohio, p. 225-236.

Hitchon, B., Bachu, S., and Underschultz, J.R., 1990, Regional subsurface hydrogeology, Peace River Arch area, Alberta and British Columbia: Bull. Can. Petroleum Geology, in press.

Hubbert, M.K., 1953, Entrapment of petroleum under hydrodynamic conditions: Am. Assoc. Petroleum Geologists Bull., v. 37, no. 8, p. 1954-2026.

Lerche, I., 1990, Basin Analysis - Quantitative Methods I: Academic Press, Inc., San Diego, 562 p.

Lytviak, A., 1987, Alberta Geological Survey well database: Provincial Geologists Journal, v. 5, p. 124-128.

Nakayama, K., and Lerche, I., 1987, Basin analysis by model simulation: effects of geologic parameters on 1D and 2D fluid flow systems with application to an oil field: Gulf Coast Assoc. Geological Societies Trans., v. 37, p. 175-184.

Porter, J.W., Price, R.A., and McCrossan, R.G., 1982, The Western Canada Sedimentary Basin: Philosophical Trans. Roy. Soc. London, Ser. A, v. 305, p. 169-193.

Steckler, M.S., and Watts, A.B., 1978, Subsidence of the Atlantic-type continental margin off New York: Earth and Planetary Sc. Lett., v. 41, no. 1, p. 1-13.

Tissot, B.P., and Welte, D.H., 1984, Petroleum Formation and Occurrence (2nd ed.): Springer-Verlag, New York, 699 p.

Tissot, B.P., Pelet, R., and Ungerer, P., 1987, Thermal history of sedimentary basins, maturation indices, and kinetics of oil and gas generation: Am. Assoc. Petroleum Geologists Bull., v. 71, no 12, p. 1445-1466.

Ungerer, P.O., Bessis, F., Chenet, P.Y., Durand, B., Nogaret, E., Chiarelli, A., Oudin, J.L., and Perrin, J.F., 1984, Geological and geochemical models in oil exploration: principles and practical examples, in Demaison, G., ed., Petroleum Geochemistry and Basin Evaluation: Am. Assoc. Petroleum Geologists Mem. 35, p. 53-77.

Welte, D.H., and Yukler, M.A., 1981, Petroleum origin and accumulation in basin evolution - a quantitative model: Am. Assoc. Petroleum Geologists Bull., v. 65, no. 8, p. 1387-1396.

WELL-LOG IMAGING AND ITS APPLICATION TO GEOLOGIC INTERPRETATION

L. Huang, D. Richers, and J. E. Robinson
Syracuse University, Syracuse, NY

ABSTRACT

Modern image-processing techniques (IP) can be useful in the display and interpretation of well-log data. By enhancing the normal presentation of log data, the application of these techniques makes it easier for geologists and other users to understand the log data. Many sophisticated image-processing and interpretation methods that have been developed and efficiently used in other sciences could be introduced successfully into the logging industry. For example, IP can be a powerful tool in log analysis by applying enhanced log images to subsurface geology, petroleum engineering, and especially in oil reservoir description.

INTRODUCTION

Since the early 1970's, image processing and interpretation techniques have progressed rapidly and are used widely in the study of the world's resources. Almost all of the major oil exploration companies have the capability to process and analyze remotely sensed image data. Thus, image techniques are a necessary tool for exploration, forest, soil, water-resource, and land-use studies. To date, however, this powerful tool and

Computerized Basin Analysis, Edited by J. Harff
and D.F. Merriam, Plenum Press, New York, 1993

its associated high-level techniques have received only limited applications in well logging.

Well logging is an important branch of exploration geophysics that spans many geologic disciplines touching almost all earth scientists. Currently, there are more than fifty different types of sensing instruments, termed logging tools, being used to measure the physical and geometric properties of the subsurface rocks encountered in drilled boreholes. Well logs provide continuous, in-situ measurements related to lithology, porosity, saturation of hydrocarbons, bed thickness, and other rock properties. Conventional well-log record data have a standard sampling rate of two samples per foot, whereas some high resolution tools sample on a much smaller interval(micrologs). The number of data values collected in a well of 10,000 ft depth, where 10 or different logging tools are used results in a fairly large digital file.

Once a well is drilled, the only economical and efficient method of evaluating the penetrated formation is the use of logging techniques. Although every well log is important, they may not be used properly and in an efficient manner. One reason might be that traditional well-log files are offered only in the form of graph with sets of curves displayed as a function of depth. In this form they may be difficult to interpret. Some casual log data users even consider well logs to be "black magic "or "voodoo science." This situation could be changed if image techniques are introduced successfully into log display and interpretation.

LOG FILES

The simplest log data file contains a single data set, displaying a single curve, such as a gamma-ray log, a neutron log, a density log or a caliper log. Each example being a sequential data set with sampling order numbers (depth index), described as a one-dimensional vector

$$x(k)=w(k \times dh) \qquad k=1,2,...,n \qquad (1)$$

where :
k	=	sampling order number or depth index
dh	=	sampling interval
w	=	well-log data at depth of (k ¥ dh)
x(k) =		digitized well-log data set
n	=	the last sampling order number, number of rows

There are some special log data files, such as the Formation Microscanner (FMS), the Borehole Televiewer Tool (BHTV), and the Well borehole Seismic, where the data file for each is a two-dimensional vector described as

$$x(j,k) \quad j = 1,2,\ldots,m; \ k=1,2,\ldots,n$$

where :

j	=	measurement number or column number
m	=	number of columns
k	=	row number
n	=	number of rows

To meet the needs of the geologist, these log data files are organized generally into a single multilog (multiband) data file. In this situation, one file typically contains all the data from one logging run or one selected set of measured or computed results. There are data from a number of different curves or logs at each sampling point corresponding to a specific depth. This results in a three-dimensional vector corresponding to a group of logs. The vector is written as follows:

$$x(i,j,k) \qquad i = 1,2\ldots,l \ ; \ j=1,2,\ldots,m \ ; \ k=1,2,\ldots,n$$

where :

i	=	log number, band number
l	=	number of bands
j	=	column number
m	=	number of columns
k	=	row number
n	=	number of rows

LOG IMAGE

An image, as displayed on an image processor display device, consists of a large number of picture elements (pixels) laid out in m columns and n rows. A pixel is displayed as a single point of light on a video screen. In a black-white image, each pixel has a specific gray-level value and can be located by its column and row numbers as an x,y coordinate. One single set of data corresponds to a set of pixels and can be displayed as a single line or a single row image that is just a straightline with different brightness or saturation values for each pixel contained on it. It is necessary for an interpretable image to have a large numbers of columns and rows to be analyzed by human vision or by computer-based methods. Typically each pixel can vary in brightness from zero (off) to 255 (full brightness) for each of the three primary colors. Thus, on a 24-bit system each pixel could be one of the 16.8 million colors depending on its properties. In a b/w system, usually only 256 levels are possible.

The volume of rocks which contributes to the log measurements is limited. This volume can be considered approximately as a column with a limited radius and height as shown in Figure 1. Thus, it is reasonable to split the column into specific m subcolumns, each of which contributes to the overall log- measurement values. Now we have an equation:

$$x(1,k) = \frac{1}{m} \sum_{j=1}^{m} x(j,k) \tag{2}$$

where:

$x(1,k)$	=	mono-log value at depth k
j	=	subcolumn number
m	=	number of subcolumns within the column
$x(j,k)$	=	the log parameter value of j-subcolumn at depth k

If the parameters of the rock are reasonably homogeneous, then $x(1,k) = x(j,k)$, and a log curve can be duplicated m times to produce a

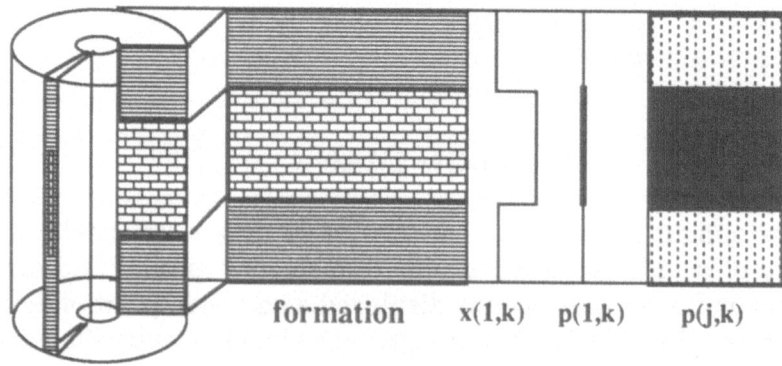

formation   x(1,k)   p(1,k)   p(j,k)

Figure 1. Log image developing.

Figure 2. Gamma-log graphical representation.

matrix with m columns and n rows, in a digital data image. The value of pixel x(j,k) represents the log measured parameter value at sampling point k corresponding depth of (k¥dh). Figure 2 is an example of a gamma-ray log image composed of well-log data from western New York State.

In color-video technology (true color or composite color), different colors are represented on the display screen by varying the intensity levels of light to their red, green, and blue 'color guns' within the video display unit. These color and intensity levels of a pixel are controlled by three variables or three sets of data. For example, the gamma-ray, neutron and density curves can be assigned to red, green, and blue guns, respectively, then a color image will be displayed on the screen. Different types of lithologies then can be shown by specific colors. In principle, combinations of three color beams with 256 brightness levels can be used to define up to 16.8 million properties of rocks. In the situation of a single log curve, a pseudocolor mode can be taken. It operates on a single image memory plane from which the pixel values are processed through the three function memories for the three color guns, and the resulting values then are displayed on the screen. This results in a " rainbow" or density slice display. A program named ROYGBIV in the ERDAS@ system, which can assign red, orange, yellow, green, indigo, and violet colors to specific areas within the display, by spreading the range of values in the file over the spectrum at equal intervals. Even so if there is only one log curve, such as a Gamma Ray, a color image can be obtained, indicating up to seven different lithologies with specific colors.

There are four different coordinate systems to work with: (1) pixel or screen coordinates; (2) file or data file coordinates; (3) database coordinates; and (4) map coordinates. File coordinates refer to the locations of pixels within the data file. A gamma image is a file. The row number of each pixel is related to the depth at which it was taken. The coordinates of the center point of a log image are used to locate the image on the specific screen coordinates. This is useful for multiwell-image display and to compile an image from a number of images. Database coordinates are used to locate a data file within a database that covers too much area to be included in a single data file. A correlation section of wells can be considered as a database with each well as a single data file. Each well-log image can be laid out within a database coordinate system to be compiled into a new large file as shown as in Figure 3. Finally the map coordinates can be used to make well locations related to a specific geographic coordinate system.

IMAGE ENHANCEMENT

Image enhancement is the process of making an image more interpretable for a particular application. After enhancement, the appearance of an image usually is improved. Two types of enhancement techniques, spectral and spatial, are available for log image processing.

Spectral Enhancement

Spectral enhancement is concerned with the adjustment of individual values of pixels in the image in which the brightness of each pixel is modified independently, not taking the values of neighboring pixels into account. Spectral enhancement may be referred to contrast enhancement or contrast stretching, because the goal of this technique is to make

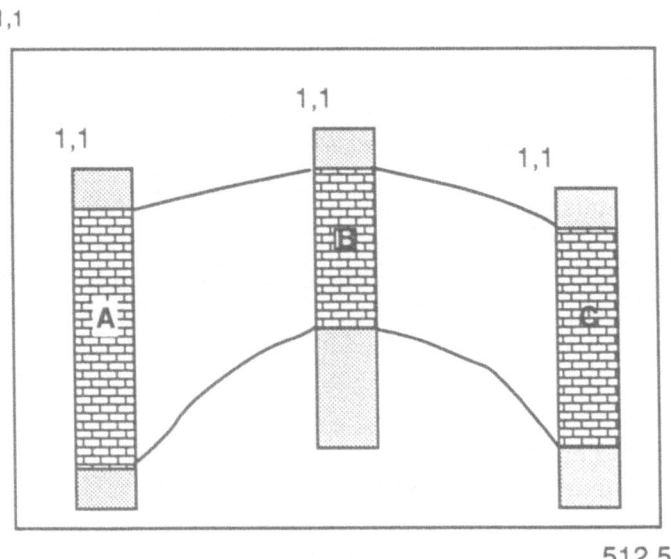

Figure 3. Database coordinate system and data file coordinate system A, B, C.

certain features more visible than old originally low contrast image by enhancing or stretching the contrast either linearly or nonlinearly.

Using a gamma-ray log as an example, suppose the range of the gamma value is 10 to 100 API throughout the entire well, but for the interval of potential reservoir rocks, the gamma intensity ranges only from 10 to 40 API. Obviously, the initial contrast is poor. To perform a linear contrast enhancement, the range of log readings will be stretched to use the entire brightness range of the display device, which covers 256 levels (0 - 255), and the transformed brightness value for one individual log reading will be computed by the equation

$$P(j,k) = int\left[\frac{x(j,k) - xmin}{xmax - xmin}\right] \times 255 \qquad (3)$$

where:
p(j,k) = pixel (j,k) brightness corresponding to x(j,k)
x(j,k) = the original log measurement, GR in API units

xmin = minimum of gamma-ray log
xmax = maximum of gamma-ray log
int   = integer function

In this example, the transform function is a straightline. According to the data distribution characters and user's goal, different forms of transform functions can be applied, as shown in Figure 4, whereas the results are given in Figure 5. In this illustration, the gamma-ray data are from Well 031-013-04154, Medina Formation, western New York State. There are six tracks in Figure 5, each displays an image enhanced by a specific method:

A – linear enhancement
B – two level slices
C – multilinear enhancement
D – double logarithmic enhancement
E – zero-logarithm-logarithm enhancement
F – histogram equalization.

Histogram equalization is one of the more likely used nonlinear contrast enhancement technique. To apply this enhancement method, the histogram of an image is derived, and the number of output gray-scale classes, into which the data are to be redistributed, are specified to assign

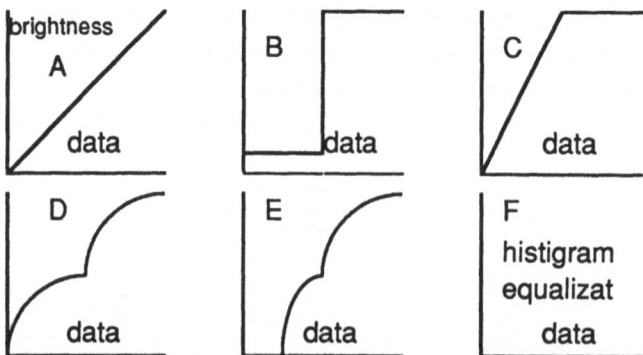

Figure 4. Spectral enhancement transformers: A, linear enhancement; B, two level enhancement; C, multilinear enhancement; D, double logarithmic enhancement; E, zero-logarithm-logarithm enhancement.

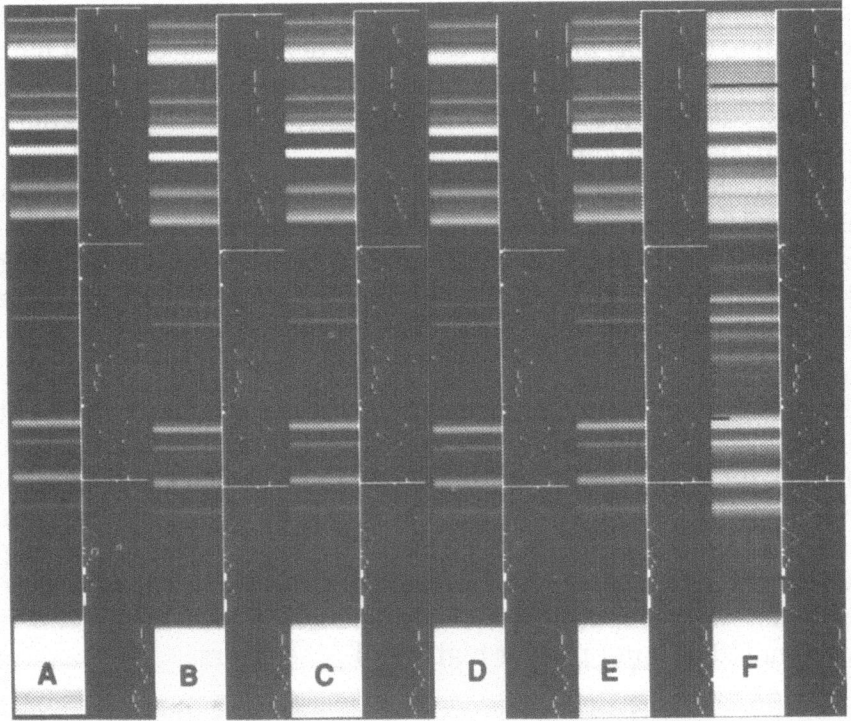

Figure 5. Spectral enhancement (gamma ray).

approximately an equal number of pixels to each of the output gray-scale bins. The result is a nearly flat histogram, so that most of the populated ranges gain a greater contrast and the pixels with a lower probability of occurrence lose their contrast. If the features we want to extract have high probability of occurrence in the image, histogram equalization is a powerful tool. Unfortunately, in most logging environments, the meaningful pixels usually have less probability than the majority of less useful points which makes histogram equalization less efficient. The multi-transformer function shown in Figure 5E seems reasonable for the log user.

Spatial Enhancement

Spatial enhancement modifies pixel values based on the values of surrounding pixels. This technique is concerned mainly with spatial frequency, defined as the number of changes in brightness values per unit distance for any particular part of an image. The image with zero spatial frequency is flat, a low spatial frequency image results in a smoothly varying image, whereas an image consisting of a checkerboard of black and white pixels has the highest spatial frequency. The basic spatial enhancement technique is spatial filtering. Convolution filtering is one of many of spatial filtering methods.

To perform convolution filtering, a matrix named a convolution kernel is used to calculate the weighted average value of each pixel with values of surrounding pixels in a particular way. The result is to change the spatial frequency of the initial image and to enhance certain features in areas of interest. The kernels frequently used are as follows:

(1)   Zero Sum Kernels are kernels in which the sum of all coefficients (elements) equals zero. When applying this type of kernels to an image, the output pixel values will be zeros in areas where all input values are equal, low values in areas with low spatial frequency, and extreme values (very high or very low) corresponding to high frequency. So it can be used as an edge detector to create a sharp contrast where spatial frequency is high. These filters can be designated as band-pass or high band-pass filters.

(2)   High-Frequency Kernels are used to enhance edges. If a high-frequency filter is applied to an image, a relatively low value surrounded by higher values will be transformed into an even lower value, conversely, a relatively higher value surrounded by lower values will upon filtering, become even higher. A high pass kernel usually is referred to as an edge enhancement or zero sum filter.

(3)   Low Frequency Kernels or low pass kernels decrease spatial frequency and make the image smoother by eliminating the high-frequency noise (such as smoother and despeckle enhancements). The sum of the coefficients in such a filtering scheme usually equals one.

The spatially filtered results are seen in Figure 6. The initial image is given in the first left track in the graph for comparison, then five different resulting images from specific enhancement methods (SMOOTH, SHARPEN, TRACE EDGE, REDUCE NOISE, and DITHER of Macintosh@ Image1.29).

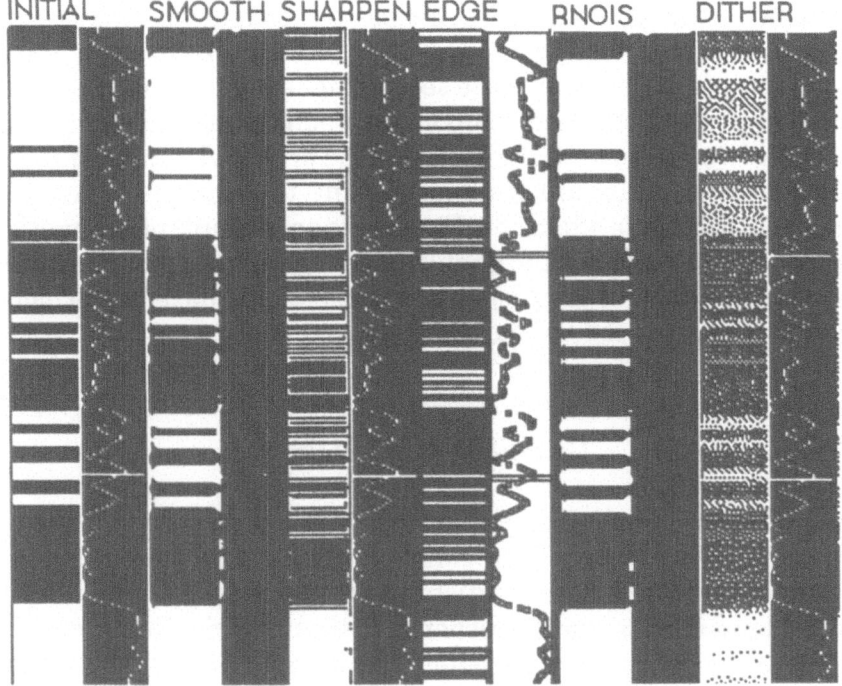

Figure 6. Spatial filtered results.

Multiband Enhancement

As mentioned, the number of log traces available to the interpreter can be large, an in fact, may exceed thirty. In order not to lose vital information methods that compress the number of bands down to less than or equal to three may be applied. Principal components analysis (PCA) is used as a band compression method to make the data more interpretable.

In a n-dimensional histogram, if the data value distributions of each input band are normal or near normal, a hyperellipsoid is formed. To perform principal components analysis, the axes of the spectral space are rotated, changing the coordinates of each pixel in spectral space. The resulting principal components are specific linear combinations of random or statistical variables (log-data sets) which have special properties in terms of variances. The first principal component has the greatest variance, measuring the highest variation within the data; the second principal component measures the second highest variation of the data, and so forth. The first few components (bands) account for a high proportion of the variance in the data, and the principal components with smaller variances often are ignored, making the number of resulting bands fewer than the original one. Figure 7 is an example of this method. In Figure 7, the left side is the original image represents a gamma-ray log, a neutron log, and a density log. The right side displays the principal components image based on four logs (plus caliper).

CLASSIFICATION

Multispectral (multiband, multilog-data) classification is the process of sorting pixels into a finite number of individual classes based on their data file values. If a pixel satisfies a certain set of criteria, the pixel is assigned to the class that corresponds to that criteria. In the situation of well logs, the purpose of classification is to determine the types of lithologies and reservoir categories based on the set of logs.

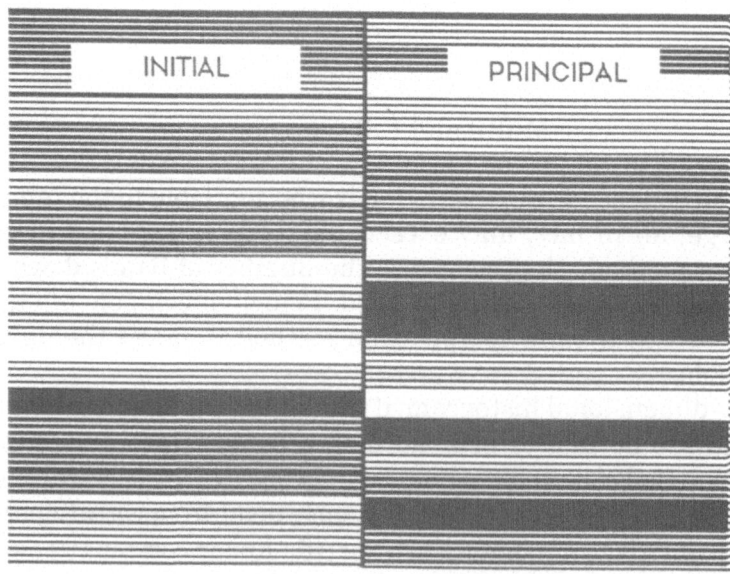

Figure 7. Principal enhancement.

To perform classification of an image, a parameter named the spectral distance is needed. There are several ways to define a spectral distance, one of them is Euclidean distance, defined as:

$$SD_{xyc} = \sqrt{\sum_{i=1}^{n} \left(m_{ci} - X_{xyi} \right)^2} \qquad (4)$$

where:

SD_{xyc} = spectral distance from pixel (x,y) to the mean of class c
n = number of bands (dimensions, number of logs)
i = a particular band, log
c = a particular class, lithology
X_{xyi} = data file value of pixel (x,y) in band i
ms_{ci} = the mean of data file values in band i for the sample for class c

Figure 8 gives the results derived from a set of logs by using different classifiers in each part of the image. The original data were gamma ray, neutron, density, and caliper through an interval of 256 feet. The explanations of the markers at the areas of the image are given as follows:

(1) MLK-MAXIMUM LIKELIHOOD/BAYESIAN

The Maximum Likelihood Decision rule is based on the probability that a pixel belongs to a particular class. The basic equation assumes that there probabilities are equal for all classes, and that the input bands have normal distributions. This procedure produces the most interpretable characteristics. When compared with other methods, MLK yields more accurate classification of the before mentioned reservoir.

(2) MIN—MINIMUM DISTANCE DECISION RULE

When the spectral distance (SD) of a pixel can computed for all possible classes, using Equation (4), the class of the candidate pixel is assigned to the class for which SD is the lowest.

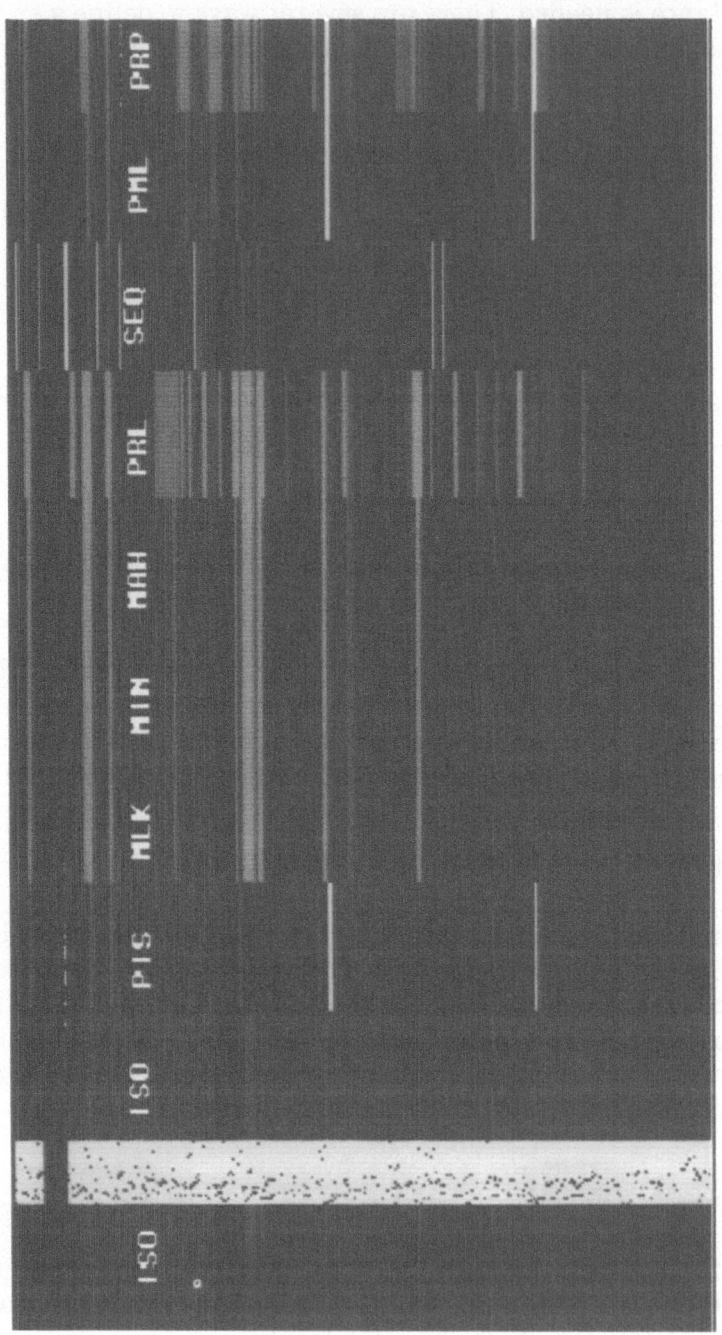

Figure 8. Classification (from GR, NT, DEN, CAL).

(3) MAH—MAHALANOBIS DISTANCE DECISION RULE

Mahalanobis distance is defined as:

$$D = \left(X - M_c\right)^T \left(Cov_c^{-1}\right) \left(X - M_c\right) \tag{5}$$

where:

D	=	Mahalonobis distance
c	=	a particular class
X	=	the measurement vector of the candidate pixel
M_c	=	the mean vector of the signature of class c
Cov_c	=	the covariance matrix of the pixels in the signature of class c
Cov_c^{-1}	=	inverse of the covariance matrix
T	=	transformation function

The pixel is assigned to the class c, for which D is the lowest. Both the results of MIN and of MAH show little differences from the MLK, the classification products are persistent.

(4) ISO—ISODATA CLUSTERING

It is an "Interactive Self-organizing Data Analysis Technique." This technique repeatedly performs an entire classification and recalculates statistics, locating clusters with minimum user input.

The ISODATA program uses minimum spectral distances to assign a cluster for each candidate pixel. Beginning with a specified number of arbitrary cluster means, the process is applied repetitively until there is little change in cluster means between iterations. The classification results for the data set in Figure 8 is listed as :

CN	1	2	3	4	5	6	7	8	9	10	11	12	13	14	15	16
%	0.2	0.0	42.0	0.0	0.0	0.9	2.7	3.5	5.4	2.6	7.0	5.6	7.2	8.3	8.3	6.2

CN – cluster number
% – percent of total pixels

The data distribution is plotted in Figure 9. According to this graphic, one can assumes that there are only two major types of litholo-

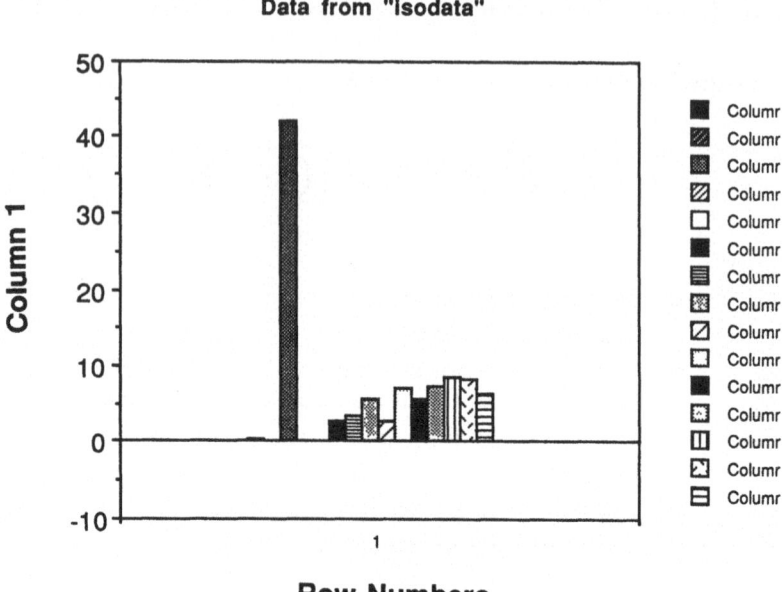

Figure 9. Distribution of classes.

gies, one of them has a percentage of 42%, the other can be divided further into certain subclasses.

(5) SEQ—SEQUENTIAL CLUSTERING

In this method, pixels are examined one at a time, beginning with the upper-left corner of the image and going left to right, line by line. The measurement vector of the first pixel becomes the mean vector of the first cluster. Then the spectral distance between the first and second pixels is calculated. If this distance is greater than R (the minimum spectral distance that will separate the means of two clusters), then the second pixel becomes the beginning of a new cluster, otherwise the second pixel is considered as a member of the same cluster with the first one, and the first cluster will have a newly calculated mean and location. Then the spectral distances between each of the analyzed pixels and the means of previously defined clusters are calculated sequentially, each pixel either joins a nearest existing cluster, or begins a new cluster, based on the spectral distances. Results are comparable with those from ISO,MLK,MIN, and MAH.

(6) PIS

The ISODATA method applied to a principal components image for the same initial data set. It gives more details of the picture.

(7) PRL—PARALLELEPIPED CLASSIFIER

To perform classification with the parallelepiped classifier, the data file values of the candidate pixel are compared to certain upper and lower limits for every signature in every band. When a pixel has values between the limits for every band in a signature, then the pixel is assigned to that class.

(8) PML

The Maximum Likelihood method applied to principal components image.

(9) PRP

The Parallelepiped method applied to principal components image.

The ERDAS@ classification programs offer an excellent "toolbox" to be used as needed. However, selection for a specific objective such as log analysis is difficult and requires practice.

APPLICATIONS

The development of image-processing techniques and their introduction to well-logging domain will increase the potential for geologists to improve geological solutions. Image methods are favorable for following applications:

(1) Lithology Identification / Litho-image Reconstruction

There are a variety of ways to display a well-log image and to perform classification. For instance, a red color image from the gamma-ray log; a green color image from the neutron log; a blue color image from the density log; a red-green color image derived from the gamma ray and neutron logs; a red-blue color image derived from the gamma ray and density logs; a green-blue color image derived from the neutron and density logs; a RGB three color image of gamma ray, neutron, and density logs. Any one of the logs can be used to construct a selected color image and can be combined with any one or more logs to produce a composite color image to display specific lithologies. For example, here the gamma-ray can be assigned to a red color, the pixels with higher red brightness will correspond to higher shaleness. When the gamma-ray log is combined with the neutron, and assigned to red and green colors respectively, both radioactivity and hydrogen content become discernible based on both color scheme and brightness. Various combinations of selected logs are available to meet different needs, such as bedding logged intervals, lithology identification, clay typing, water/oil or gas contact locating.

(2) Deposition Facies Study

In a study of depositional facies, an image gives the facies transient characteristics. Figure 10 shows a set of deposit models showing the "bell", "box", and "funnel" imagery displays. Geological phenomena, such as transgression, regression, unconformity, and overlap, may be extracted from corresponding images.

(3) Well Wall Image

At present there are two types of logging tools that can obtain well wall images. They are the Formation MicroScanner (FMS) and the Borehole Televiever Tool (BHTV) as mentioned previously.

The FMS tool obtains oriented, two-dimensional, high-resolution imagery of microresistivity variations recorded by four imaging arrays, each of the four has 27 small electrodes apart 0.1 inch from each other in lateral.

(4) Well-Log Correlation

Correlating the similar patterns in a set of log traces from well to well when the wells are a long distance apart is difficult, image techniques can make it a little easier. Figure 11 gives an example with 15 well-images of the Silurian Medina Formation widely distributed in western New York State. All of the gamma traces are normalized. Figure 12 is a location map of these wells.

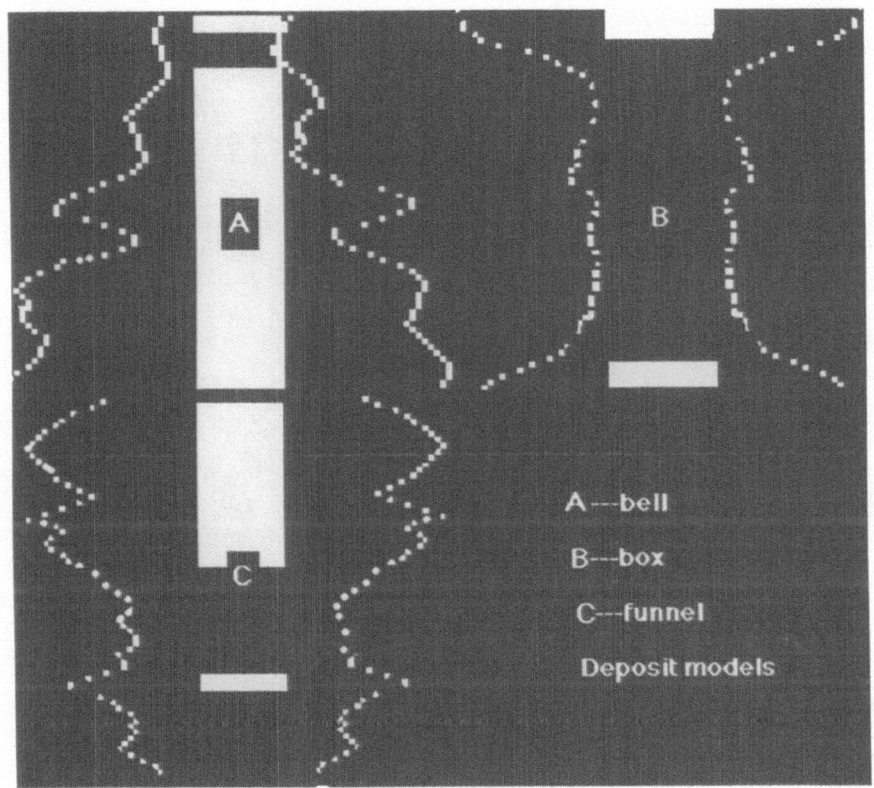

Figure 10. Gamma-ray depositional models.

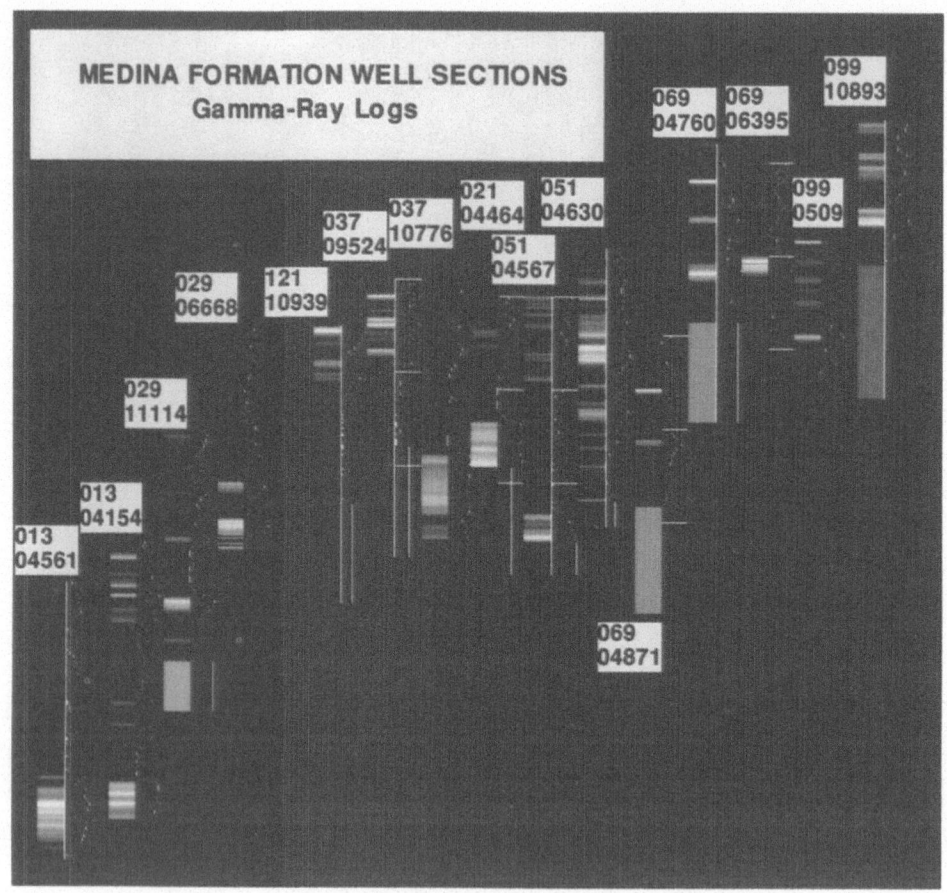

Figure 11. Medina Formation well sections, gamma-ray logs.

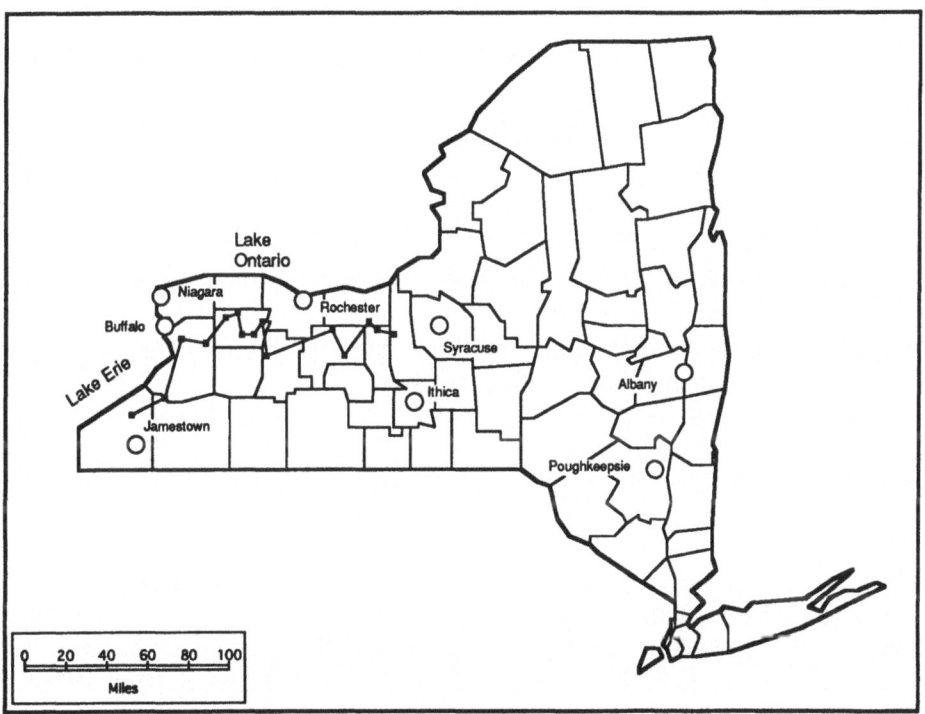

Figure 12. Well location map for gamma ray (Fig. 11).

(5) Reservoir Description

In reservoir description, the problems most difficult to resolve are those related to data matching, such as between core-data and log-data depths and log and seismic data pattern matching. In considering depth matching of cores and logs, it may be difficult because of missing sections or physical deterioration of the core. Introducing image-processing techniques to help correlate the core-image and log-image can facilitate this task through the use of histogram matching and image compiling techniques.

CONCLUSIONS

(1)   It is possible and useful to introduce image techniques into log data processing and interpretation.
(2)   Image visualization techniques can make log information easier for geologists to understand.
(3)   Image methods are ideal for rock typing, facies study, section correlation, and reservoir description.

REFERENCES

Ellis, D.V., 1987, Well logging for earth scientists: Elsevier Science Publ. Co. Inc., New York, 532 p.

Ekstrom, M.P., 1986, Formation imaging with microelectrical scanning arrays: Trans. SPWLA Annual Logging Symposium, p. BB1 - BB21.

ERDAS Field Guide,1990, v. 7.4, ERDAS, Inc., Atlanta, Georgia, 410 p.

Etnyre, L.M., 1988, Finding oil and gas from well logs: Van Nostrum Reinhold, New York, 305 p.

Hord, R.M., 1986, Remote sensing/methods and applications: John Wiley & Sons, New York, 362 p.

Schlumberger Corporation, 1987, Log interpretation principles/ applications: Schlumberger Ed. Services., Houston, 198 p.

Tissot B.P., and D.H.Welt, 1978, Petroleum formation and occurrence (2nd ed.): Springer-Verlag Publ. Co., Berlin, 699 p.

Walker, R.G., 1984, Facies models, (2nd ed.), Geoscience Canada Reprint Series 1, Geol. Assoc. Canada, Toronto, 317 p.

INTERACTIVE DEPOSIT MODELING WITHIN LIGNITE EXPLORATION

K.-D. Hanemann and K.O. Zeissler
Geological Survey of Saxony, Freiberg, Germany

ABSTRACT

The exploration of lignite deposits in eastern Germany is based on c. 2000 - 5000 boreholes as well as results of geophysical exploration such as gravity, seismic, and well-log measurements. The resulting amount of data can only be interpreted by computers using modern database systems, geostatistical methods, and interactive computer graphics. The Geological Survey of Saxony in Freiberg developed a computer-based technology for this purpose, demonstrated at a lignite exploration project with 2400 boreholes. Methods for the evaluation of the information density and the distribution of geological horizons or characteristics are presented. Furthermore, possibilities of graphical representation and interactive modeling of the deposit are discussed.

GEOLOGICAL SITUATION IN THE INVESTIGATED AREA

The Tertiary lignite deposit Delitzsch-Süd is located in the Middle-German lignite district some kilometers northeast of the town Leipzig (Fig. 1). The total area of geological investigations is about 225 km². From the geological viewpoint, this area belongs to the North-Saxonian anticlinorium.

Computerized Basin Analysis, Edited by J. Harff
and D.F. Merriam, Plenum Press, New York, 1993

The pre-Tertiary basement consists of Proterozoic graywacke, lower Cambrian sandstones, and Cambrian and Ordovician granodiorites. In some regions Permian rhyolites occur. The upper part of the pre-Tertiary is deeply eroded (Fig. 2).

The Tertiary sedimentation began in this area during the upper Eocene. In this time, sands, silt, clay, and some lignite layers were deposited. Among the four lignite layers occurring in the investigated area only the lignite bed BiO1 (Bitterfelder Oberbank 1) is of economic interest. Its mean thickness is 4.7 m. It ranges from 1.2 m to 10.8 m. The mean depth to the lignite bed is about 50 m, so the proportion between the thickness of the overburden and the thickness of the lignite bed is about 10:1.

OBJECT OF GEOLOGICAL INVESTIGATION

The detailed objects of investigation have been the following:

- base of the Quaternary
- distribution, structure, and thickness of the lignite bed

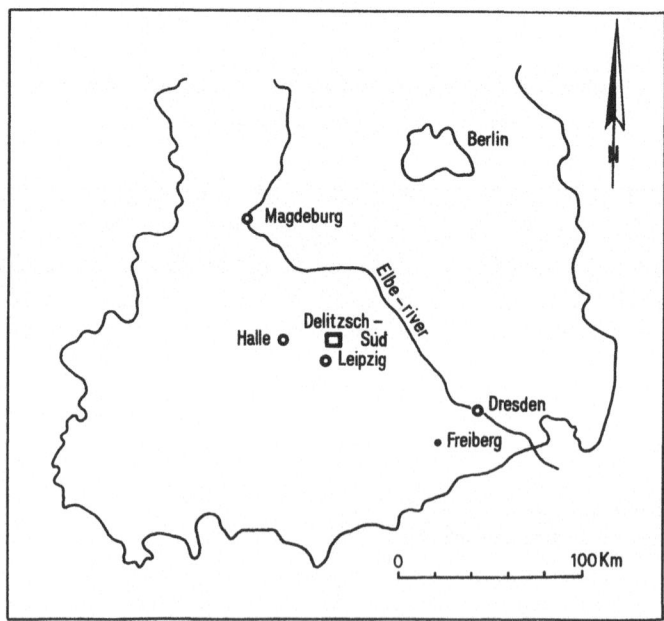

Figure 1. Location of Delitzsch-Süd-lignite deposit.

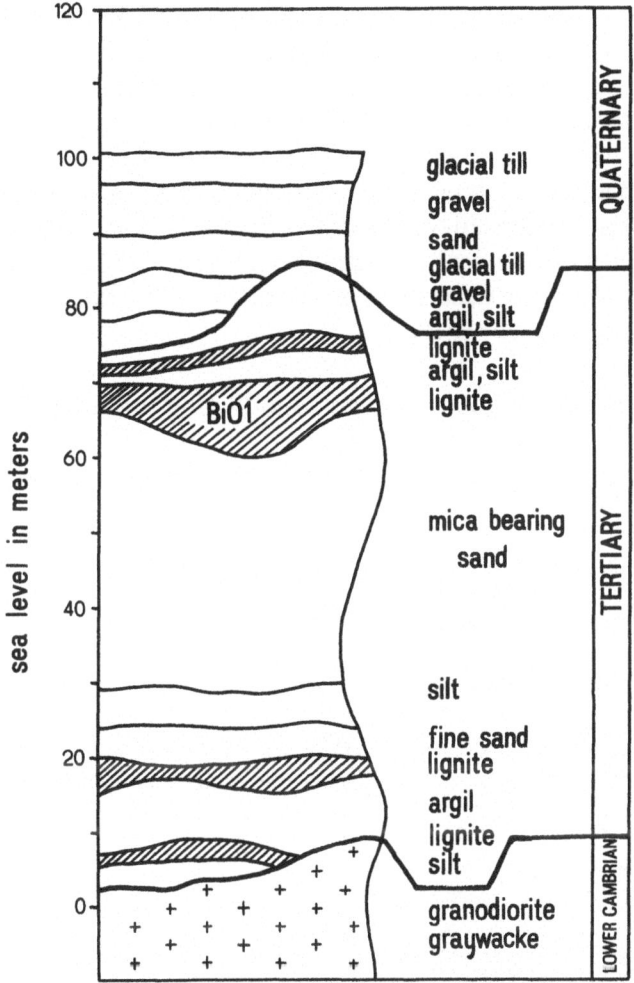

Figure 2. Generalized stratigraphic column of investigated area.

- relief of the top of the mica-bearing sand (which is equal to the bottom of the lignite bed, if this occurs)
- top of the pre-Tertiary.

DATABASE

The basis of the deposit modeling is the database SVZ (Schichtenverzeichnisse) including all the lithological and stratigraphical information about the boreholes, that is all geological information about the investigated area (Seifert, Kleinstaeuber, and Hanemann, 1991).

Storage and processing of the data are oriented spatially, but represented two dimensionally. The main graphic applications are map construction, display of point distribution, and profile construction. In layer oriented areas such as lignite deposits, we may use the top of the layer, its bottom, or the thickness of the layer for the description of the deposit. Besides numerical parameters of the strata description, we defined a new one, that describes the condition of the borehole in relation to the investigated layer (Fig. 3).

The symbols in Figure 3 have the following meaning:

" - ": The layer is present in the borehole and it was penetrated by this borehole.

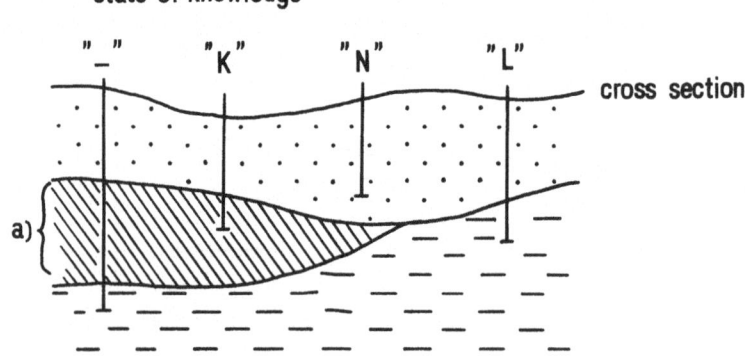

Definition of the term

"state of knowledge"

the geological unit a) is:

"–" present and drilled

"K" present but not drilled

"N" unreliable

"L" not present

Figure 3. Definition of term "state of knowledge".

" K ": The layer is present in the borehole, but the borehole reached only the top of the layer, not the bottom.

"N ": There is no information about the presence of the layer; the borehole did not reach the stratigraphic horizon.

" L ": The layer is not present in the borehole. The borehole reached the stratigraphic horizon but did not prove the layer.

These parameters have a great influence on the graphical presentation of strata parameters. In the following we will demonstrate this by some selected examples of the Delitzsch-Süd lignite deposit.

RESULTS

We will discuss the information about the geological objects shown in Table 1.

The total number of boreholes, that can be used for the description of the deposit is 2414. Because only boreholes with the parameter "-", "K", and "L" (column 1, 2, and 4) describe exactly the stratum, one can see, how the number of boreholes decreases with the depth of the stratum. For the general description of this dependence we defined the parameter "knowledge-accuracy", that is the number of boreholes that can be used for the description of a strata divided by the total number of boreholes.

The following figures show the knowledge-state of the discussed horizons (Fig. 4).

For Figure 5 the boreholes with the parameter "N" (no information about the presence of the layer) have been ignored.

Figure 6 shows the complete state of knowledge, including boreholes with the parameter "N". Generally the mica-bearing sand represents the underlying horizon of the lignite bed. The high amount of boreholes, that have been finished within the mica-bearing sand results from the object of geological investigation, that is the exploration of the lignite bed. Boreholes with the knowledge-state "K" (present, but not run through) may be used of course for the description of the top of a layer, but not for the modeling of the thickness or the base of the layer.

Figure 7 shows, that the pre-Tertiary is present everywhere. But it shows also, that only 15% of the boreholes reached the top of the pre-Tertiary.

Table 1. Data amount and state of knowledge in investigated area

Formation	- (1)	K (2)	N (3)	L (4)	knowledge-accuracy
base of the Quaternary	2123	289	0	2	1.00
thickness of the lignite bed	1820	93	340	161	0.86
top of the mica-bearing sand	345	1573	451	45	0.81
top of the pre-Tertiary	0	356	2058	0	0.15

$$\text{knowledge-accuracy} = \frac{(1) + (2) + (4)}{(1) + (2) + (3) + (4)}$$

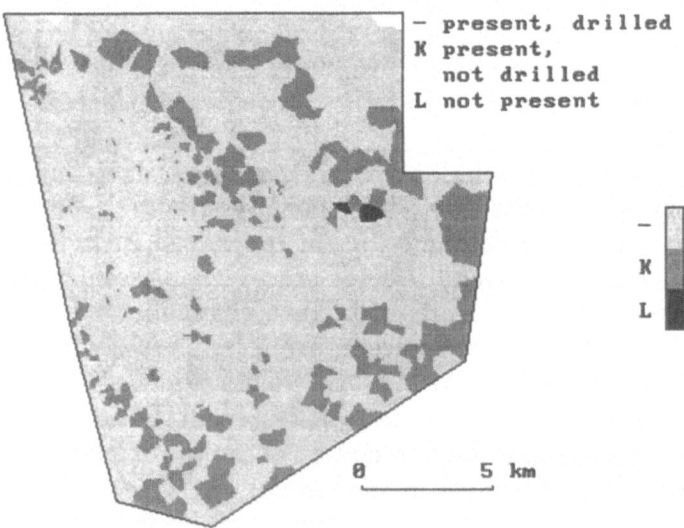

Figure 4. Knowledge-state of Quaternary rocks.

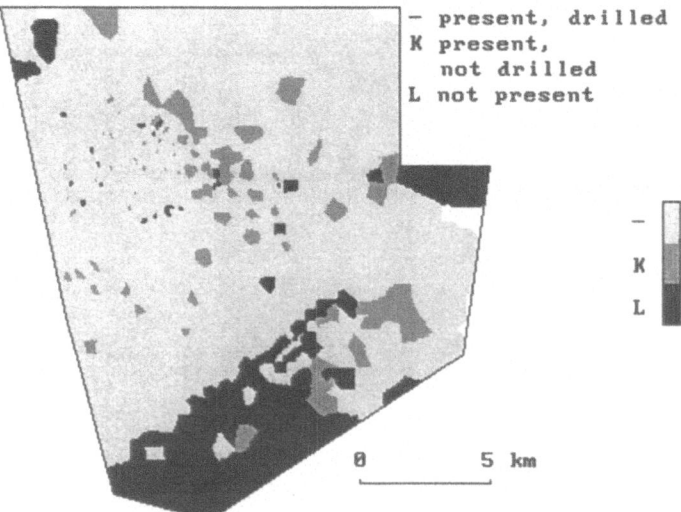

Figure 5. Knowledge-state of lignite layer.

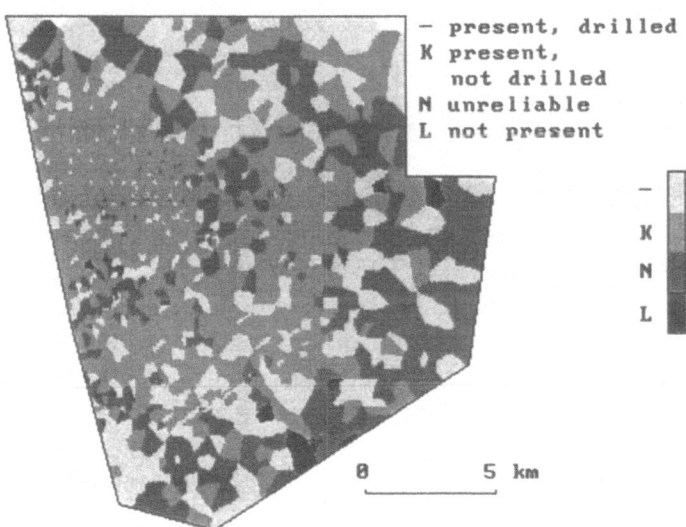

Figure 6. Knowledge-state of mica-bearing sand.

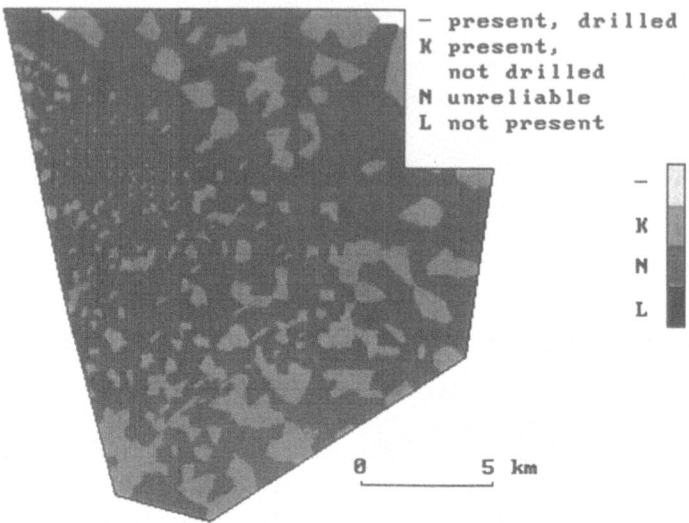

Figure 7. Knowledge-state of pre-Tertiary.

TOP, BASE, AND THICKNESS OF A LAYER

After the discussion of the state of knowledge and the distribution of
the main interesting horizons, we will show the geological models of these
horizons.

The bottom of the Quaternary (Fig. 8) is determined by runway
structures. They are caused by glacial processes. In Figure 9 one can see
the thickness of the lignite bed. The boundary of the lignite bed is visible
clearly in the south and southeast of the picture. Remarkable are some
NE-SW-striking structures of low lignite thickness. These structures are
caused essentially by the mica-bearing sands, underlying the lignite bed.
Figure 10 shows the top of the mica-bearing sand.

Figure 11 shows the general uplift of the pre-Tertiary from north to
south. The white areas in the picture indicate large borehole distances.
They are caused by the method of map-construction from single-point
map (point growing processes). Thus the white areas indicate a low state
of knowledge.

Lignite Deposit   DELITZSCH
base of the Quaternary

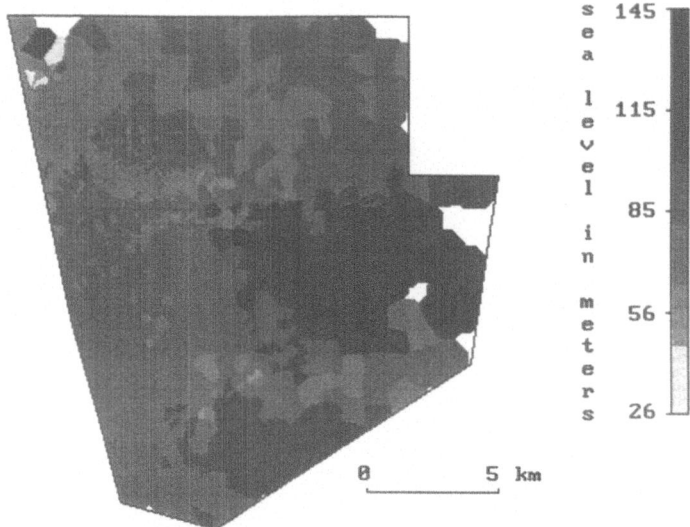

Figure 8. Base of Quaternary.

Lignite Deposit DELITZSCH
thickness of the lignite layer

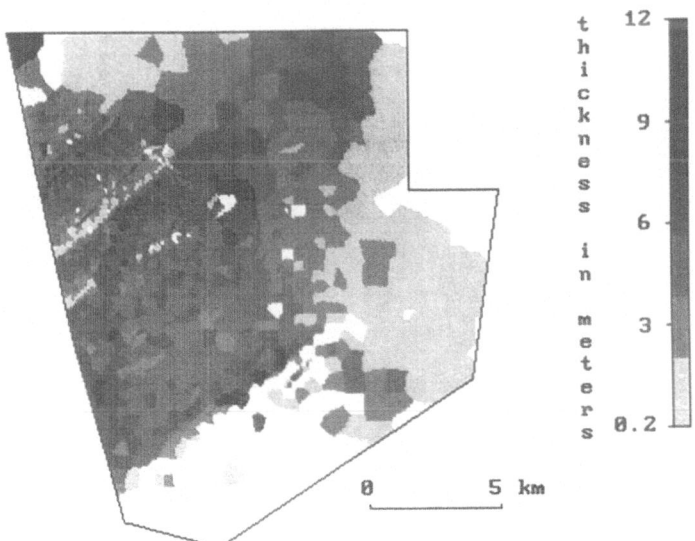

Figure 9. Thickness of lignite layer.

Lignite Deposit DELITZSCH
top of the mica-bearing sand

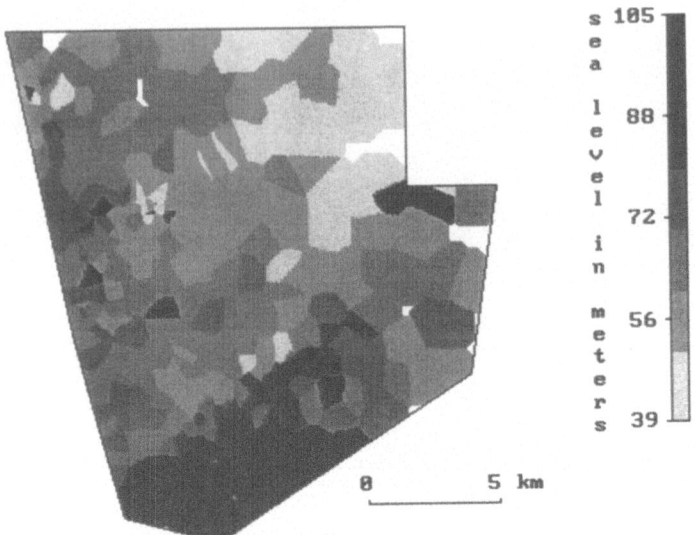

Figure 10. Top of mica-bearing sand.

Lignite Deposit DELITZSCH
top of the pre - Tertiary

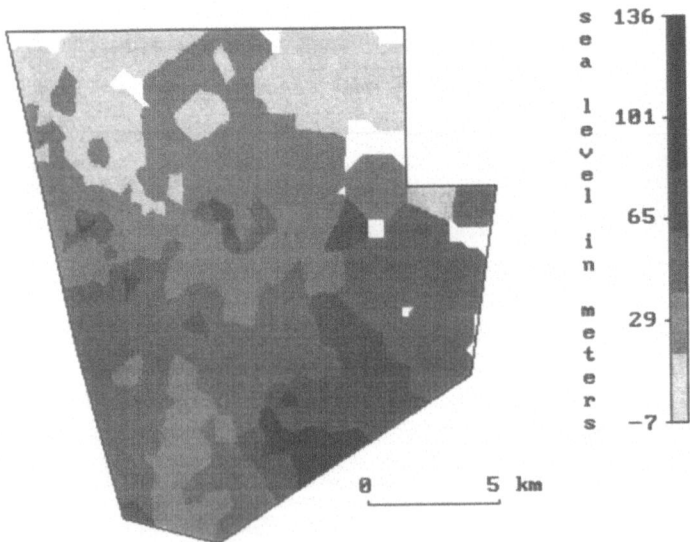

Figure 11. Top of pre-Tertiary.

Table 2. Comparison between state of knowledge and geometrical deposit data.

Type of graphical	state of knowledge			
representation	-	K	N	L
state of knowledge	*	*	*	*
distribution of a horizon	*	*		*
top of a layer	*	*		*
base of a layer	*			*
thickness of a layer	*			*

Table 3. Boreholes usable for graphical representation in Delitzsch-Süd-lignite deposit.

graphical representation	no. of boreholes	%
Quaternary, base	2125	88
Quaternary, distribution	2414	100
lignite bed BiO2, thickness	1981	82
lignite bed BiO2, distribution	2074	86
mica-bearing sand, top	1963	81
mica-bearing sand, knowledge-state	2414	100
pre-Tertiary, top	356	15
pre-Tertiary, knowledge-state	2414	100

COMPARISON BETWEEN STATE OF KNOWLEDGE
AND GEOMETRICAL DATA

If we ask, what boreholes can be used in connection to their knowledge-state to what graphical representation, we get Table 2.

From Table 2 we can use the number of boreholes shown in Table 3 for the graphical representation of the Delitzsch-Süd-lignite deposit.

Taking into consideration the state of knowledge of the boreholes we can guarantee a maximum of accuracy in the construction of geological maps.

SUMMARY

Summarizing these experiences we would say, that the use of computer-aided technologies for lignite exploration made it necessary to introduce a new parameter evaluating the information about a strata in the borehole. We term it "state of knowledge".

This parameter plays a central role in our data-processing system. During the entire process of modeling and utilization we make use of this parameter. The pictures shown in this paper represent only a small part of the entire computer-aided technology for lignite exploration. Utilization of geodetic data, chemical, and mechanical analysis and geophysical measurements are integral parts of this technology.

REFERENCE

Seifert, A., Kleinstaeuber, G., Hanemann, K.-D. ,1991, The computer aided construction of geological cross sections and maps within the system of computer aided geological exploration in the German Democratic Republic: Geol. Jahrbuch, A122, Hannover 1992, p. 243-256.

AN INTEGRATED APPROACH TO BASIN ANALYSIS AND MINERAL EXPLORATION

D. F. Merriam, B.A. Fuhr

*Stratigraphic Studies Group, Wichita State University,
Wichita, Kansas*

and

U. C. Herzfeld

Scripps Institution of Oceanography, La Jolla, California

ABSTRACT

An integrated approach to mineral exploration in basin analysis involves taking into consideration geological, geophysical, geochemical, and topographical data. Different spatial parameters such as structure, unit thickness, porosity, gravity, magnetics, paleotopography, etc., are important in locating potential areas of interest for mineral exploration. A complete analysis in precomputer times was accomplished using hand-prepared maps, many of which took considerable effort and time to construct. Now with the availability of databases and plotting/contouring software these tasks can be completed in a matter of minutes. Different combinations of maps can be compared or integrated to determine which data sets are the most informative, by comparing the results to a target map—one showing the location of the known mineral-type of interest. Most data have to be pretreated (normalized, standardized, transformed, etc.) before analysis, which is accomplished by (1) pairwise comparison, (2) stacking and weighting, or (3) combining. This approach then can be used to determine the optimum combination of data sets and weights for use in exploration. A mature oil-producing area in south-

Computerized Basin Analysis, Edited by J. Harff
and D.F. Merriam, Plenum Press, New York, 1993

central Kansas, which includes part of the Pratt Anticline, is used as an example of this integrated, automated approach. The area has a thin cover of Paleozoic sediments over the crystalline Precambrian basement. Most of the structures are the result of vertical movement of, and compaction of sediments over, basement features where oil/gas accumulations are present in structural and stratigraphic traps.

INTRODUCTION

An integrated approach to basin analysis requires that different areal properties be compared and integrated. Thematic maps concerning the geology, geophysics, topography (and paleotopography), and geochemical attributes need to be considered simultaneously in order to evaluate properly which are important in locating resources and interpreting the geologic history of the area.

Prior to the use of computers, analyses of this type were accomplished by visually comparing map properties. Perhaps, the geologist overlaid the maps in different combinations two at a time and interpreted the relationship of different spatial parameters such as structure, unit thickness, porosity, gravity, magnetics, paleotopography, etc. These maps were prepared manually, and because of the time involved in the preparation only a few were constructed. This meant that the geologist had to preselect those parameters deemed most important in the analysis.

Now, with the availability of computers and digitized databases, it is relatively easy to make many maps for study and to manipulate them in many ways. These comparisons can be accomplished by (1) pairwise comparison, (2) stacking and weighting, or (3) combining maps. This approach allows different combinations of maps to be compared or integrated to determine which data sets are the most informative.

Several methods may be used to pretreat the data. It is necessary, of course, to standardize the data if they are measured in different units. It also is necessary to grid the data before analysis just as it is with many quantitative analyses.

There are many problems with quantitative analyses, and most of those will not be discussed here. However, it is necessary to point out that data distribution, data interdependence, transformation of the data, and data reliability are important. The statistical technique used and the possibility of obtaining different results with slightly different data sets also should be considered.

The objective of this study is to apply the algebraic map comparison algorithm developed by Herzfeld and Merriam (1990) and Herzfeld and Sondergard (1988) to integrate a geological data set in a spatial model map that indicates favorable exploration sites. By searching for geologically meaningful weights for the individual data sets and comparing the result visually to a map showing known mineral deposits in a basin setting, it is possible to determine which combination of variables and in what weight they can be used to outline the favorable areas for future and more detailed study.

The area selected for study was a 9-township (324 sq mi, 840 km^2) area in south-central Kansas in Pratt County. In all, 900 wells were used in constructing the subsurface maps; published maps on the geophysics of the area also were considered. This area is a mature oil province and the data, therefore, are relatively easy to obtain. A thin veneer of Paleozoic sedimentary rocks covers a crystalline Precambrian basement. Rocks range in age from Cambro-Ordovician to Permian. The major structural feature is the north-south trending, south-plunging Pratt Anticline. The smaller subsidiary structures, trending north-northeast subparallel to the major features, are mainly the result of episodic vertical movements and compaction of sediments over basement features.

GEOLOGICAL SETTING

The major structural feature of the study area is the Pratt Anticline, located in northern Barber County, much of Pratt County, and extending northward into southern Stafford County (Fig. 1). The Pratt Anticline is a southward-plunging extension of the Central Kansas Uplift, bounded on the west, south, and east by elements of the Anadarko Basin. The southern limit of the Pratt Anticline is on the northern shelf of the Anadarko Basin of Oklahoma. The Hugoton Embayment lies to the west and the Sedgwick Basin lies to the east. Deformation in early Paleozoic time and again in pre-Desmoinesian post-Mississippian time formed the Pratt Anticline. On the crest of the anticline is a large area where Mississippian rocks are absent and Pennsylvanian beds directly overlie Ordovician rocks. The Pratt Anticline is reflected weakly on Lansing structure, but strongly reflected on Ordovician structure. Post-Paleozoic activity produced mainly tilting—first to the west and then to the east. Merriam (1963) described the Pratt Anticline as the smallest of the major structural features in Kansas.

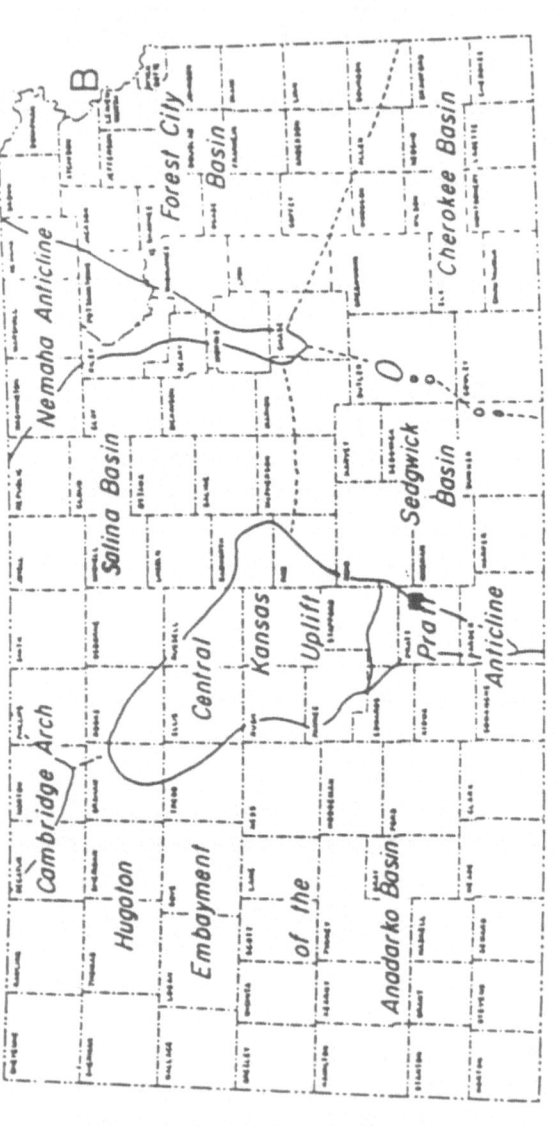

Figure 1. Major pre-Pennsylvanian post Mississippian structural features of Kansas and location of study area in Pratt County (after Merriam 1963).

The Cambro-Ordovician section is comprised by the Arbuckle Group, Simpson Group, and Viola Formation, in ascending order. The Arbuckle is mostly carbonate—dolomite and limestone—and up to about 1100 ft (352 m) thick. The Simpson unconformably overlies the Arbuckle Group and, in turn, is uncomfortably overlain by the Viola. The Simpson is mostly clastic, and in this area mostly sandstone about 40 ft thick. The Viola Formation, composed almost entirely of limestone, dolomite, and dolomitic limestone, is from 30 to 80 ft thick.

The Mississippian rocks, unconformably overlying the Cambro-Ordovician sequence, are mostly carbonates in this area ranging up to 250 ft thick. Their age is late Mississippian and they overlie the Chattanooga Shale of Devonian age. On top of the Mississippian carbonate section is a unit of residual chert up to 150 ft thick formed from weathering and erosion of Mississippian rocks.

The alternating limestone, sandstone, and shale of the Pennsylvanian unconformably overlies the Mississippian and ranges in age from middle to late Pennsylvanian. Thickness of these units is up to about 1600 ft.

The Permian, ranging up to about 1800 ft in thickness, forms the upper part of the section in this area and crops out locally from under Tertiary and Quaternary units. Most of the Permian is comprised of redbeds, sand and siltstone, micaceous shale, and limestone.

In the study area, the highly weathered upper portion of the Arbuckle is a prolific producer of oil and gas, as is the Simpson and Viola. The residual chert where permeable on top of the Mississippian carbonate section also serves as the reservoirs for many oil and gas fields in the Pratt Anticline area. Several units in the Pennsylvanian section have the potential for oil and gas production and collectively the zones form prolific reservoirs in the study area. The Permian produces primarily gas as shallow as 1800 ft.

DATA SOURCE AND PREPARATION

Most of the sources from which data were collected for this study are available from the Kansas Geological Society Library and the Kansas Geological Survey. In addition to well logs, another important source was the completion cards published by Petroleum Information, Inc. The cards were used to obtain formation tops, legal locations, and other information

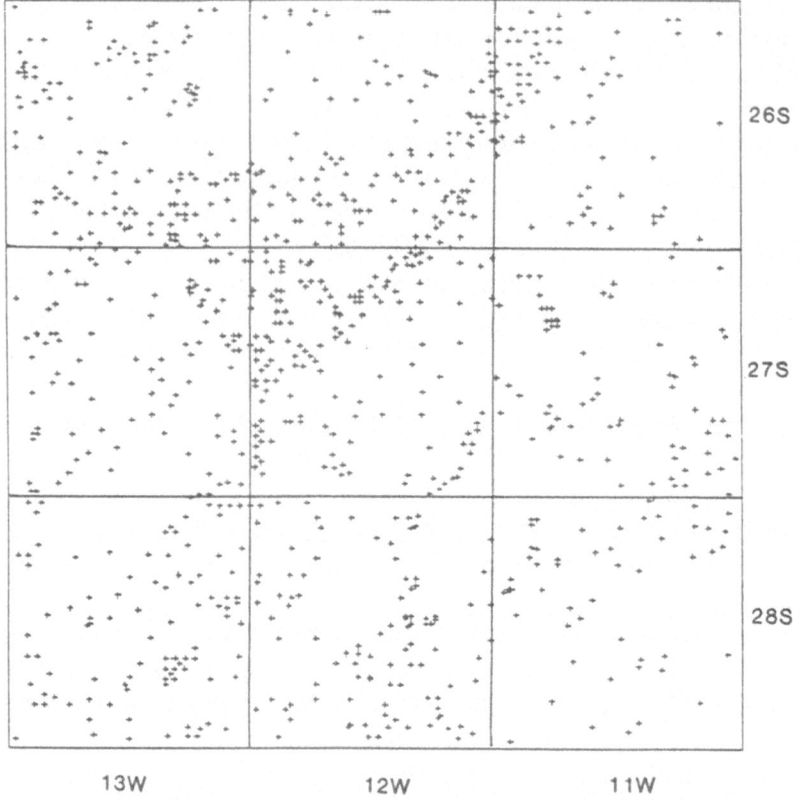

Figure 2. Distribution of data in study area, Pratt County, Kansas.

not obtainable through the KGS Library. Approximately 900 wells, both dryholes and producing wells, were selected to build the database (Fig. 2).

Once the set of wells was selected, data sources of each well were examined at the Kansas Geological Library. Formation tops were determined from available logs. The formation tops used in this study include the Lansing-Kansas City, Mississippian chert, Mississippian limestone, Viola, and Arbuckle. Where Mississippian chert is present, an average porosity value was calculated.

After the data have been gathered, each well with its location, formation tops, and porosity value (if available) was stored in a data file. Missing or unknown values were assigned the number 999.

As the original data localities are given as a legal location based on the traditional township and range description, the original data first are transformed into an orthogonal xy coordinate system using internal coordinates with the size of the grid imposed on the study area of 50 rows by 50 columns. Gridding was performed using (mostly default parameters in) SURFACE III (Sampson, 1988).

The values at the grid nodes were calculated using an average distance weighted averaging method (scaled 1/D2 function). Any data points lying outside the maximum distance range were flagged. After this stage of data preparation, each geological input map is given as a digital variable model (DVM), characterized by underlying grid and a value (or the missing value flag) at each grid node. The DVMs were contoured using SURFACE III. For the data integration problem then, the DVMs can be used as input to the algebraic map comparison method. Maps were hand contoured to check the reliability of the structure and isopachous computer-prepared maps utilizing SURFACE III.

ALGEBRAIC MAP COMPARISON METHOD

Methods for integration of several data sets with disparate units termed map-comparison methods have a tradition in mathematical geology and in resource assessment (an overview of quantitative map-comparison methods is given in Merriam and Jewett, 1989). The algebraic map-comparison algorithm used here is based on the calculation of weighted averages between standardized values of any pair of input maps, for each point of the map area (in practice, for each node of a common underlying grid). For a detailed account of the algorithm see Herzfeld and Merriam (1990), the software is described in Herzfeld and Sondergard (1988).

Because the individual data sets (depth of formation tops, thicknesses, porosity) are recorded in noncommensurable units, they have to be standardized and nondimensionalized prior to comparison. For each digital variable model (DVM), this is achieved by a linear transformation into the interval [0, 1]: If a and b are the minimal and maximal values observed and y [a, b], then proportion of range standardization zp and inverse proportion of range standardization zp- are defined by:

$$S_{ij} = \frac{b - M_{ij}}{b - a} \qquad (1)$$

where the latter is used if one variable decreases whereas most of the others increase.

Let M denote the map area, n the number of input maps, and M_1,..., M_n the input maps with $M_k(x)$ the (standardized) value of map M_k at a grid node x in the map area M. Then the comparison map F is defined by:

$$F(i,j) = \frac{\sum_{t=1,s<1}^{m} W_s W_t \left| m_{sij} - m_{tij} \right|}{\sum_{t=1,s<1}^{n} W_s W_t} \qquad (2)$$

where w_k is a (positive) weight assigned to map M_k, (k=1,..., n). At the point x in the comparison map F, a weighted average of differences of standardized values in all pairwise combinations of input maps is calculated. The weights account for the fact that one indicator variable might be considered more important as a predictor of the target than another. Note that the denominator reduces to the number of comparisons, if all maps are given equal weight $W_k=1$, for all k. By computing F for each grid node, a DVM of the comparison map F is obtained. Entries in F range between 0 and 1 because of standardization and restriction to nonnegative weights. Because the similarity calculated is a distance measure (a seminorm in a space of dimension of the number of comparisons possible), low values indicate high similarity between the input maps, whereas high values are interpreted as poor relationships between the input variables.

GEOLOGICAL INPUT MAPS

A series of geological maps were constructed for the comparison study as described in the previous section.

Lansing Structure Map. The Lansing-Kansas City map (Fig. 3A) was contoured on top of the Lansing-Kansas City, and probably is the most reliable map because most wells penetrated these units. Because there is not as much structural relief on the Lansing-Kansas City map as on the Mississippian limestone or Viola structure maps, the main fault is reflected only weakly on Lansing structure. The Cunningham Anticline is readily apparent, as well as the Brehm Anticline. Both of these features are prolific Lansing producers.

Mississippian Chert Structure Map. The Mississippian chert map (Fig. 3B) is probably the least reliable of the maps, because chert is not present in all of the wells. The fault which extends northeast through the center of the study area serves as a trap for a Mississippian chert reservoir. The greatest permeability was created near the fault, and this is where the chert is most productive.

Mississippian Limestone Structure Map. The Mississippian limestone is not a producing unit, but is a good marker for mapping structural features (Fig. 3C). On the crest of the anticlines in the area, the limestone is not present, but is is present almost everywhere else.

The fault is apparent on the Mississippian limestone structure map. As stated earlier, the major movement occurred in post-Mississippian pre-Pennsylvanian time, so many structures are prominent on the Mississippian and older structure and thickness maps.

Viola Structure Map. The fault once again is outlined by the Viola structure map (Fig. 3D) as well as the Cunningham Anticline. The Viola also is a prolific producer on the crest of the Pratt and Cunningham Anticlines, as well locally on the upthrown side of the fault. This is a fairly reliable map, although many of the wells did not penetrate the Viola.

Arbuckle Structure Map. This map (Fig. 3E) shows the major fault, a faint outline of the Cunningham trend, and the Brehm Pool, although many wells did not penetrate the Arbuckle.

Mississippian Chert to Viola Isopachous Map. A thickness map was generated to show the thickness between the Mississippian chert and Viola structure maps (Fig. 3F). This map indicates thinning and thickening in this interval. A thin interval indicates little chert present and possibly a high Viola datum, and a thick interval may indicate a thick chert section. Chert wells produce best where there is an interval greater than 50 ft. For this reason, this map is important in prospecting for Mississippian or Viola reservoirs.

Lansing to Arbuckle Isopachous Map. This map (Fig. 3G) shows the thickness between the Lansing-Kansas City and Arbuckle. This map is somewhat less reliable, once again caused by the lack of data on the Arbuckle.

Mississippian Chert Porosity Map. A porosity map was generated for the Mississippian chert (Fig. 3H). An important factor in a Mississippian chert well is the development of porosity. A high porosity value in the range of 14% to 25% is required for commercial production. As stated

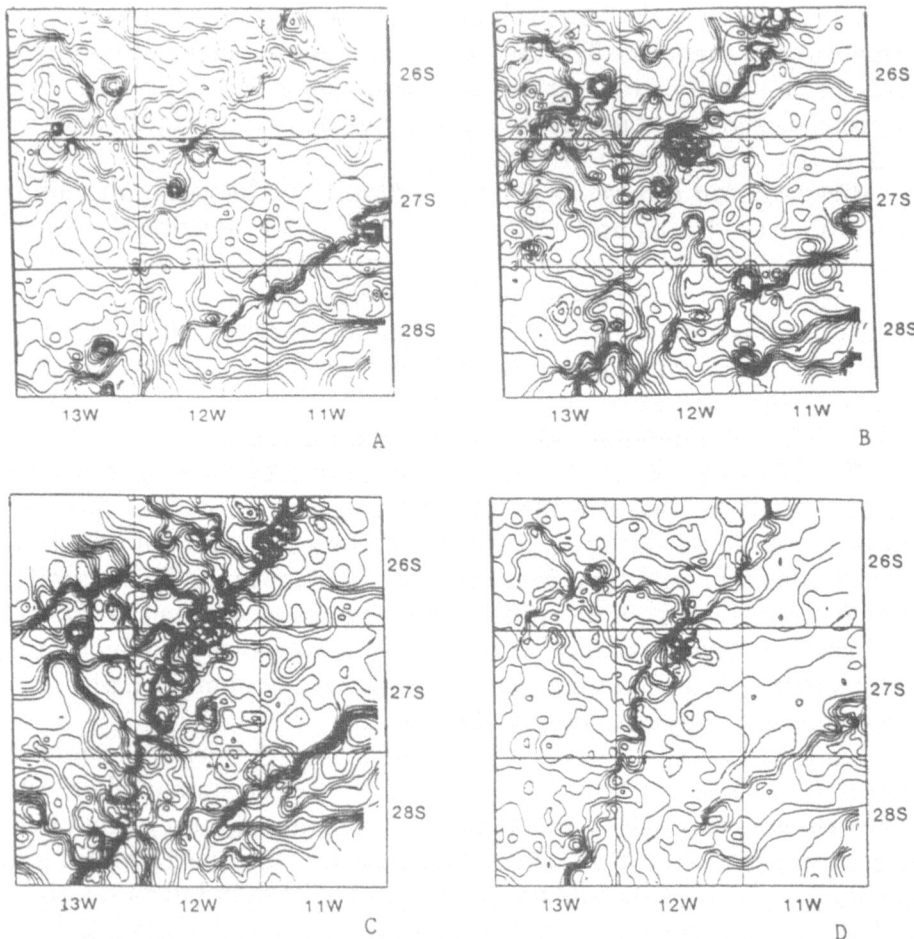

Figure 3. Geological input maps for study area in Pratt County, Kansas. Data from electric logs and well logs from Kansas Geological Society Library and Petroleum information (Fuhr, 1990). Contouring and gridding by SURFACE III (Sampson, 1988), coordinates indicate legal locations, area includes townships, (approximately 18 mi by 18 mi, 324 square miles, 1 mi = 1.609 km). Accounting for latitude effect (townships are not exactly squared), grid spacing is 0.366 mi (589 m) in east-west direction and 0.380 mi (612 m) in north-south direction, 50 by 50 grid node in area. Structure maps (A-E) show depth of formation top in feet (1 ft = 0.32 m). Continued.

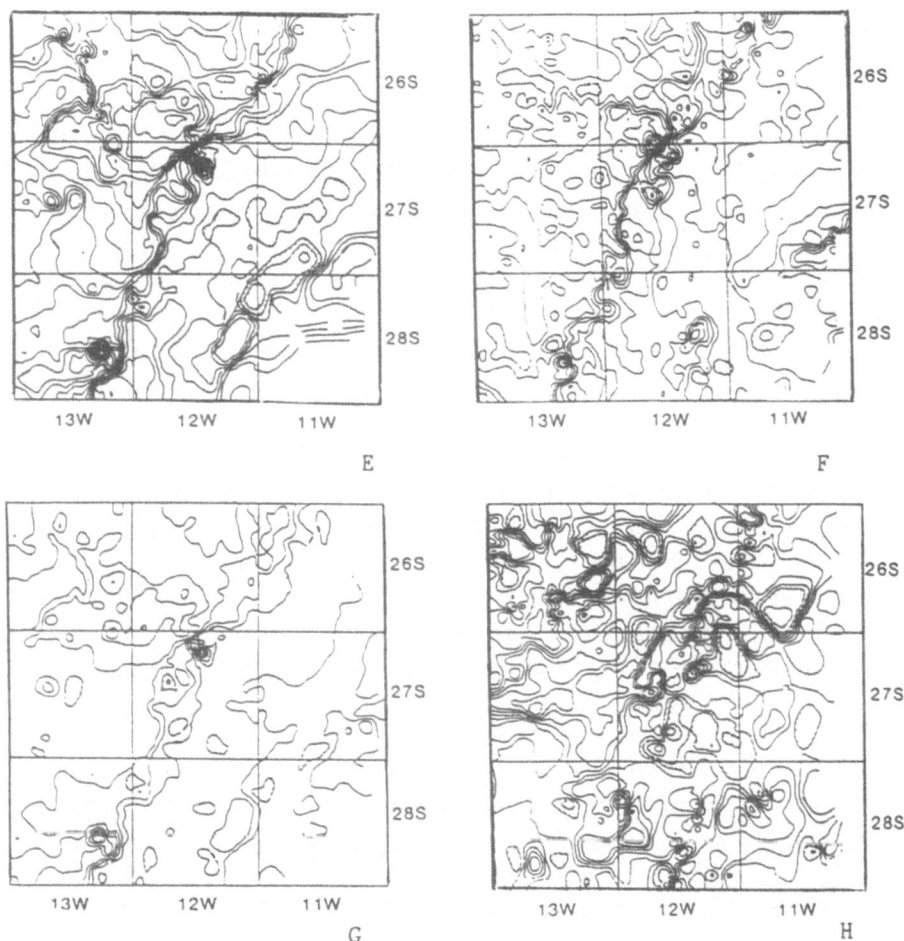

Figure 3 Continued. Isopachous maps (G,H) show thickness of intermediate layer. A. Lansing structure map, contour interval = 10 ft. B. Mississippian chert structure map, contour interval = 10 ft. C. Mississippian limestone structure map, contour interval = 10 ft. D. Viola structure map, contour interval = 25 ft. E. Arbuckle structure map, contour interval = 25 ft. F. Mississippian chert to Viola isopachous map (difference of B and D) contour interval = 50 ft. G. Lansing to Arbuckle isopachous map (difference of A and E), contour interval - 50 ft. H.Mississippian chert porosity map, contour interval = 2%.

earlier, there are many areas where the chert is not present. This map also is unreliable in areas where no chert is present.

RESULTS

The goal in map comparison and integration is to use the similarity map resulting from the algebraic map comparison as the basis for predicting future exploration sites in the basin analysis. The key to successful usage is to determine the "best" combination of variables and their weight to produce a similarity map which will reflect the location of known mineral deposits on the target map. A target map shows the information known on the target variable (Fig. 4). The similarity map and the target map are compared visually then to determine the site of future and detailed exploration. This favorability map is a model of the distribution of the target variable.

The map data sets are selected from those data available to the investigator. The weights are selected by the investigator based on the geological knowledge of the area and the constraints under which the study is made. The resulting similarity map then is compared visually with the target map to determine if the particular set of variables and weights reflect the distribution of known mineral deposits. If not, then another trial is made using a different set of variables or different weights. This procedure is repeated until the mineral deposits are defined on the similarity map and the similarity values are considered to reflect the conditions favorable for mineral accumulation.

A subarea of the main study was used as a trial and error example of combinations of variables and weights. Again the variables used were those readily available and the geologist weighted each variable according to a preconceived idea as to the importance of each in the entrapment of petroleum in the area.

A full-scaled data set of 900 wells as described was used to evaluate the application of the algebraic map comparison method. Figure 4 is a map showing location of known Mississippian chert production. Figure 5 is a similarity map resulting from an unweighted comparison of the 10 input maps for the entire study area (Table 1). Using the results from the prestudy, the Mississippian chert and porosity were weighted 10, all the other variables 1. Comparing Figure 5 and Figure 6, it is apparent that the structures are more pronounced on Figure 6 and known producing areas can be identified better.

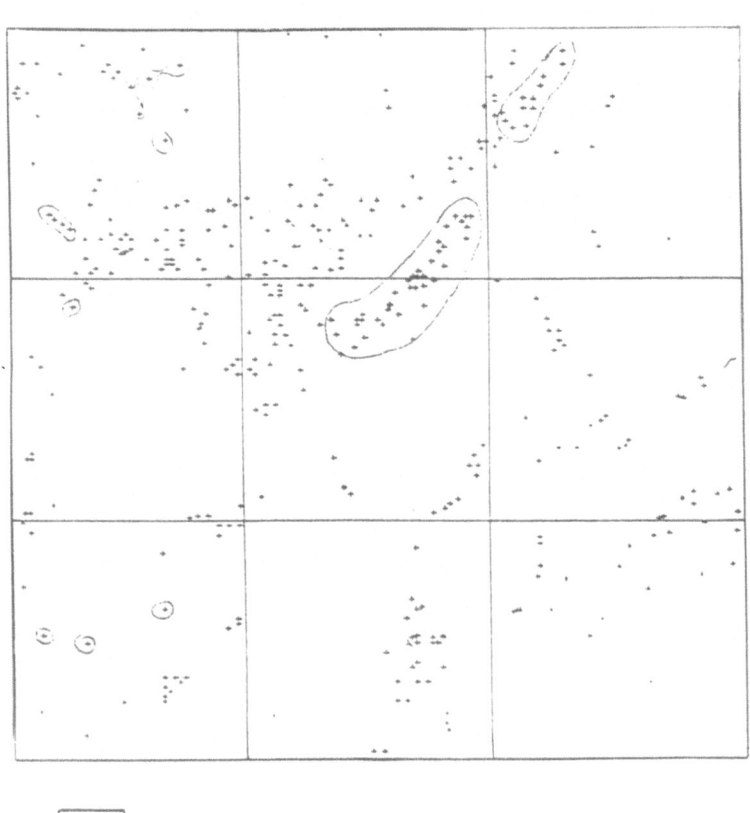

MISSISSIPPIAN CHERT PRODUCTION

Figure 4. Mississippian chert production map (Target Map). Outline of major fields and location of Mississippian chert production.

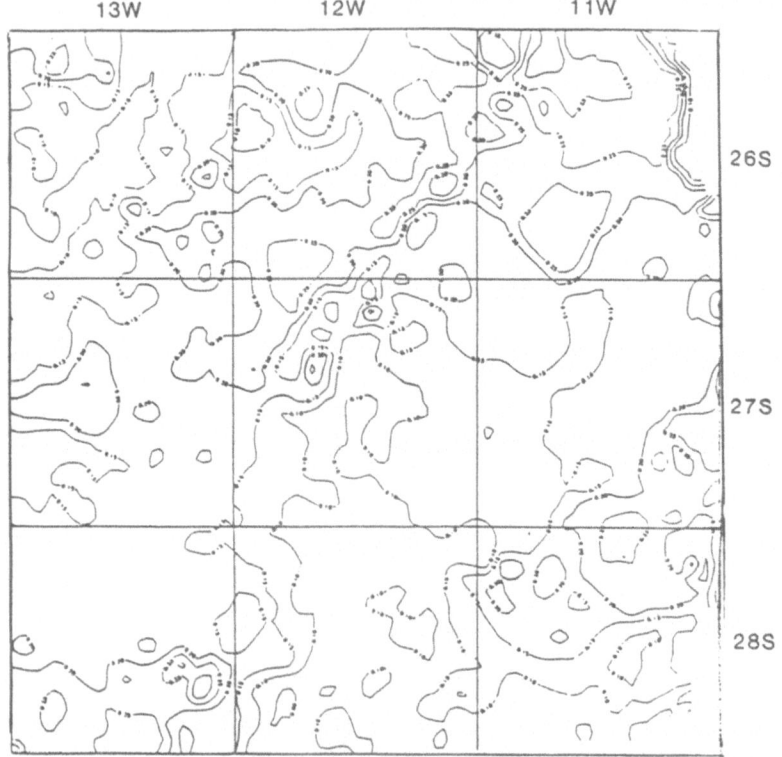

Figure 5. Similarity map for study area in Pratt County, Kansas, resulting from unweighted map comparison. Input weights are given in Table 1. Low similarity values indicate high similarity, all values are in interval [0, 1].

Table 1. Maps and weights used in study area, Pratt County, Kansas (study area based on Townships)

Map	Structure						Thickness			Porosity	Figure
	Lansing	Mississippian Chert	Mississippian Limestone	Viola	Arbuckle		Mississippian Chert to Viola	Lansing to Arbuckle			
Weight	1	1	1	1	1		1	1		1	5
	1	10	1	1	1		1	1		10	6

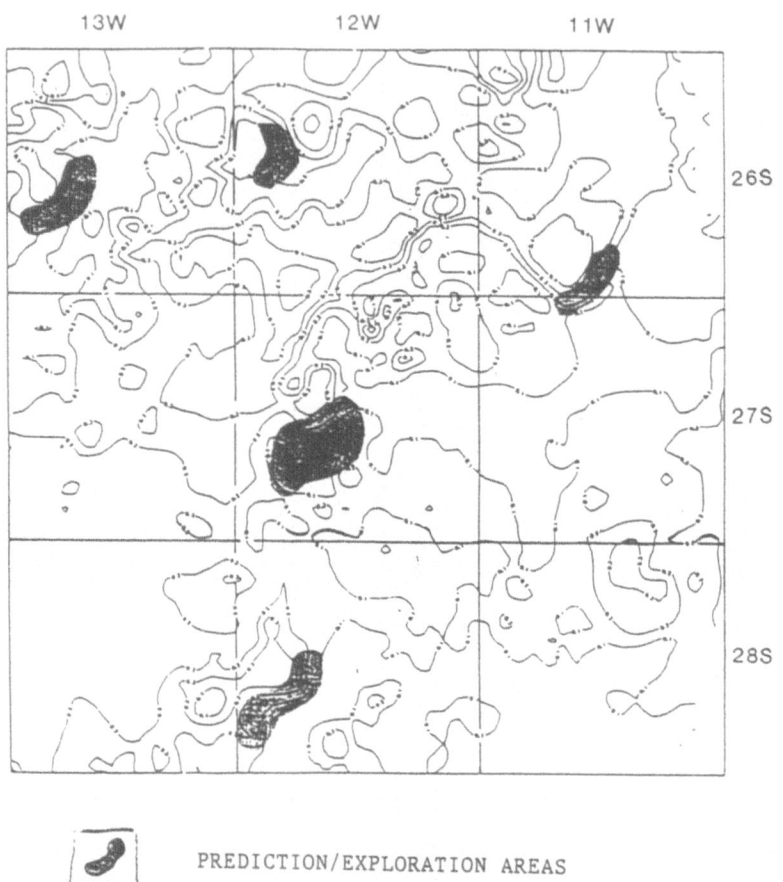

PREDICTION/EXPLORATION AREAS

Figure 6. Similarity map for study area in Pratt County, Kansas, resulting from map comparison with Mississippian chert structure top and porosity weighted 10, and favorability map overlay. Contours: similarity map, weights given in Table 1 (cf. Fig. 5). Comparison with target map (Fig. 4) led to result shaded on that map. Shaded Areas: proposed areas favorable for future exploration for Mississippian chert reservoirs.

The previously discovered production seems to be confined between the 0.1 and 0.2 contour lines with a few fields near or in the 0.3 line. The larger structures lie in the 0.3 to 0.5 contour interval. Because the Mississippian is an important producing unit in this area, the Mississippian chert structure and porosity were weighted heavily in relation to the other variables.

The chert seemingly accumulated on the flanks of the structures represented by the 0.1 to 0.2 contour interval and that is where exploration should begin. Those areas of interest deemed favorable for oil accumulation are shaded on Figure 6.

SUMMARY AND CONCLUSIONS

The use of integrated and combined maps of different themes can be another guide in the exploration of petroleum. Once the optimum combination of variables is ascertained, that combination will help locate additional deposits. In addition to structure, structural development (thickness maps), and porosity, other variables that could be important are gravity, magnetics, geochemical surveys, and paleotopographic maps. Studies currently are being conducted to evaluate further this approach as a valid exploration tool in basin analysis.

We have outlined five areas which may be of interest in looking for additional petroleum deposits contained in the Mississippian chert. The combination of 8 variables where two of them—the Mississippian chert structure and porosity—were weighted heavily by the geologist who believes these two variables are most important in the entrapment of petroleum in these units. Only further drilling in the area to test our prediction will give proof of the value of the technique as an aid in an integrated approach to basin analysis and mineral exploration as we propose.

ACKNOWLEDGMENTS

We would like to thank Mark Sondergard, formerly of Wichita State University, for help in processing some of the data. Rick Brownrigg at the Kansas Geological Survey assisted with running SURFACE III and wrote the conversion program from township and range to x-y coordinates. Peter G. Sutterlin of the Department of Geology at Wichita State

University and W. Lynn Watney of the Kansas Geological Survey read a preliminary version of the manuscript and helped clarify several concepts. Partial funding for this project was supplied by Conoco Oil Company in Ponca City, Oklahoma. U. C. Herzfeld was supported by the Alexander von Humboldt Foundation through a Feder Lymen Fellowship and by a grant from the American Chemical Society Petroleum Research Fund.

REFERENCES

Fuhr, B.A., 1990, An integrated approach to petroleum exploration in Pratt County, Kansas: unpubl. masters thesis, Wichita State University, 87 p.

Herzfeld, U.C., and Merriam, D.F., 1990, A map-comparison technique utilizing weighted input parameters, computer applications in resource estimation and assessment for metals and petroleum: Pergamon Press, Oxford, p. 43-52.

Herzfeld, U.C., and Sondergard, M.A., 1988, MAPCOMP—a FORTRAN program for weighted thematic map comparison: Computers & Geosciences, v. 14, no. 5, p. 699-713.

Merriam, D.F., 1963, The geologic history of Kansas: Kansas Geol. Survey Bull. 162, 217 p.

Merriam, D.F., and Jewett, D.G., 1989, Methods of thematic map comparison, *in* Current advances in mathematical geology: Plenum Press, New York, p. 9-18.

Sampson, R.J., 1988, SURFACE III: Kansas Geol. Survey, 277 p.

CONTROLS ON PETROLEUM ACCUMULATION IN UPPER PENNSYLVANIAN CYCLIC SHELF CARBONATES IN WESTERN KANSAS, USA, INTERPRETED BY SPACE MODELING AND MULTIVARIATE TECHNIQUES

W. Lynn Watney, Jan Harff, John C. Davis,
Geoff Bohling, and J.C. Wong
Kansas Geological Survey, Lawrence, KS

ABSTRACT

Techniques of "regionalized classification" have been applied to problems of oil exploration in Paleozoic strata of the western Kansas shelf. Well locations, each described by a set of geological data, are hierarchically classified. Kriging is used to map the results of well classification across the entire area of investigation, subdividing it into homogeneous subareas termed "regions." The reliability of the regionalization is expressed by a Bayesian probability as a function of the plane. Applying this methodology to stratigraphic data from western Kansas results in a regionalization of the Paleozoic (Pennsylvanian to Permian) sequences. These results are compared to oil productivity and its regional separation. The Bayesian probability of correct allocation to each class can be used as well for the detection of zones of tectonic instability and for assessing exploration risk because the classes reflect oil productivity.

Computerized Basin Analysis, Edited by J. Harff
and D.F. Merriam, Plenum Press, New York, 1993

INTRODUCTION

Predicting additional petroleum discoveries in a mature petroleum province becomes increasingly difficult because the remaining pools tend to be smaller and more subtly controlled. New approaches are needed to evaluate the risks of exploration and development drilling and to create new opportunities for exploration by departing from conventional strategies.

An extensive database of borehole information on Upper Pennsylvanian (Kasimovian) cyclic shelf carbonates of the Lansing and Kansas City Groups of western Kansas, USA, has been interpreted through space modeling and multivariate techniques to determine the controls on petroleum accumulation and to identify new exploratory targets. These strata were deposited on a broad shelf along the actively, but episodically, subsiding northern margin of the Anadarko Basin. Hundreds of oil fields already have been discovered in limestone reservoirs that apparently are controlled by subtle stratigraphic, diagenetic, and structural features. The database includes structural elevations of key horizons, isopachs of critical intervals, petrophysical measurements, and relative differences in thicknesses measured in approximately 1700 wells over an area of 70,000 km^2.

Regionalization is an objective, quantitative procedure for subdividing an area into spatially homogeneous regions, based on geologic variables measured at specific locations such as boreholes. Observations are clustered to form compact groups in variable space that simultaneously delineate contiguous geographic areas. The nature of the regionalized units depends upon the variables; results may not correspond to conventional classifications. However, variables can be weighted optimally so units reflect a desired characteristic, such as high initial oil production.

Regionalization based on structural elevation of formation tops delineates areas of similar structure. Regionalization based on isopachs delineates areas having similar depositional histories. Optimizing a regionalization to reflect initial oil-production rates and lease performance defines areas where discoveries and field extensions have been prolific and identifies stratigraphic and structural variables associated with high production. In addition to maps of the regions, the method produces maps of the probability of correct classification into regions that can aid risk assessment. An improved understanding of the controls and mechanisms for reservoir development and petroleum accumulation also may be obtained from this analysis.

REGIONALIZED CLASSIFICATION MODEL

The model and the corresponding procedure for regionalized classification are based on a generalization (Harff and Davis, 1990) of the concept of the "classification of geological objects" by Rodionov (1981) and coworkers in the Soviet Union. Well records, b, are extracted as elements of a set B. Each of the wells is assigned to a point (vector) r of the region R to be investigated. The observed values form vectors x of an n-dimensional variable X and are allocated to the well records. X may be regarded as a spatially dependent (regionalized) random variable

$$X(b(r)) = m(b(r)) + Y(b(r)), \; A \; r \in R.$$

Here m(b(r)) represents the multivariate expected value function and Y(b(r)) denotes the n-dimensional random variate. The region of investigation is assumed to be composed of homogeneous subregions $R_i \subset R$, $i \in \{1, ..., H\}$. These subregions are characterized by parameter functions which are constant. Thus, the expected value function,

$$E\{X(b(r))\} = m_i, \; \forall r_i \in R_i, \; \forall i \in I, \; I = \{1,...,K\}, \; K \leq H,$$

assumes value vectors m_i as elements of a finite set M.

The intraclass covariance matrix Σ_i is assumed constant within each of the homogeneous regions,

$$E\{(X(b(r)) - (X)(b(r)) - m_i)'\} = \Sigma_i, \; \forall r_i \in R_i, \; \forall i \in I.$$

The taxonomic distance between the elements of M is described by Mahalanobis' generalized distance,

$$d(m_i, m_j) = (m_i - m_j)' \Sigma^{-1} (m_i - m_j),$$

where Σ denotes the between group covariance matrix.

The following conditions are assumed to hold for the elements of the set of distances M:

$$M = \{m_i\}, \; i \in I, \; m_i \neq m_j, \; d(m_i, m_j) \geq \bar{d}, \; i \neq j.$$

The number K of distinguishable vectors m, as well as the number H of subregions R, are dependent upon the threshold $\bar{d}$. Changing this threshold will lead to a hierarchical structure of the set M and the Region R.

A random vector $\Delta(r)$ now is introduced:

$$\Delta(r) = (d_i(r)), \quad \forall i \in I, \forall r \in R,$$

$$d_i(r) = (X(b(r)) - m_i)' \Sigma_i^{-1} (X(b(r)) - m_i).$$

The model implies the following relation,

$$E\left\{ d_i(r_i) \right\} \overset{\min}{=} j \ E\left\{ d_j(r_i) \right\}, j, i \in I \qquad (1)$$

REGIONALIZED CLASSIFICATION

The method of regionalized classification described by Harff and Davis (1990) is based on a random sample B' of drillhole records $b \in B' \subset B$ in which a corresponding number of measurement vectors of an n-dimensional variable $x(b(r))$, $j \in \{1, ..., N\}$ is assigned. The procedure is split into two steps:

(a) Typification by classification
   Using a numerical classification procedure, a number of classes

$$B_i = \left\{ b(r_j) \right\}, j \in \{1, ..., N_i\}, i \in I * (\bar{d})$$

are determined. The classes form a division of the random sample $Z(\bar{d}) = \{B_j\}, i \in I * (\bar{d})$, where the number of classes depends upon the taxonomical distance threshold $\bar{d}$ between the centroid m_i^* in the n-dimensional Euclidean space. The relationships between the number of classes, their experimental expected value vectors m_i^* (centroids), and the threshold $\bar{d}$ can be explored by hierarchical cluster analysis. At each stage in an experimental subdivision of set B', estimates

$$M^*(d) = \left\{ m_i^* \right\}, \; i \in I^*(\bar{d})$$

can be determined by statistical methods.

(b) Regionalization by interpolation

For points $r_t \in R''$, $R'' \subset R$ that are not represented by exploratory holes such as the nodes of a regular grid covering the entire area of investigation, the distance vector

$$\Delta^*(r_t), \forall r_t \in R'' \qquad (2)$$

can be estimated by interpolation methods.

Initially, distance vectors for elements of the random sample are estimated by

$$\Delta^*(r_j) = \left(d_i^*(r_j) \right), \; \forall i \in I^*(\bar{d}), \; \forall j \in \{1,...,N\},$$

$$d_i^*(r_j) = \left(x(b(r_j)) - m_i^* \right)' S_i^{-1} \left(x(b(r_j)) - m_i^* \right) \qquad (3)$$

where S_i^{-1} is the experimental covariance matrix.

Using the estimates defined in Equation (3), the interpolation procedure is used to estimate values at the grid nodes r_t

$$d_i^*(r_t) = \sum_{j \in J_t} \lambda_j \cdot d_i^*(r_j), \; i \in I^*(\bar{d}).$$

Optimal weights λ_j can be determined by kriging as suggested by Harff and Davis (1990).

The probability of allocating a well record $b (r_t)$ to a class B_i can be calculated using Bayes' equation:

$$p\left\{ b(r_t) \in B_i \mid x(b(r_t)) \right\} = \frac{P_i \cdot p\left\{ x(b(r_t)) \mid b(r_t) \in B_i \right\}}{\displaystyle\sum_{j \in I^*(\bar{d})} P_j \cdot p\left\{ x(b(r_t)) \mid b(r_t) \in B_j \right\}} \qquad (4)$$

Here, is the a priori probability of the occurrence of class B_i, and

$$p\left\{x(b(r_t)) \mid b(r_t) \in B_i\right\} = (2\pi)^{-n/2} e^{-\hat{d}_i^*(r_t)/2} dx_1, \ldots, dx_n$$

with $\hat{d}_i^*(r_t) = d_i^*(r_t) + \ln |S_i|$ is the probability that a randomly drawn member of class B_i will have a combined measured value between $(x_{1i}, x_{2i}, \ldots, x_{ni})$ and $(x_{1i} + dx_1, x_{2i} + dx_2, \ldots, x_{ni} + dx_n)$ (Tatsuoka, 1977).

After interpolating to each of the grid nodes $b(r_t)$, $rt \in R''$, they are assigned to the class $Z(\bar{d})$ determined in step (a), according to the discriminant rule given by Equation (1)

$$b(r_t) \in B_i \Leftrightarrow p\left\{b(r_t) \in B_i \mid x(b(r_t))\right\} \stackrel{\max}{=} j \; p\left\{b(r_t) \in B_j \mid x(b(r_t))\right\}.$$

The reliability of the resulting regionalization can be expressed by the Bayesian misclassification probability

$$p\left\{b(r_t) \to Z(\bar{d})\right\} \stackrel{\max}{=} i \; p\left\{b(r_t) \in B_i \mid x(b(r_t))\right\} \qquad (5)$$

CASE STUDY

Kansas is located on the northern shelf extension (Fig. 1) of the Anadarko Basin (Lidiak, 1982). The basin was active during the late Paleozoic and was preceded during the early Paleozoic by the Oklahoma Basin (Johnson, 1990). Development of the Anadarko Basin is linked closely to the Ouachita orogeny and associated thrusting along the Wichita-Arbuckle Uplift (Ham and Wilson, 1967; Brewer and others, 1983; Kluth and Coney, 1981a, 1981b).

The configuration of the Precambrian surface in Kansas (Fig. 2) illustrates broad basins with two localized areas of uplift, the NE–SW trending Nemaha Uplift in eastern Kansas and the NW–SE trending Central Kansas Uplift, diagonally crossing the western half of the State (Baars and Watney, in press). A small structural saddle divides the Central Kansas Uplift, forming the Cambridge Arch at the north. Major movement along the uplifts during the late Mississippian to early Penn-

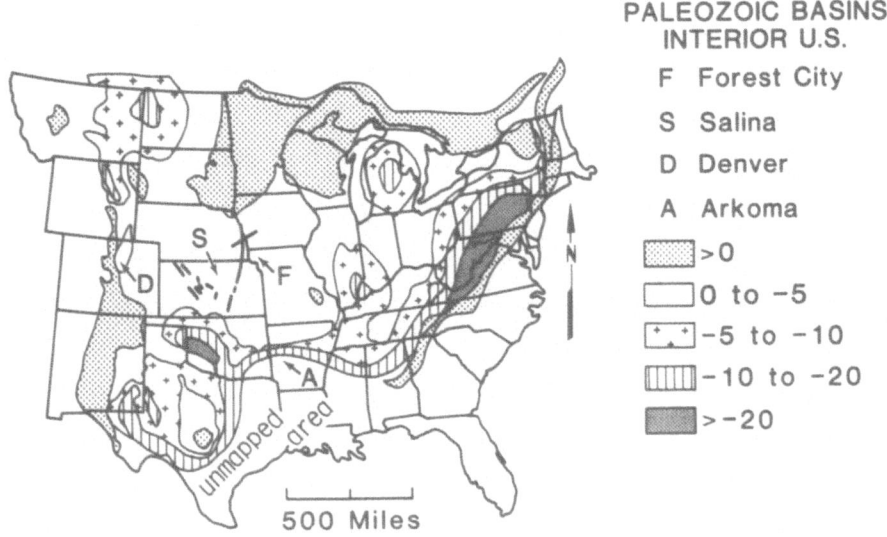

Figure 1. Depth to basement in Paleozoic basins of interior of U.S. including foreland basins bordering northern edge of Appalachian-Ouachita-Marmaton orogenic systems. Depths shown in thousands of feet. Modified from Lidiak (1982).

sylvanian coincides with tectonic activity along other cratonic uplifts and basins in the region (Kluth, 1986; Kluth and Coney, 1981a). The general area of the Central Kansas and Nemaha Uplifts experienced earlier episodes of deformation (Merriam, 1963). Recurrent deformation along basement heterogeneities including faults, fractures, and juxtaposition of contrasting rock types occurred in conjunction with laterally transmitted tectonic stresses (Watney and Wilson, 1983; Watney, 1985a, 1985b).

Reactivation of basement structures differed as the magnitude, direction, and duration of tectonic stresses changed, as during the Appalachian orogeny (Quinlan and Beaumont, 1984). The Precambrian basement essentially provided the template for Phanerozoic deformation.

The cyclic carbonate reservoirs in the Upper Pennsylvanian Lansing and Kansas City Groups have produced large quantities of oil and gas in Kansas during the last 70 years. These widespread strata consist of relatively thin marine limestones stacked between thin marine and nonmarine shales that were deposited in response to major fluctuations in sealevel (Wanless and Shepherd, 1936; Heckel, 1986; Ross and Ross, 1987; Veevers and Powell, 1987). Cyclicity dominated sedimentation worldwide from the late Mississippian through the early Permian (Veevers

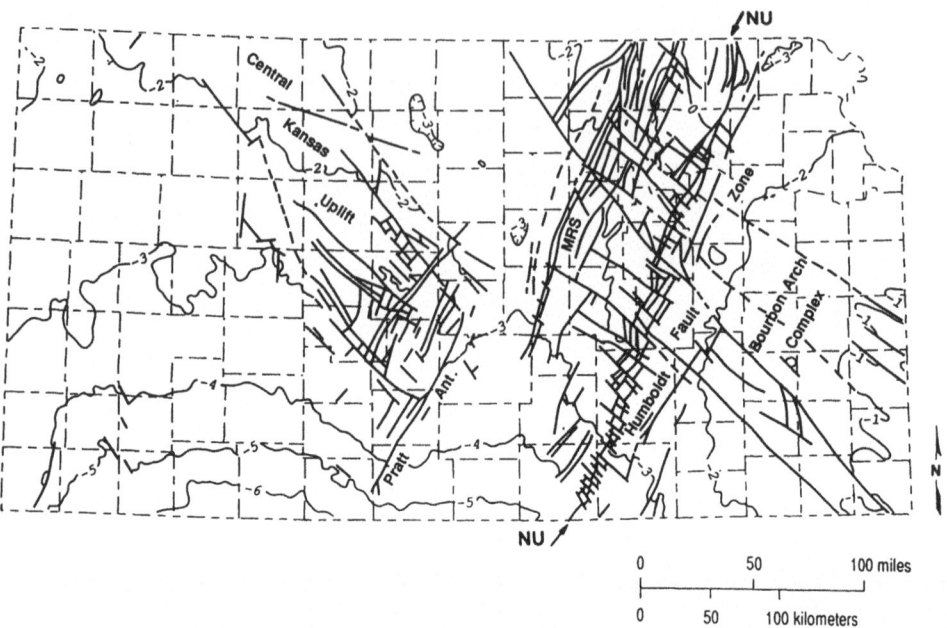

Figure 2. Configuration of Precambrian surface in Kansas (Baars and Watney, in press). Contours in thousands of feet below to sealevel. NE–SW-trending fault system in eastern Kansas is location of Nemaha Uplift. NW–SE-trending faults in central Kansas denote location of Central Kansas Uplift.

and Powell, 1987; Ross and Ross, 1987). The evolving structural configuration of the western Kansas shelf significantly influenced the nature and location of reservoir rock within these cycles in the Lansing–Kansas City interval. Both depositional setting and early diagenesis seem to have been affected strongly by differing shelf position (Watney, 1985a, 1985b). Global sealevel changes were sufficient to move the shoreline hundreds of kilometers across this shelf, but the structural framework of the shelf controlled both the deposition of high-energy grainstones at areas of increased depositional slopes and the location of intense early diagenesis in more positive areas of the shelf. Early diagenesis resulted in extensive carbonate dissolution or cementation (Watney, 1980; Watney, Wong, and French, 1989).

A NW–SE structural cross sections along the Central Kansas Uplift (Fig. 3) and a NE–SW section extending across the Central Kansas Uplift and onto the lower shelf (Fig. 4), illustrate the changes in structure and thickness of the stratigraphic interval between the basal Pennsylvanian

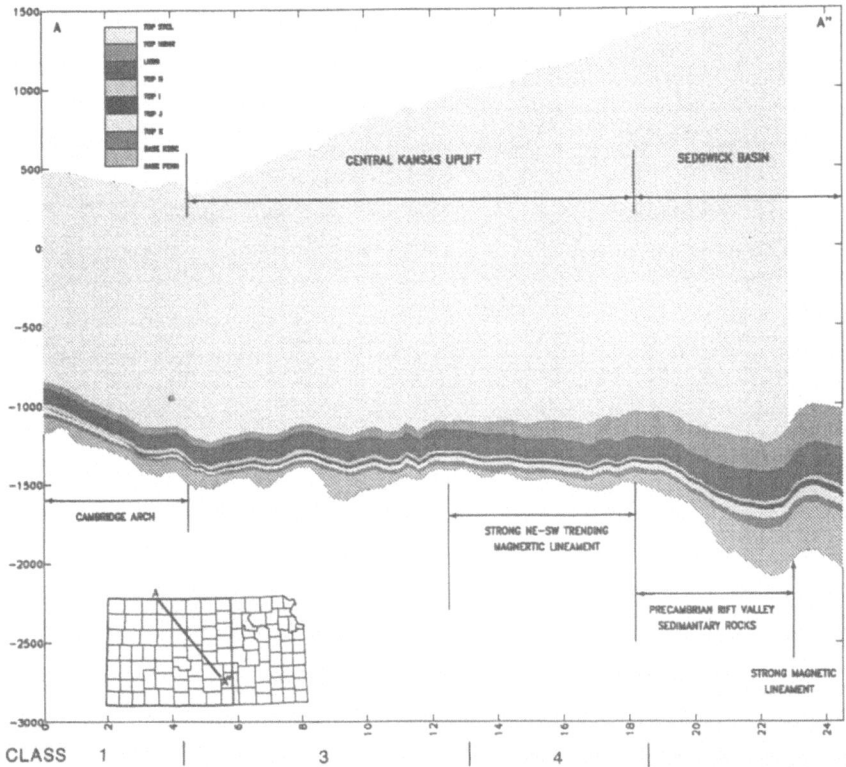

Figure 3. NW–SE structural cross section along crest of Central Kansas Uplift. Gray tones indicate nine stratigraphic intervals used for regionalization. Major basement structures are indicated and location of regions are identified at bottom.

unconformity and the top of the Lower Permian Stone Corral Formation which represents nearly 60 Ma of geologic history. Precambrian basement structures, including boundaries between areas of significantly different ages or rock compositions, are identified on the cross sections. Thicknesses of the Permo-Pennsylvanian strata abruptly change across these boundaries and correspond to many of the class boundaries computed in this study.

The data used in this study are derived from cores, wireline logs, and completion reports on 2112 wells in western Kansas (Fig. 5). All wells penetrate the Lansing and Kansas City Groups and include producing

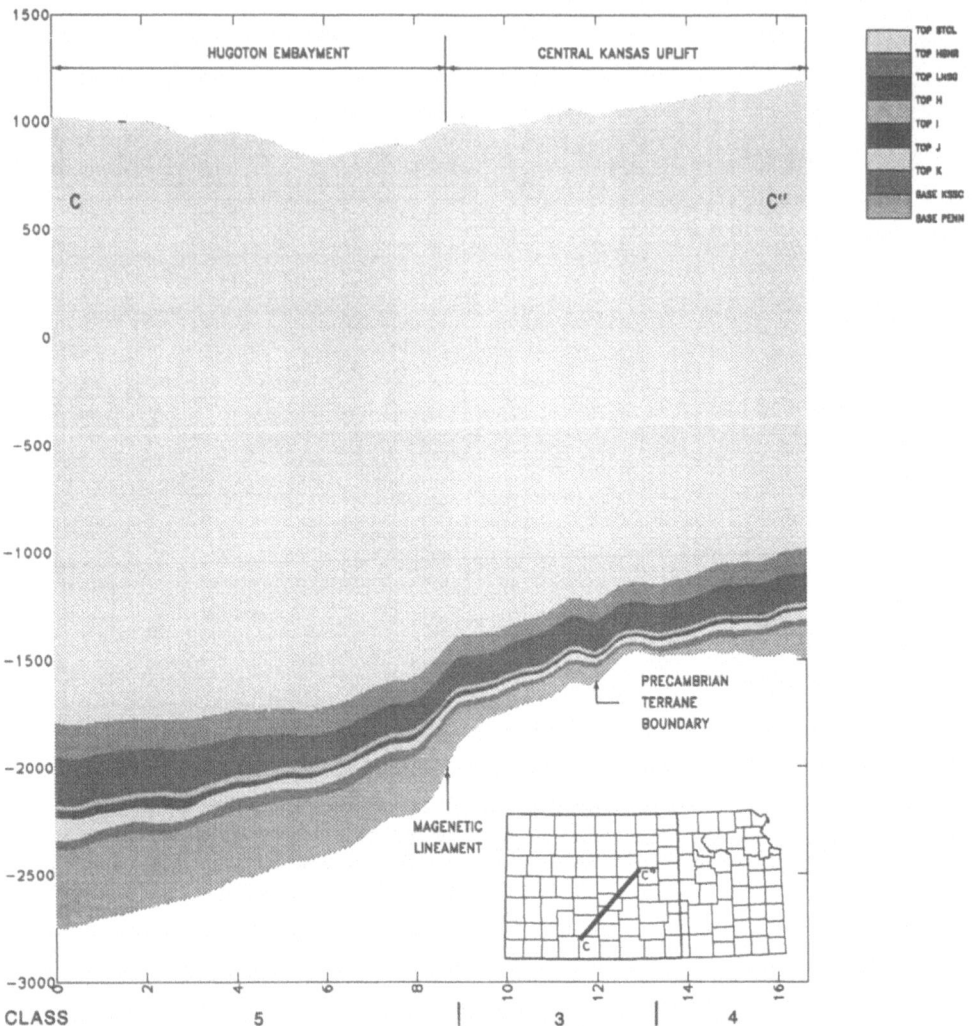

Figure 4. NE–SW structural cross section across southern part of Central Kansas Uplift, extending into lower shelf of Hugoton Embayment. Same format as in Figure 3.

wells from 350 different fields and nearby dryholes. Each well is represented by 28 stratigraphic and petrophysical variables.

The regionalization described here only thickness data from nine stratigraphic intervals; results reflect the combined effects of differential subsidence and sealevel changes. The thicknesses of large stratigraphic intervals are expressed in Variables 1–5:

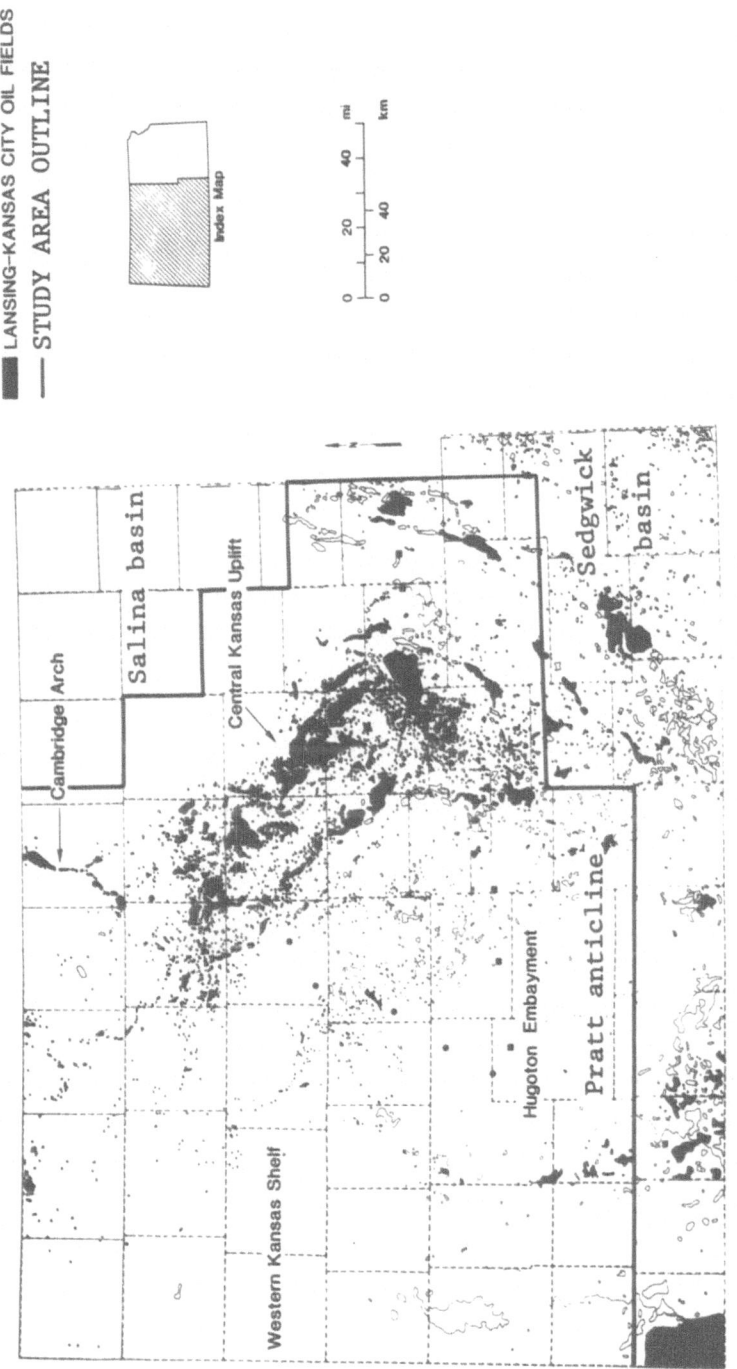

Figure 5. Outlines of oil and gas fields in western Kansas showing fields that produce from Lansing–Kansas City. Index map shows area included in following maps.

Variable	Interval
1	Top of Stone Corral Formation (Lower Permian) to top of Heebner Shale (Upper Pennsylvanian)
2	Top of Heebner Shale to top of Lansing Group
3	Top of Lansing Group to base of Iola Limestone (G-zone)
4	Base of Iola limestone (G-zone) to base of the Swope Limestone (K-zone)
5	Base of Swope Limestone (K-zone) to basal Pennsylvanian unconformity

The thicknesses of individual depositional sequences in the Upper Pennsylvanian Kansas City Group (marine-nonmarine cycles, as described by Watney, Wong, and French, 1989) are expressed in Variables 6–9.

Variable	Interval
6	Thickness of Dewey Limestone (H-zone)
7	Thickness of Cherryvale Formation (I-zone)
8	Thickness of Dennis Limestone (J-zone)
9	Thickness of Swope Limestone (K-zone)

Variables 6–9 are separate depositional sequences in the Kansas City Group and consist of a basal flooding unit ending with a subaerially weathered limestone or a siliciclastic paleosol (Watney, Wong, and French, 1989). These cycles are widespread and reflect major sealevel changes. Using conodonts (Lambert and others, 1991), many individual depositional sequences in a long core from western Kansas have been correlated with rocks outcropping almost 600 km to the east. Similarly, correlations of these same sequences also have been extended into Texas, some 800 km to the south (Boardman and Heckel, 1988), and into the

eastern United States at a distance of about 1600 km (Heckel and others, 1991).

The 2112 wells were grouped using the hierarchical clustering and regionalization techniques described by Harff and Davis (1990). An initial classification into three clusters distinguishes major broad, homogeneous structure elements of the shelf, including: the Central Kansas Uplift; the northern part of the shelf that is farthest from the Anadarko Basin; and the shelf closest to the Anadarko Basin. A six-class regionalization subdivides these three major structural regions. The differences in thicknesses of the nine stratigraphic intervals between these regions can be thought of as differences in subsidence or sediment accommodation space. Locations classified within the same region underwent similar changes in sediment accommodation space. Figure 6 is a map showing the regionalization produced by subdividing the area into six subregions on the basis of differential subsidence.

The changes in accommodation space within each of the six regions can be represented by subsidence curves (age vs. depth), constructed using the estimated means of the nine thickness variables for each class (Fig. 7). Because of the differences in scale, subsidence curves must be constructed separately for large interval variables and individual depositional sequences.

Pennsylvanian sediments were deposited first on the surface represented by basal Pennsylvanian unconformity in Regions 6 and 5, located in the south adjacent to the Anadarko Basin. Regions 5 and 6 exhibit the most sustained subsidence during the time from the basal Pennsylvanian to Stone Corral deposition, spanning some 60 Ma. This subsidence can be attributed to proximity to the Anadarko Basin. During deposition of the Kansas City Group, subsidence was greatest in Region 6 adjoining the western portion of the Anadarko Basin. Significant subsidence in the western Anadarko Basin during the upper Pennsylvanian is suggested by: evidence of sediment starvation; deep water (>1100 ft) with considerable shelf relief; and turbidite deposition in the western part of the Anadarko Basin (Dickinson and Yarborough, 1976; Kumar and Slatt, 1984). However, the subsidence rate in Region 6 diminished below that in several other regions during the time of deposition of the Heebner to Stone Corral interval. Nevertheless, cumulative subsidence was greatest in Region 6. A cratonic salt basin developed in central and southern Kansas over the locations of Regions 3, 4, and 5 during the Leonardian Permian, immediately beneath the Stone Corral datum, reflecting the eastward displacement of subsidence at this time (Watney , Berg, and Paul, 1988).

Region 2 is a broad area in west-central Kansas immediately west of the Central Kansas Uplift. The area initially was a low shelf that became filled with older Pennsylvanian strata. Subsidence in Region 2 prior to deposition of the Kansas City Group exceeded that is areas on the Central Kansas Uplift, including Regions 1, 3, and 4. However, subsidence rates in Region 2 were similar to that on the Central Kansas Uplift and upper shelf during deposition of the Kansas City Group. Region 2 experienced lower subsidence rates during the time of the youngest Heebner to Stone Corral interval, resulting in a cumulative subsidence intermediate between the upper and lower shelf areas.

Region 1, the upper shelf and the Cambridge Arch in the northern portion of the Central Kansas Uplift, has undergone the least amount of subsidence. Pennsylvanian sedimentation began earlier in Region 1 than

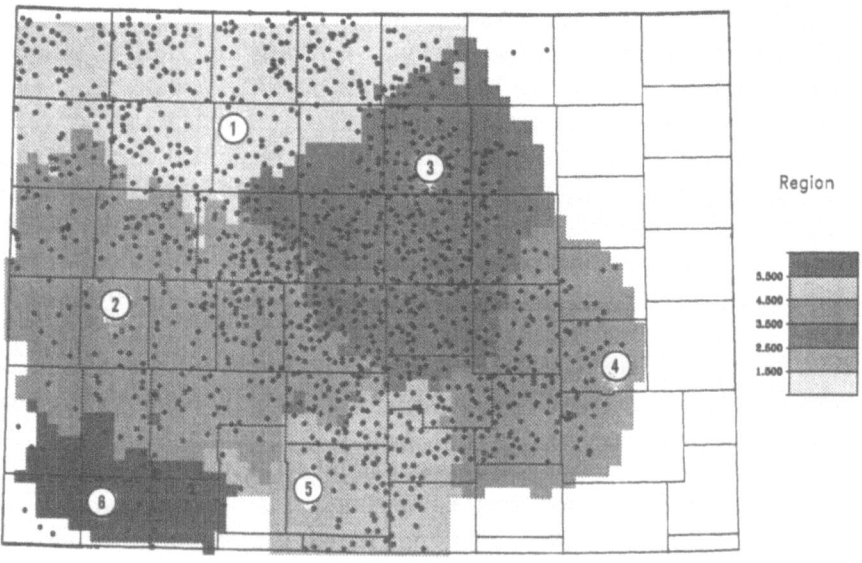

6-Category Regionalization
Thickness of 9 Pennsylvanian Stratigraphic Intervals
Regionalization of the Western Kansas Shelf

Figure 6. Six-class regionalization based on nine interval thicknesses. Dots indicate well locations. Numbers indicate regions identified in text.

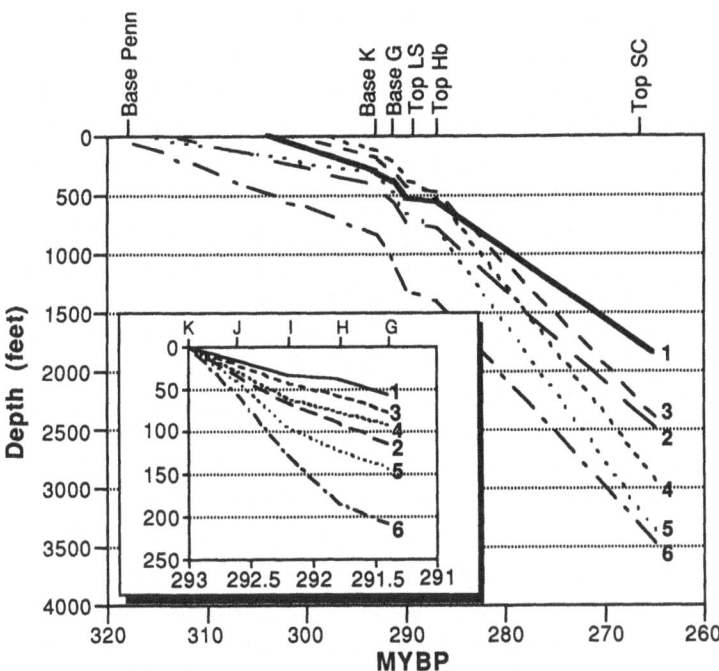

Figure 7. Subsidence plot of geologic time vs. depth for mean thicknesses of nine intervals in each of six regions.

in Regions 3 and 4 on the Central Kansas Uplift, reflecting an early western downwarp associated with a long, narrow embayment that developed in westernmost Kansas and adjoining Colorado during early Pennsylvanian Morrowan and Atokan time. Minimal subsidence in Region 1 is consistent with its position as most distant from the Anadarko Basin. Region 3, located on the northern portion of the Central Kansas Uplift, is characterized by only slightly greater subsidence rates than Region 1 during the deposition of the interval from base of the Pennsylvanian to top of the Heebner Shale. However, subsidence in Region 3 accelerated significantly during the time of Heebner to Stone Corral deposition.

Thicknesses of individual depositional sequences in the Kansas City Group (Fig. 7, insert) represent short intervals of geologic time (~300,000 to 500,000 years). Thicknesses of strata deposited during these shorter time intervals probably were a function of changes in accommodation space caused by both subsidence and global sealevel changes (Watney, Wong, and French, in press). Sediment accommodation space generally increased from the northern part of the uplift region and upper shelf that is most distant from the Anadarko Basin (Region 1) to the southern portion of the Central Kansas Uplift (Region 4). This trend continues southwestward into Region 6 toward the western part of the Anadarko Basin, which was the most tectonically active.

The probability of correct classification of the wells was calculated as a space-dependent Bayesian probability [Eq. (4)]. Figure 8 shows the probability map for Region 4 as an example. The map for the probability of correct classification of all six regions [given in Eq. (5)] is displayed as Figure 9. The central parts of the homogeneous regions are marked by high probabilities, reflecting more definitive classifications than in the zones of transition from one homogeneous region to another. In these zones, the probability of correct classification drops rapidly, forming "probability valleys." These may be interpreted as regions of tectonic instability characterized by fracturing, faulting, and folding as seen on the basement configuration map (Fig. 2) and as inferred from gravity and magnetic potential field maps (Yarger, 1983; Baars and Watney, in press). Cross sections (Figs. 3 and 4) illustrate the changing structural styles in the different regions.

Most of the boundaries and locations of the six regions correspond to recognizable elements of the Precambrian basement; these are especially apparent on the cross sections (Figs. 3 and 4):

(1) The boundary between Regions 3 and 4, which divide he Central Kansas Uplift, corresponds to the western edge of a basement area crossed by faults associated with the Keweenawan (1.1 billion year old) Midcontinent rift system.

(2) The boundary between Regions 4 and 5 corresponds to the southwestern end of an area of rifted basement rocks.

(3) The northernmost boundaries along Regions 5 and 2 correspond to a Precambrian terrain boundary defined by potential field mapping and differences in basement geochronology (Yarger, 1983). Regions 2, 5, and 6 are south of this boundary and underwent greater subsidence during deposition of the Kansas City Group than did the area north of the boundary. This boundary also is the axis of an inferred southward flexure of the shelf, characterized by abruptly thickened

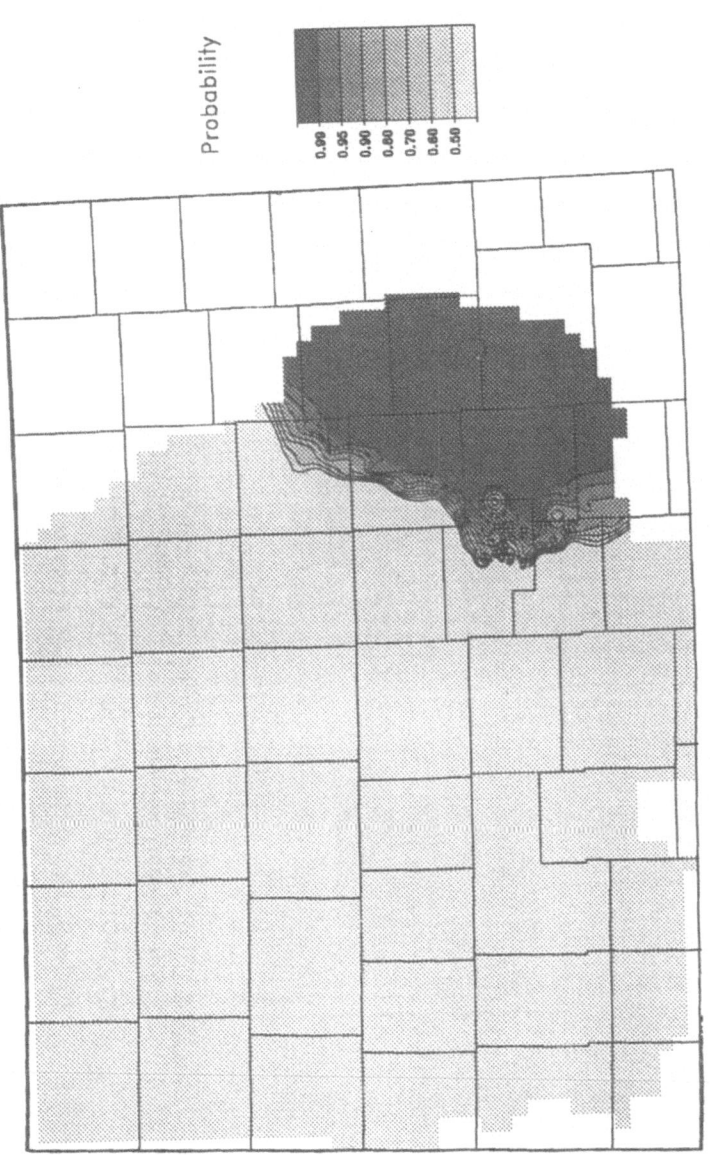

Probability

0.90
0.95
0.90
0.80
0.70
0.60
0.50

Probability of Correct Classification
Class #4 Regionalization Based on Thickness

Regionalization of the Western Kansas Shelf

Figure 8. Map showing probability of correct classification in Region 4.

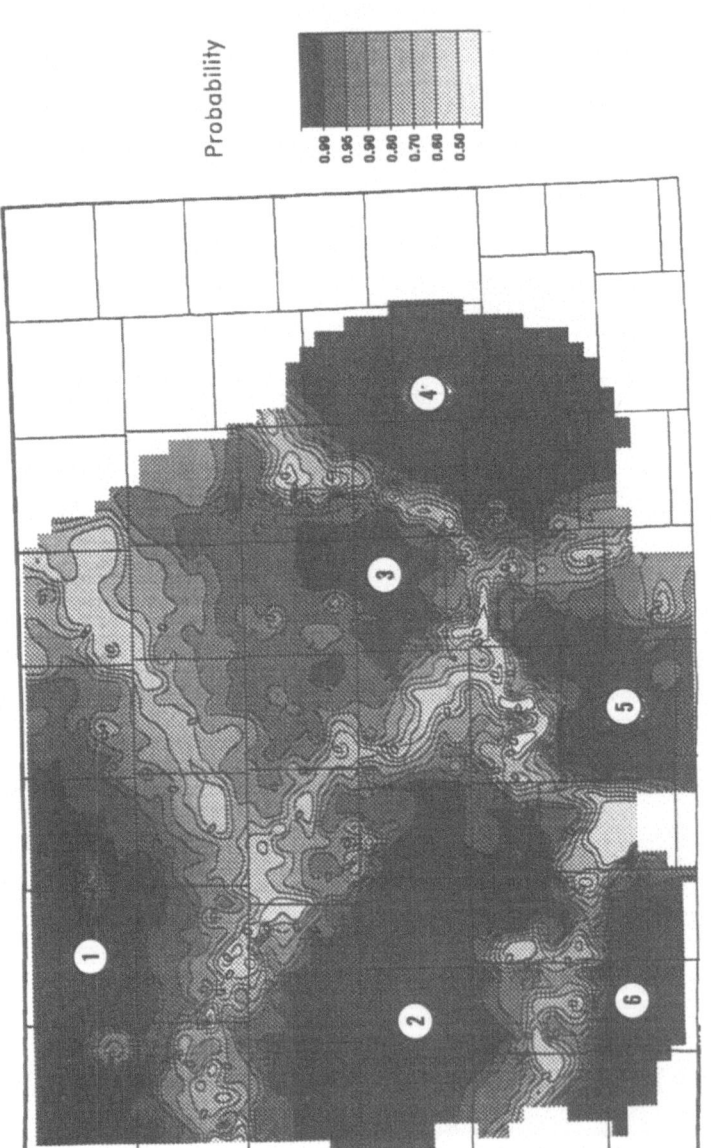

Probability of Correct Classification
6-Category Regionalization Based on Thickness
Regionalization of the Western Kansas Shelf

Figure 9. Map showing probability of correct classification in all six regions.

trends in oolitic grainstone in sequences of the Kansas City Group (Watney, 1985a, 1985b).

(4) The boundary between Regions 2 and 6 is not well defined from basement mapping because of the absence of subsurface control.

Fundamental heterogeneities in the basement rocks consist of lateral contrasts in age, composition, and structure. The characteristics of the variables that define the six regions can be attributed to a combination of differential structural deformation between the basement elements and lateral heterogeneities.

Region 4 contains the highest percentage of productive oil wells (20% of the wells in the database) of all the regions. The percentage is 7% in Region 1, 8% in Region 2, 14% in Region 3, 2% in Region 5, and 10% in Region 6. Region 4 probably has a higher percentage of productive wells because:

(1) The area was predominantly a sustained positive site (Fig. 5) during deposition of Upper Pennsylvanian strata, leading to subaerial exposure during the later stages of each depositional sequence in the Lansing–Kansas City. Criteria diagnostic of subaerial exposure are abundant in sequences in Region 4, accompanied by nontectonic fracturing and dissolution of carbonate.

(2) Region 4 includes the southern part of the Central Kansas Uplift and represents the first major uplift beyond the northern edge of the Anadarko Basin, the probable source of much of the oil that accumulated on the western Kansas shelf (Schmoker, 1986). Oil migrating out of the basin would encounter this uplift as one of the first major structural traps.

(3) Structural flexure associated with development of the southern margin of the Central Kansas Uplift was concurrent with sedimentation of Kansas City strata, providing favorable sites for high-energy carbonate accumulation (Watney, 1984, 1985). Updip pinchouts of primary intergranular porosity associated with these grainstones added to the potential for trapping hydrocarbons.

(4) The Precambrian basement beneath the area of Region 4 is characterized by numerous intersecting conjugate faults at the intersection of the NW–SE-trending Central Kansas Uplift and the NE–SW-trending Midcontinent rift system (Baars and Watney, in press). Changing stress patterns from southern orogenic systems would likely lead to formation of fractures in overlying strata during the intermittent reactivation of faults in this area. Fracturing facilitates oil and gas migration and formation of reservoirs. The NW–SE structure cross section (Fig. 3)

shows the clear association of thickness and structural changes in this zone of basement faulting.

(5) The positive structural location resulted in thinning of shales deposited prior to Lansing–Kansas City time, causing a closer proximity or juxtaposition of successive limestones and unconformities. This configuration produced greater opportunities for vertical oil migration. Most of the oil can be correlated geochemically with source rocks in the underlying Mississippian-Devonian Chattanooga (Woodford) Shale.

CONCLUSIONS

Regionalization can be regarded as a mathematical expression and computer implementation of traditional methods of geological mapping and interpretation applied to the structural component of basin analysis. The subsidence behavior of subunits in a basin during different time steps can be constructed from data on stratigraphic thicknesses. These multivariate data, derived from well records, are used for regionalization. Thicknesses of large stratigraphic intervals are assumed to be proportional to subsidence. Thickness variations in nine successive Pennsylvanian and Permian stratigraphic intervals are sufficient to define six coherent regions, forming geologically distinctive areas that reflect basement heterogeneity, anisotropy, and proximity to orogenic activity. The pattern of subsidence within each region is regular but distinctive from adjoining regions, reflecting changes in rates and locations of directed tectonic stress that originated primarily in the south as the Anadarko Basin developed.

Typification of wells based on these variables distinguishes different characteristic subsidence (sealevel rise) curves. Regionalization of western Kansas based on this typification expresses the regional differences in subsidence history and delineates major structures. Each of the homogeneous regions also can be judged on the basis of favorability for petroleum accumulation, reflecting the formation of traps, reservoirs, and hydrocarbon migration pathways. The map of probability of correct classification (regionalization) shows linear zones of transitions from one homogeneous region to another. These transitions are marked by low classification probabilities that may be due to local instability, corresponding to faults and folds. The regions reflect oil productivity and can be used for risk assessment in exploration.

REFERENCES

Baars, D.L., and Watney, W.L., in press, Paleotectonic control of reservoir facies, *in* Franseen, E.K., Watney, W.L., Kendall, C.J., Ross, eds., Sedimentary modeling: computer simulations and methods for improved parameter definition: Kansas Geological Survey Bull.

Boardman, D.R., II, and Heckel, P.H., 1988, Glacial-eustatic sea-level curve for early Late Pennsylvanian sequence in north-central Texas and biostratigraphic correlation with curve for midcontinent North America; Geology, v. 17, no. 9, p. 802-805.

Brewer, J.A., Good, R., Oliver, J.E., Brown, L.D., and Haufman, S., 1983, COCORP profiling across the southern Oklahoma aulacogen-overthrusting of the Wichita Mountains and compression within the Anadarko Basin: Geology, v. 11, no. 2, p. 109-114.

Dickinson, W.R., and Yarborough, H., 1979, Plate tectonics and hydrocarbon accumulation: Am. Assoc. Petroleum Geologists Continuing Education Course Notes Series 1, 150 p.

Ham, W.E., and Wilson, J.L., 1967, Paleozoic epeirogeny and orogeny in the central United States: Am. Jour. Science, v. 265, no. 5, p. 332-407.

Harff, J., and Davis, J., 1990, Regionalization in geology by multivariate classification: Jour. Math. Geology, v. 22, no. 5, p. 573-588.

Heckel, P.H., 1986, Sea-level curve for Pennsylvanian eustatic marine transgressive-regressive depositional cycles along Midcontinent outcrop belt, North America: Geology, v. 14, no. 4, p. 330-334.

Heckel, P.H., Barrick, J.E., Boardman, D.R., II, Lambert, L.I., Watney, W.L., and Weibel, C.P., 1991, Biostratigraphic correlation of eustatic cyclothems (basic Pennsylvanian sequence units) from Midcontinent to Texas and Illinois (abst.): Am. Assoc. Petroleum Geologists Bull., v. 75, no. 3, p. 592-593.

Johnson, K.S., ed., 1990, Anadarko Basin symposium: Oklahoma Geol. Survey Circ. 90, 289 p.

Kluth, C.F., 1986, Plate tectonics of the Ancestral Rocky Mountains, in Peterson, J.A., ed., Paleotectonics and sedimentation *in* the Rocky Mountain Region, United States: Am. Assoc. Petroleum Geologists Mem. 41, p. 353-369.

Kluth, C.F., and Coney, P.J., 1981a, Plate tectonics of the ancestral Rocky Mountains: Geology, v. 9, no. 1, p. 10-15.

Kluth, C.F., and Coney, P.J., 1981b, Reply to comments on plate tectonics of the Ancestral Rocky Mountains: Geology, v. 9, no. 9, p. 388-389.

Kumar, Naresh, and Slatt, R.M., 1984, Submarine-fan and slope facies of Tonkawa (Missourian-Virgilian) Sandstone in deep Anadarko Basin: Am. Assoc. Petroleum Geologists Bull., v. 68, no. 8, p. 1,839-1,856.

Lambert, L., Heckel, P.H., Watney, W.L., Stevenson, G.M., Barrick, J.E., and Boardman, D.R., II, 1991, Precision correlation of Kansas Pennsylvanian cyclothems emphasizing wireline logs and conodonts in a sequence stratigraphic framework (abst.): Am. Assoc. Petroleum Geologists Bull., v. 75, no. 3, p. 616.

Lidiak, E.G., 1982, Basement rocks of the main interior basins of the Midcontinent, in Proctor, P.D., and Koenig, J.W., eds., Selected structural basins of the Midcontinent, United States of America: Univ. of Missouri-Rolla Jour., no. 3, p. 5-24.

Merriam, D.F., 1963, The geologic history of Kansas: Kansas Geol. Survey Bull. 162, 317 p.

Quinlan G.M., and Beaumont, C., 1984, Appalachian thrusting, lithospheric flexure and the Paleozoic stratigraphy of the eastern interior of North America: Can. Jour. Earth Sciences, v. 21, no.9, p. 973- 996.

Rodionov, D., 1981, Statisticeskie resenija v geologii: Izd. "Nedra", Moskva, 231 p.

Ross, C.A., and Ross, J.R.P., 1987, Late Paleozoic sea levels and depositional sequences: Cushman Foundation for Foraminiferal Research, Spec. Publ. 24, p. 137-168.

Schmoker, J.W., 1986, Oil generation in the Anadarko Basin, Oklahoma and Texas: modeling using Lopatin's Method: Oklahoma Geol. Survey Spec. Publ. 86-3, 40 p.

Tatsuoka, M.M., 1977, Multivariate analysis: techniques for educational and psychological research: John Wiley & Sons, New York, 310 p.

Veevers, J.J., and Powell, C.McA., 1987, Late Paleozoic glacial episodes in Gondwanaland reflected in transgressive-regressive depositional sequences in Euramerica: Geol. Soc. America Bull., v. 98, no. 4, p. 475-487.

Wanless, H.R., and Shepard, F.P., 1936, Sea level and climatic changes related to late Paleozoic cycles: Geol. Soc. America Bull., v. 47, no. 8, p. 1177-1206.

Watney, W.L., 1980, Cyclic sedimentation of the Lansing and Kansas City Groups (Missourian) in northwestern Kansas and southwestern Nebraska - a guide for petroleum exploration: Kansas Geol. Survey Bull. 220, 72 p.

Watney, W.L., 1984, Recognition of favorable reservoir trends in Upper Pennsylvanian cyclic carbonates in western Kansas, in Hyne, N.J., ed., Limestones of the Midcontinent: Tulsa Geol. Society, Spec. Publ. No. 2, p. 201-246.

Watney, W.L., 1985a, Evaluation of the significance of tectonic sedimentary control versus eustatic control of Upper Pennsylvanian cyclothems in the western Midcontinent, *in* Watney, W.L., Kaesler, R.L., and Newell, K.D., eds., Recent interpretations of Late Paleozoic cyclothems: Proc. Third Annual Meeting, Mid-Continent Section, SEPM, p. 105-140.

Watney, W.L., 1985b, Resolving controls on epeiric sedimentation using trend surface analysis: Jour. Math. Geology, v. 17, no. 4, p. 427-454.

Watney, W.L., Berg, J.A., and Paul, S., 1988, Origin and distribution of the Hutchinson Salt (Lower Leonardian) in Kansas, *in* Morgan, W., and Babcock, J., eds., Permian rocks of the Midcontinent: Midcontinent Section of SEPM, Spec. Publ. 1, p. 113-135.

Watney, W.L., and Wilson, F.W., 1983, Relationship of epeirogeny and sedimentation in Kansas (abst.): Am. Assoc. Petroleum Geologists Bull., v. 67, no. 8, p. 1328.

Watney, W.L., Wong, J.C., and French, J., 1989, Computer simulation of coastal marine sedimentary sequences influenced by glacial-eustatic sea-level fluctuations: 28th Intern. Geological Congress, Washington, D.C., Abstracts, v. 3, p. 3-336 – 3-337.

Watney, W.L., Wong, J.C., and French, J.A., in press, Computer simulation of Upper Pennsylvanian (Missourian) carbonate-dominated cycles in western Kansas (United States), *in* Sedimentary modeling: computer simulations and methods for improved parameter definition: Kansas Geol. Survey Bull.

Yarger, H.L., 1983, Regional interpretation of Kansas aeromagnetic data: Kansas Geol. Survey, Geophysics Ser. 1, 35 p.

COMPUTER-ASSISTED BASIN ANALYSIS FOR THE SEARCH OF OIL AND GAS IN THE NORTH GERMAN BASIN

J. Harff, P. Hoth,
Central Institute for Physics of the Earth,Telegrafenberg, Germany

and

W. Eiserbeck
Erdöl-Erdgas-Geotechnologie Gommern GmbH
Magdeburger Chaussee, Germany

ABSTRACT

The prediction of oil and gas is a complex task involving assessment of source-rock potential, alteration of organic matter, migration, and accumulation of hydrocarbons.

In addition to information as well on recent geological structure as on petrological and petrophysical features of the rocks data on the geological history of the basin have to be included in the investigation. Drilling and geophysical survey only provides measuring data. Paleodata have to be derived from numerical modeling of the basin's formation. Additional expert appraisals on structure and geological history of the basin play an important role also in the process of computerized basin analysis.

PC software was developed for structural modeling as interpolation of stratigraphic core data from drills, back stripping and process modeling involving compaction, heat transfer and maturation of hydrocarbons.

The technology includes an interpretation tool for the integration of different types of assessment criteria (measuring data, results of structure, process modeling, and expert appraisals). The data interpretation

Computerized Basin Analysis, Edited by J. Harff
and D.F. Merriam, Plenum Press, New York, 1993

process is split in a step for the assessment of hydrocarbon generation potential and a step for the prediction of reservoir rocks providing both regionalization schemes of the area to be investigated. The map of hydrocarbon generation capability can be linked genetically with the map of reservoir favorability by paleostructural maps expressing migration paths of hydrocarbons.

The method is demonstrated by the research on gas in the Upper Rotliegend sediments of the northeastern part of the North German Basin.

INTRODUCTION

For the investigation of deep sedimentary basins it is necessary to interpret all available information for the elaboration of an investigation and search strategy before executing expansive geophysical measuring and drilling. Mathematical modeling and information processing can help the geologist in decision-making for exploration on the basis of minimization of risk. For complex data interpretation in the Russian schools of mathematical geology specific methods in particular for the prediction of ore reserves have been developed. Harff and Davis (1990) generalized the theory and method of "classification of geological objects" (Rodionov, 1981) for geological regionalization and showed possibilities for an application of this method to the regional prediction ("prognosis") of hydrocarbons (HC) in the frame of basin analysis.

First attempts were undertaken to use this method in the northeastern part of the North German Basin for gas prediction in the Rotliegend (Lower Permian) sediments (Harff and others, 1990).

Additional results of this study for this area of research are discussed here. The assessment of HC generation potential in the Carboniferous sequences of the basin after the sealing of the Permo-Carboniferous sequences by halitic rocks in Upper Permian time and the prediction of reservoir facies in Upper Rotliegend sediments is distinguished. Both results are demonstrated as regionalization schemes for the area of investigation showing more or less favorable subregions as well for gas generation as for the occurrence of reservoir rocks.

For the discovery of the relations between that subregions having been identified as favorable for post-Rotliegend gas generation with those being favorable for the formation of reservoirs one have to reconstruct possible migration paths for HC in post-Rotliegend time. In the investigation area the Permo-Carboniferous volcanic rocks are believed to have guided the gas migration. A paleomodel of the Carboniferous subsurface (the subsurface immediately under the volcanites) at the beginning of late Permian time indicates migration paths from the basin's center (northwestern part of investigation area) to the north and south (Altmark region). The result can serve as a guideline for the elaboration of a strategy for future exploration.

BACKGROUND/METHOD

Time-space dimensions of the structure and the formation of sedimentary basins and require a specific method for information extraction and interpretation (data integration).

Figure 1 shows as a scheme the method of computerized basin analysis used here. For HC prediction as well-assessment criteria derived from the analysis of the recent basins structure as from the entire process of subsidence, structure formation,generation, migration, and accumulation of HC have to be taken into consideration.

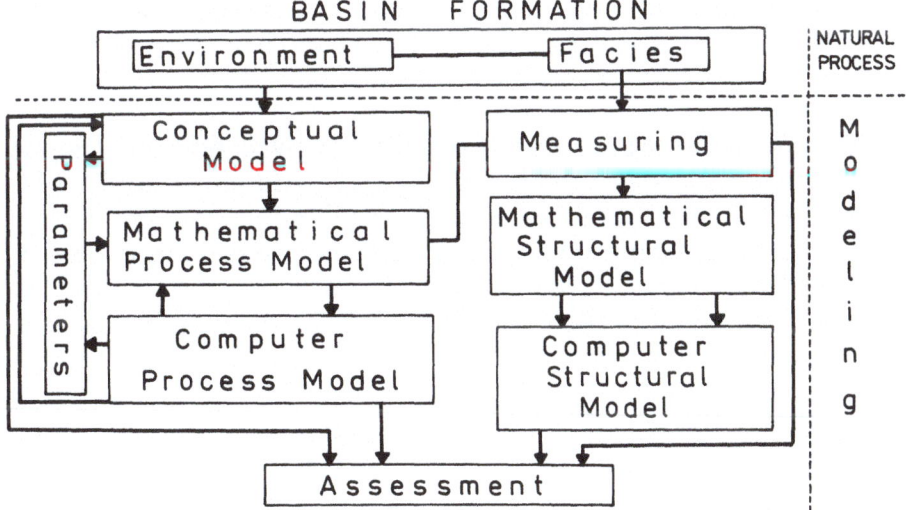

Figure 1. Schematic description of computerized basin analysis.

Four different types of assessment criteria are distinguished:

(1) Measuring data derived from geochemical, petrological, and petrophysical core sample analysis, wire-line logs, and drill data. These data describe the facies of sedimentary sequences and are allocated at the one hand to the stratigraphic units and on the other hand to the coordinates of drill points in the area of research.

(2) Results of structural modeling of the recent structure as the result of interpolation between stratigraphic records allocated to drill points. Models of subsurfaces of stratigraphic units are constructed by using the optimized IDW-method (Watson and Philip, 1985) for the interpolation between the observations in drill profiles.

(3) Results of process modeling describing the basins reconstructed status in the geological past. For the prediction of HC not only the recent structure and facies but also assumptions on the formation of the basin play an important role for the evaluation of source rock conditions, migration, alteration, and accumulation of HC. As the process is not observable a conceptional model have to be constructed by geologist. On the base of this model compaction, pore fluid movements, geothermal processes, and organic diagenesis have to be described by general physical and chemical laws as well as their mathematical expressions (differential equations). By numerical methods and parameters (initial and boundary conditions) computer process models are constructed.

Because measured rock data are available only for drill profiles, one-dimensional models are used here for a hybrid procedure between forward simulation and backstripping. Wireline logs, stratigraphy as well as petrography, and petrophysics provide parameters for subsidence modeling (reconstructed by backstripping), compaction (Harff and others, 1989), heat transfer (Berthold and Galushkin, 1986), and HC generation (Tissot and Welte, 1978). By feedback procedure parameters (material parameters, initial and boundary conditions), mathematical model and conceptual model are changed step by step in order to fit the models optimal with observations.

Figure 2 shows the relation between measuring data, processes to be modeled and results of one-dimensional modelings. These results for

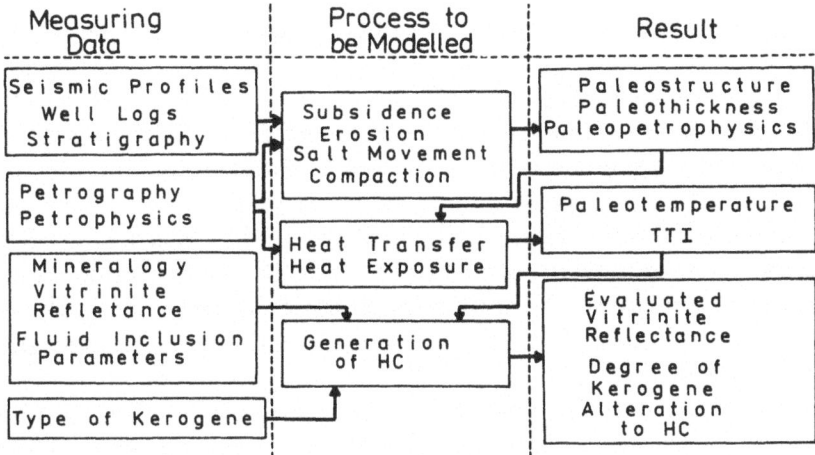

Figure 2. Relation between measuring data, geological processes to be modeled, and modeling results.

drill profiles are connected spatially by interpolation procedures providing two- and three-dimensional space-time models represented graphically as isoline maps, perspective graphs, and block diagrams.

(4) Expert appraisals. In every basin study a set of data needed for modeling cannot be supplied by measuring and observation. For instance, data from deeper parts of the basin having not been reached by drills (type and thickness of HC source rocks, depth and temperature of the basement, etc.) Therefore measuring data have to be completed by expert appraisals on the basis of long-term experiences of geologists.

For complex interpretation of these four types of assessment criteria methods of data integration have been developed using basic ideas of "classification of geological objects" (Rodionov, 1981). Figure 3 shows the principle of the regionalization method described in detail by Harff and Davis (1990).

For each drill profile investigated by current research values of the assessment criteria form an n-dimensional vector. By a classificator the set of profiles is subdivided into classes of different favorability for HC. Each class is described by the experimental mean vector of assessment criteria, representing a specific type of favorability. The favorability according to the current task (source- or reservoir-evaluation) have to be assessed leading to a rank order of these classes. At each drill point the Mahalanobis distance between the corresponding values of assessment criteria and the mean vector is computed. Distance maps for each class are constructed by interpolation and stored as gridded values. By determining the minimum distance at the grid points the class is identified to which the corresponding point at the area of investigation belongs (discriminant analysis). The mapped typification forms a regionalization scheme. This regionalization scheme subdivides the area of investigation in subregions of different favorability serving as a guideline for the researcher. Maps of the Bayes- probability for regionalizations reliability can be used for the assessment of the exploration risk (Harff and Davis, 1990).

CASE STUDY

As a case study serves an example from the investigation of the Upper Carboniferous and Lower Permian rocks in the northeastern part of the North German Basin. Figure 4 represents a three-dimensional structural model of the area of research as a view from the northwest (Hamburg) to southeast (Berlin).

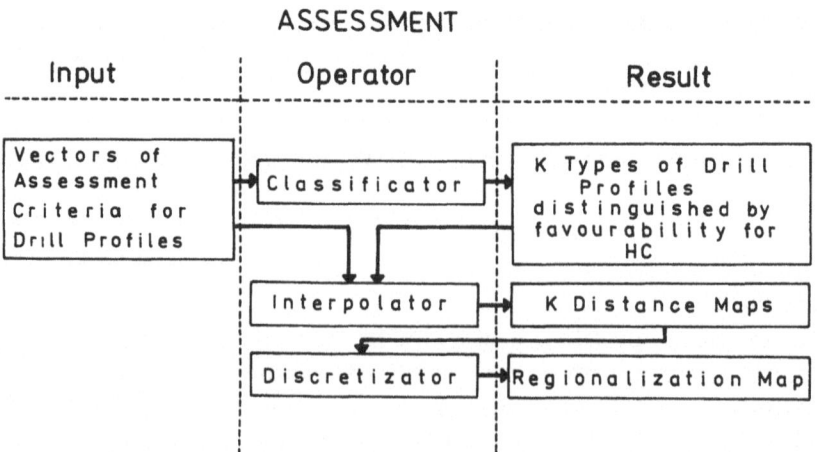

Figure 3. Schematic description of regionalization procedure for HC prediction.

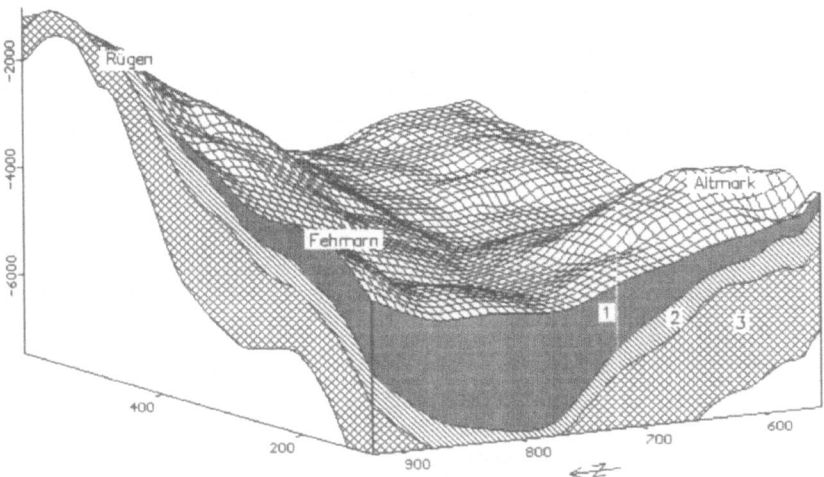

Figure 4. Three-dimensional model of Permo-Carboniferous complex - northeastern part of North German Basin. 1- Rotliegend (Lower Permian) sediments, 2- Permo-Carboniferous volcanic rocks, 3-Upper Carboniferous sediments.

The stratigraphic unit favorable for gas deposits is the Lower Permian sedimentary sequence (Rotliegendes). These clastic sediments are covered by 2 to 5 km sediments from Upper Permian salts to Quarternary sediments. The gas is believed to have been generated by coal-bearing Upper Carboniferous sequences and is believed to have migrated under and through Permo-Carboniferous volcanic rocks resting between Carboniferous source rocks and Rotliegend sediments. Figure 5 shows a north - south directed cross-section through the entire basin from the Danish coast to the Altmark region. Data from 43 drilling profiles distributed in the area of research have been involved in the analysis. The research is split into two main steps: The favorability investigation of gas generation and the research aiming to predict reservoir capabilities.

For the assessment of the gas generation after sealing of Rotliegend rocks by salts during Zechstein time, three assessment criteria have been involved in the analysis. Figure 6 shows as the first assessment criterion the potential of source rocks. It is ranked from 0 - that denotes the coaly sequences are not present - to 7 standing for the best source rock conditions drilled in the area. If a drill has not reached the corresponding sequences, missing observations are replaced by expert appraisals on the base of paleogeographical ideas and seismic records. In the map the coal-bearing belt linking the Northwest Mecklenburg area with the Rügen/Northwest Poland region is clearly outlined.

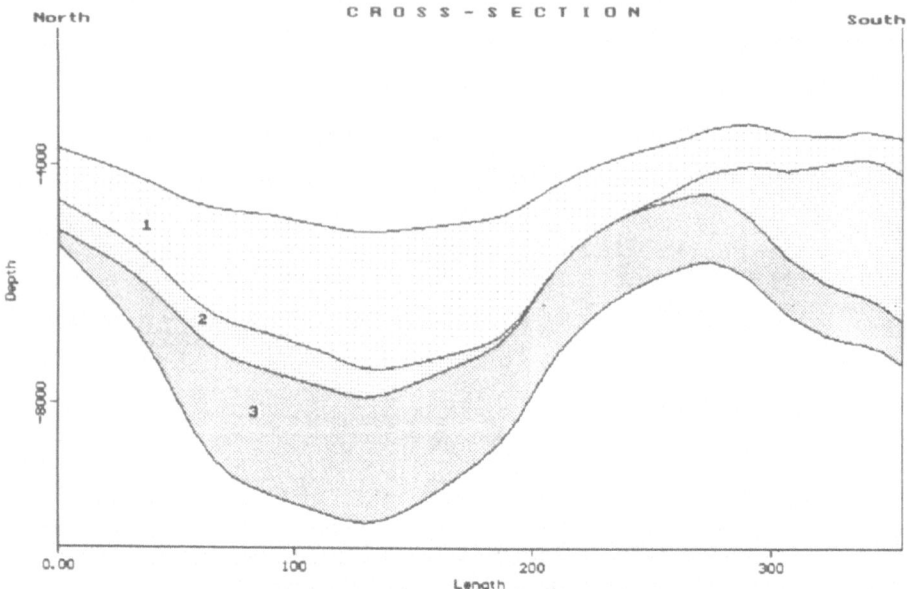

Figure 5. North-South directed cross section through basin.

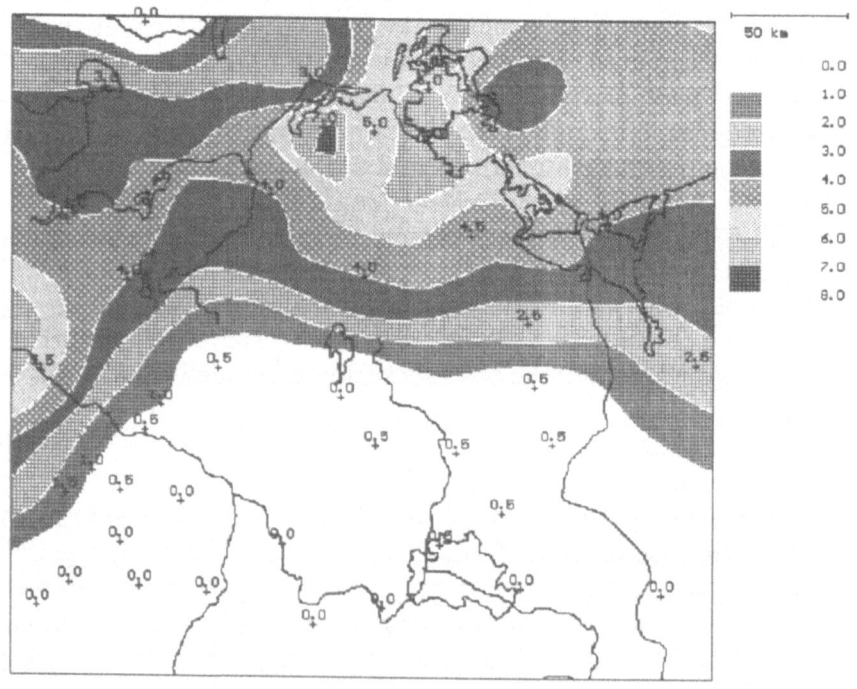

Figure 6. Map of Upper Carboniferous source-rock potential.

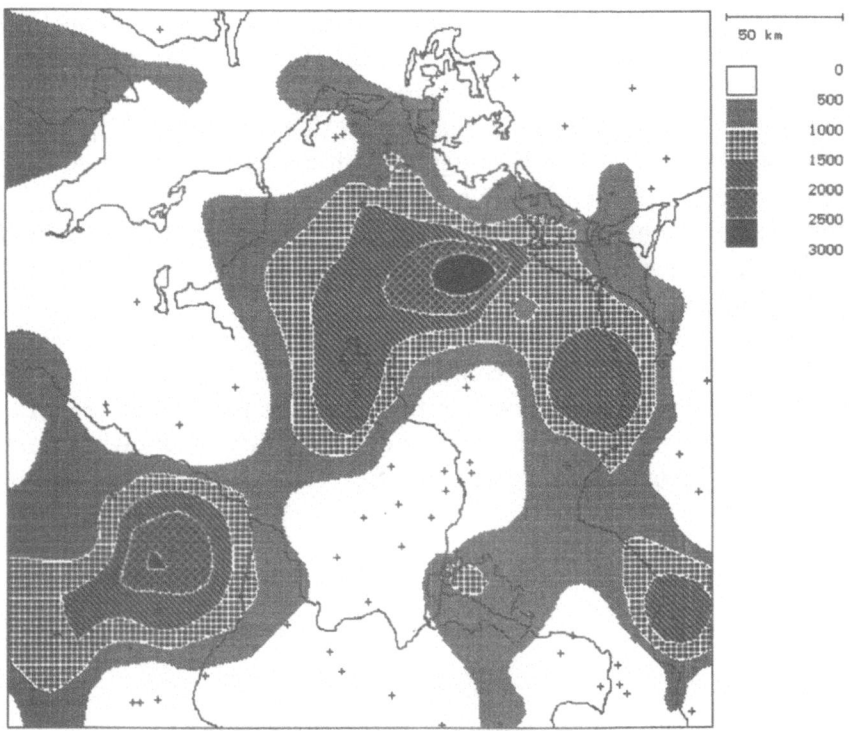

Figure 7. Thickness map of Permo-Carboniferous volcanic rocks.

Figure 7 presents the thickness of Permo-Carboniferous volcanic rocks as indicator for an thermal exposure and subsidence of source rocks (second assessment criterion). The shape of the map is determined by a southwest-northeast-trending zone with the maximum thickness of the volcanites. By process modeling the third criterion, the Temperature Time Index (TTI) at the subsurface of Carboniferous rocks at the beginning of the Zechstein salt sedimentation (the main sealing horizon), have been determined (Fig. 8). The TTI-values are determined mainly by the geothermal regime during the period of strong volcanic activity and the partly strong and rapid subsidence in Rotliegend time.

The drilling profiles have been classified (hierarchical cluster analysis) into five classes. Methods of assessment criteria are given by Table 1. The classes are ordered downward expressing decreasing favorability for post-Rotliegend gas generation. Classes 2 and 1 only represent good conditions for the generation of gas after the beginning of Zechstein salt sedimentation.

By interpolation and discretization the regionalization scheme (Fig. 9) was received. The favorable classes 2 and 1 surround the center of the investigation area similar to a horseshoe.

The second step leads to a regionalization corresponding to the reservoir capabilities in the Rotliegend sedimentary sequences. Four criteria have been involved in the analysis.

The reconstructed paleothickness of Zechstein salts (first criterion) stands for the possibility of sealing traps in the Upper Rotliegend sediments (Fig. 10).

As the second criterion serves the reconstructed paleorelief of the Permo-Carboniferous volcanic complex giving an impression on the distribution of the sedimentary environment in the paleobasin (Fig. 11). One is looking for regions with intermediate elevations showing energetic conditions favorable for sand formation (Harff and others, 1989).

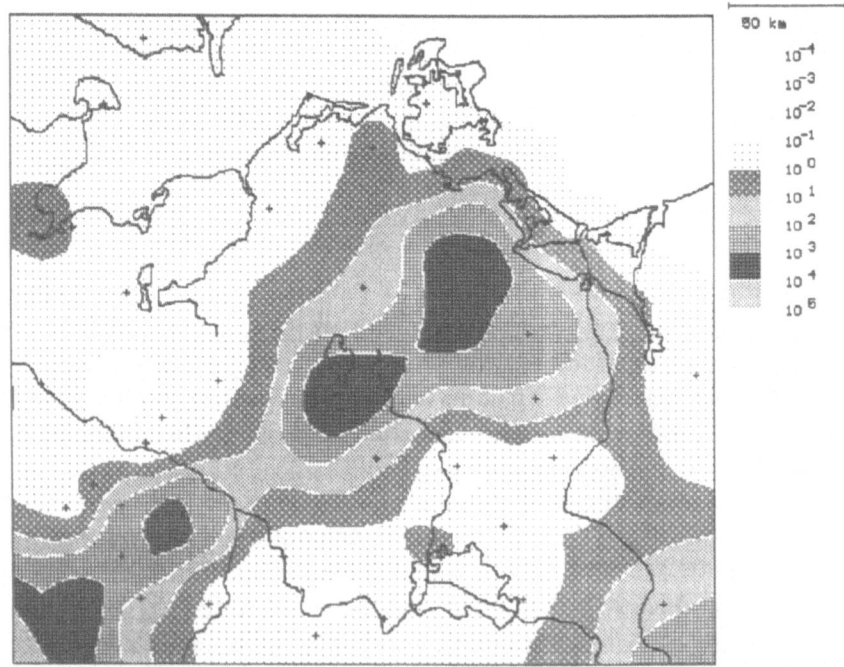

Figure 8. TTI-values for top of Upper Carboniferous sediments at beginning of Zechstein sedimentation.

Table 1. Methods of assessment criteria for gas generation.

classes	source rock potential	thickness of Permocarb. Volcanic rocks	TTI (Beginning of Zechstein)
2	5.929	490.286	0.700
1	3.563	341.625	0.450
3	1.375	2088.750	1432.625
4	0.455	269.909	1.964
5	0.333	1255.333	34.667

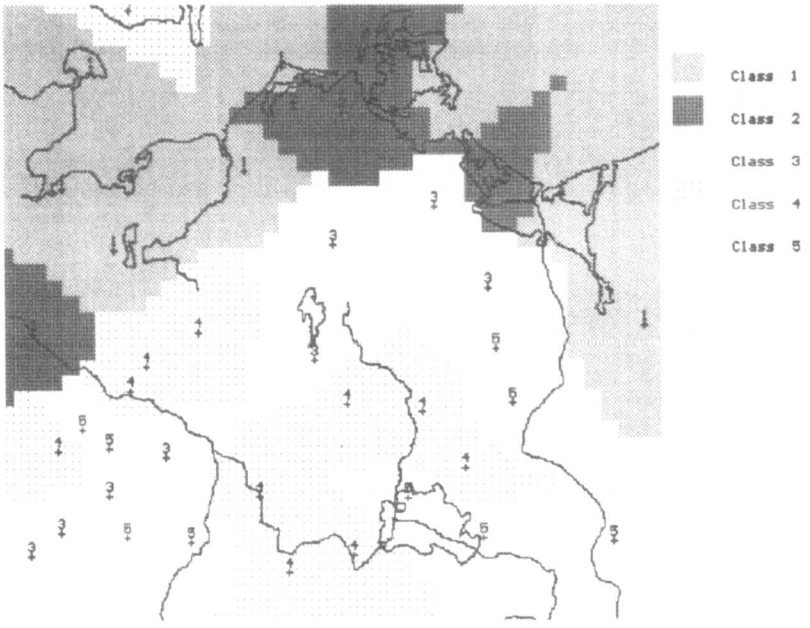

Figure 9. Regionalization scheme for favorability of gas generation after Rotliegend sedimentation.

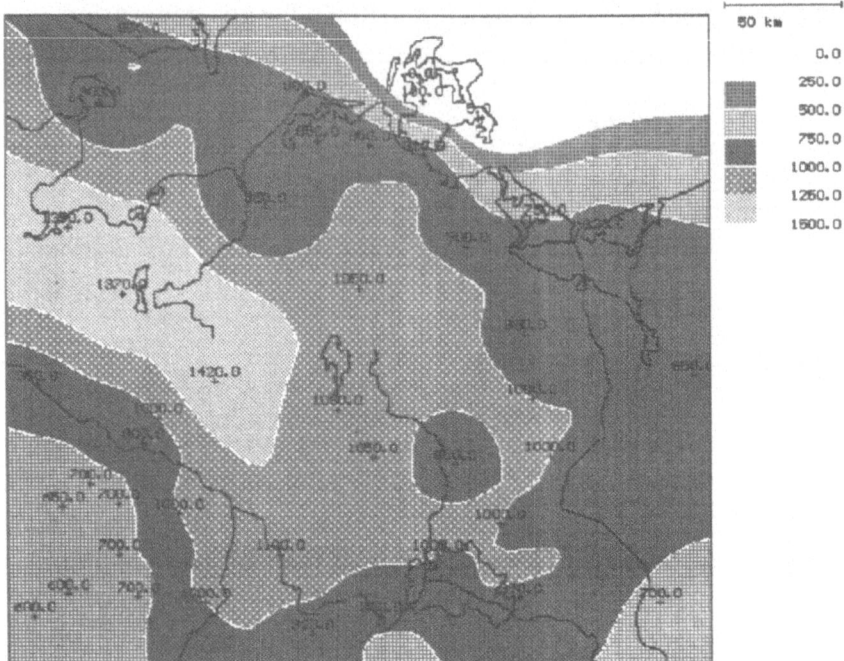

Figure 10. Map of reconstructed paleothickness of Zechstein salts.

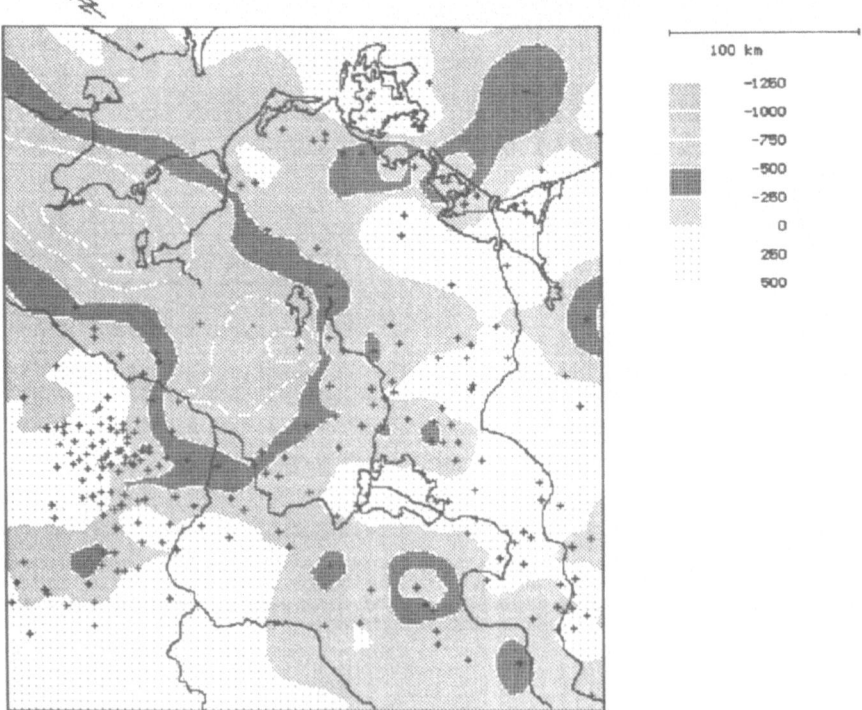

Figure 11. Isoline map of reconstructed paleorelief of Permo-Carboniferous volcanic complex.

Relative and absolute sand content in the drilled Upper Rotliegend sedimentary sequences provide the third and fourth assessment criterion.

Seven different types of drilling profiles corresponding to the reservoir capability have been determined.

Table 2 shows the mean values of assessment criteria for the different classes which are ordered downward from most favorable to most unfavorable ones. The regional distribution of the classes is shown by Figure 12. The most favorable regions are located in the southern part of the investigated area, medium conditions are connected with the northwestern part, whereas the RHgen area shows a negative result in connection with reservoir capabilities.

For the prediction of HC deposits the map of HC generation favorability can be linked with the prediction map of reservoir capability. This can be done for example by the isoline map of paleosubsurface of pre-Permian rocks at the beginning of Zechstein time (Fig. 13) and the corresponding three-dimensional model (Fig. 14). This paleostructure model can be used for the reconstruction of migration paths of HC from the Carboniferous source rocks guided at the base of the volcanites in the direction from deep subsided areas to higher positions. Gas having been generated in the central part of the investigated area can move on the base of this assumption mainly in two general directions: To the south-

Table 2. Methods of assessment criteria for reservoir potential

classes	Paleothickness of Zechstein salts	relative sand content of Saxonian Sediments	absolute sand content of Saxonian Sediments	Paleo-relief
2	1033.333	53.000	322.667	-387.000
4	708.333	72.167	175.833	69.167
5	910.909	35.636	80.455	74.455
7	904.545	12.455	56.000	-122.818
1	1380.000	3.667	64.333	-960.333
6	820.000	2.500	12.833	-231.000
3	35.000	0.000	0.000	196.667

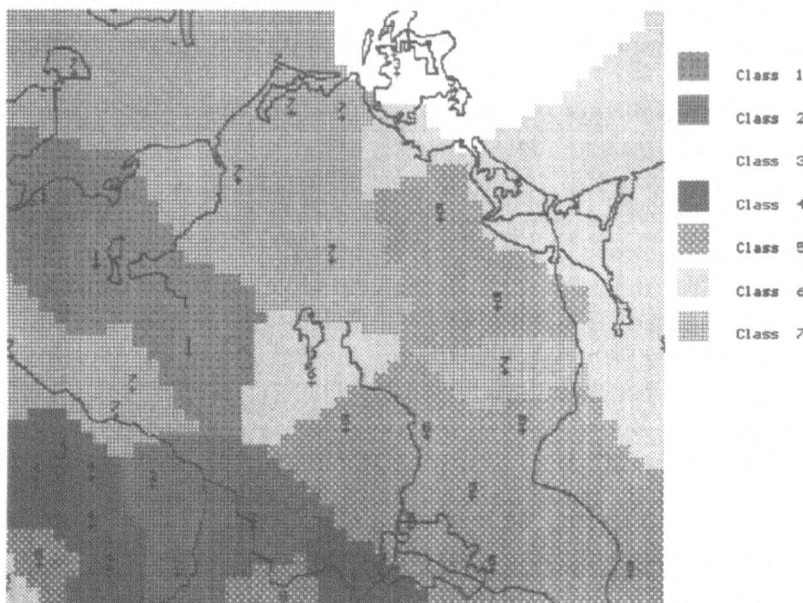

Figure 12. Regionalization scheme for favorability of reservoir potential.

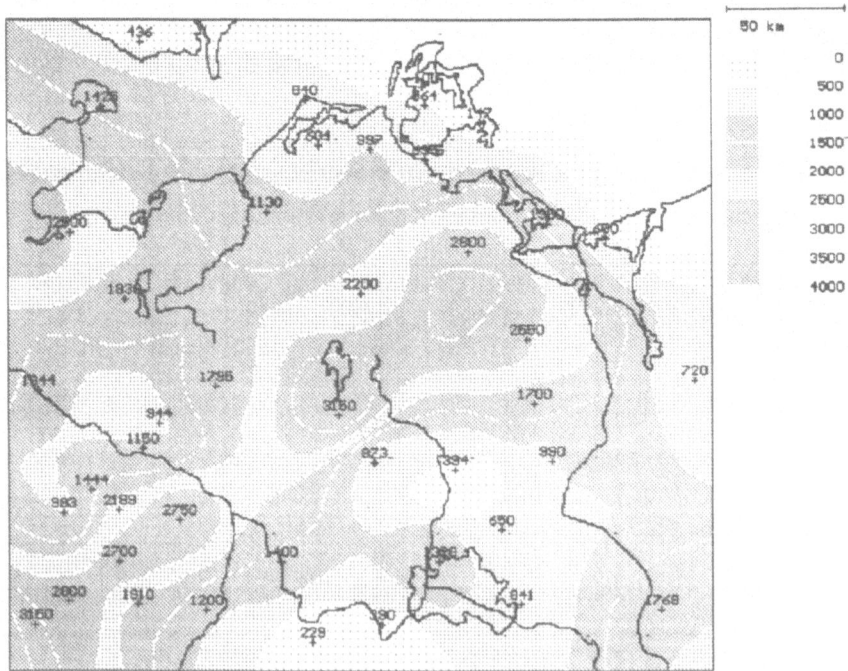

Figure 13. Isoline map of paleosubsurface of Carboniferous rocks at beginning of Zechstein time.

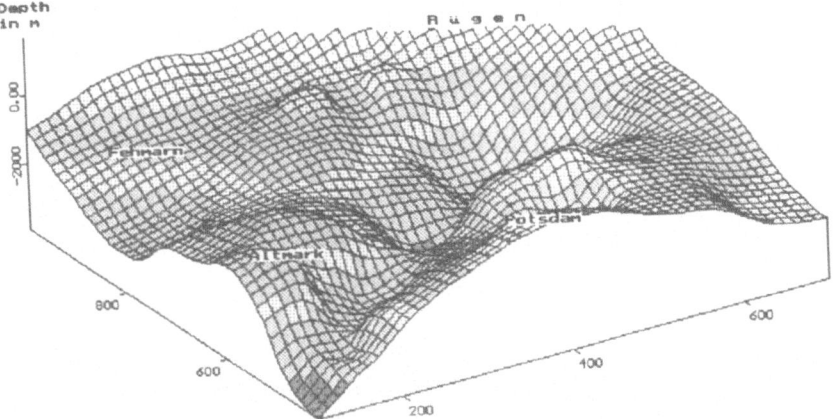

Figure 14. Three-dimensional model corresponding to Figure 13.

west where the Altmark gas deposit is located and secondly to the northeast. The southwest-northeast-trending negative anomaly connected with thick volcanites is to understand as a migration barrier for gas during the whole of late Permian time. The investigation of the development of this structure during post-Permian time seems to be important for the analysis of later gas migration.

The summarized proposal for additional exploration is to focus activities to a region in the northwestern part of the investigation area, where good conditions for HC generation and medium conditions for reservoir rocks exist. This area also is linked by possible migration paths with gas-generating areas in the basin's center similar to the well-known Altmark-gas-district.

SUMMARY

A computerized method for the complex interpretation of assessment criteria for oil and gas deposits is introduced. As assessment criteria measuring values, results of structural and process modeling as well as expert appraisals can be involved in the analysis. Structure and process modeling is carried out by a PC program using well data. The data integration is accomplished by the generalization of the "classification of geological objects" theory developed by D.A. Rodionov and others in Russia. It is recommended to split the research in two steps: the prediction of favorability for HC generation, and the prediction of reservoir capabilities expressed both by regionalization schemes. These two maps can be connected by paleostructural maps as indicators for migration pathways of HC. An application of the method to the northeastern part of the North German Basin indicates a new favorable area for the search for gas nearby the Baltic Sea coastline.

REFERENCES

Berthold, A., and Galushkin, J.I., 1986, Mathematische Modellierung der Senkenbildung am Beispiel der Norddeutsch- Polnischen Senke: Z.angew.Geol., v.32, no. 10, p. 262 - 267.

Harff, J., Schretzenmayr, S., Springer, J., Hoth, P., Eiserbeck, W., and Süssmuth, S., 1989, Mathematisch-numerische Modellierung regionaler bis lokaler Einheiten in sediment ren Becken an einem Beispiel für die Kohlenwasserstofferkundung im Rotliegenden: Z.geol.Wiss., v. 17, no. 7., p. 747 - 759.

Harff, J., and Davis, J.C., 1990, Regionalization in geology by multivariate classification: Jour. Math. Geology, v. 22, no. 5, p.573-588.

Harff, J., Eiserbeck, W., Hoth, K., and Springer, J., 1990, Computer assisted basin analysis and regionalization aid the search for oil and gas: Geobyte, v. 5, no. 4, p. 11 - 15.

Rodionov, D.A., 1981, Statisticeskie resenija v geologii: Nedra, Moskva, 231 p.

Springer, J., Lewerenz, B., Bobertz, B., Harff, J., Hoth, P., and Sübmuth, S., 1990, BASIN - a software system for the computer aided modeling of sedimentary basins - software description (diskette available from ZIPE Potsdam).

Tissot, B., and Welte, D.H., 1978, Petroleum formation and occurrence: Springer, New York, 699 p.

Watson, D.F., and Philip, G.M., 1985, A refinement of inverse distance weighting: Geoprocessing, v. 2, no. 8, p. 315-327.

THE STATIC MODEL OF THE
RUDOLSTADT BASIN AND ITS
METALLOGENETIC INTERPRETATION

H. Thiergärtner and J. Rentzsch
Central Geological Institute, Berlin, Germany

ABSTRACT

Metal content of the Kupferschiefer unit (Zechstein subformation) as well as the Rotliegendes thickness in the Permo-Silesian Rudolstadt Basin, northern foreland of the Thuringian Forest, Germany, were studied using a computer-aided image-processing technology consisting of

(1) the generation of an area-covering information from the original data obtained in borehole points with the aid of an interpolation by inverse distance weighting of all data points that are relevant with reference to a Gabriel graph selection procedure;

(2) a gray value transformation of the interpolated data using the dialog regime for a geological control of the results; and

(3) a multispectral classification of the data by a self-learning nonhierarchical method.

The interpretation of the images obtained shows a predominance of the paleotectonic control of the copper distribution in the Rudolstadt Basin area against the paleogeographical environment. Local occurrences of the copper-type Kupferschiefer are concerned with the adjacency of marginal faults of the basin which supported inflows of copper-bearing ascending solutions.

Computerized Basin Analysis, Edited by J. Harff
and D.F. Merriam, Plenum Press, New York, 1993

GEOLOGICAL POSITION

The Rudolstadt Basin in the northern foreland of the Thuringian Forest (Germany) covers about 15 km in NW-SE and 7 km in NE-SW direction. It belongs to the Variscan local isolated intramountainous depressions in Central Europe. Within it, there are deposited Permo-Silesian coarse-clastic red sediments up to 90 m thickness which are in a discordant position to folded Paleozoic rock formations. The SW and the NE boundaries of the basin as well as an interior NW-SE-striking structure of the basin are controlled by pre-Permian NW - SE striking faults (Seifert, 1972). The subsidence along the faults began in the early Lower Rotliegendes (Upper Silesian? - early Lower Permian). From this point of view, the Rudolstadt Basin represents a "cross-depression" in relation to the system of NE-SW-oriented ridges and depressions (Fig. 1). It differs, with respect to its fracture tectonic conditioning, "by its size, orientation, and genesis clearly from the more extended, NE-SW oriented and epirogenic caused parageosynclines of the Rotliegendes and Silesian formations" (Seifert, 1972). The original area of the detrital material are the near Thuringian Slate Mountains adjacent in the south. The sediments of the Rotliegendes wedge out at the basin margins. By boreholes, only the northwestern part of this basin is explored (Fig. 2).

Together with the origin of the German depression by a NW tipping of the area, the Rudolstadt Basin extended from the beginning of the Zechstein transgression. Now, in addition to the pre-existing NW-SE-striking structures, NE-SW-oriented elements belong to the tectonic system (formation of the NNE-SSW-oriented Rudolstadt channel) and the coast line moves towards the units of the crystalline basement complex folded in the Paleozoic. Thus, continually new parts of the up to now continent are flooded. The deposition leads to a concordant sequence of beds [see stratigraphic section (Table 1) of the complete overburden series in the Rudolstadt area; after Jungwirth and Seifert, 1966; Jungwirth and Seidel, 1968].

The Kupferschiefer (copper shale) shows an increasing carbonate content in connection with a decreasing silicate proportion and an increasing thickness at the basin margins. The content of carbonates is proportional directly to the thickness of the marlstone, but the contents of silicic material and the organic carbon, however, indirectly (Rentzsch and others, 1973a, 1973b). The character of the Kupferschiefer is controlled by the geomorphology, consequently. This paleogeographic influence is evident in the Zechstein limestone, too.

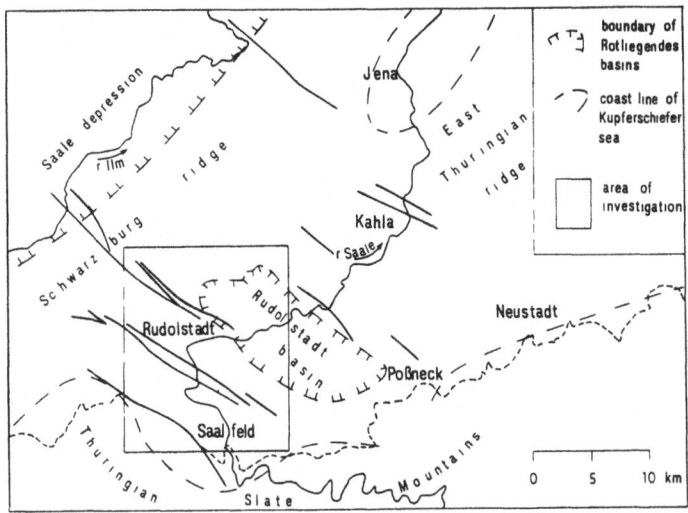

Figure 1. General Geological Position of Rudolstadt Basin (after Seifert, 1972, Fig. 1, and Hoppe and Seidel, 1974, Fig. 79).

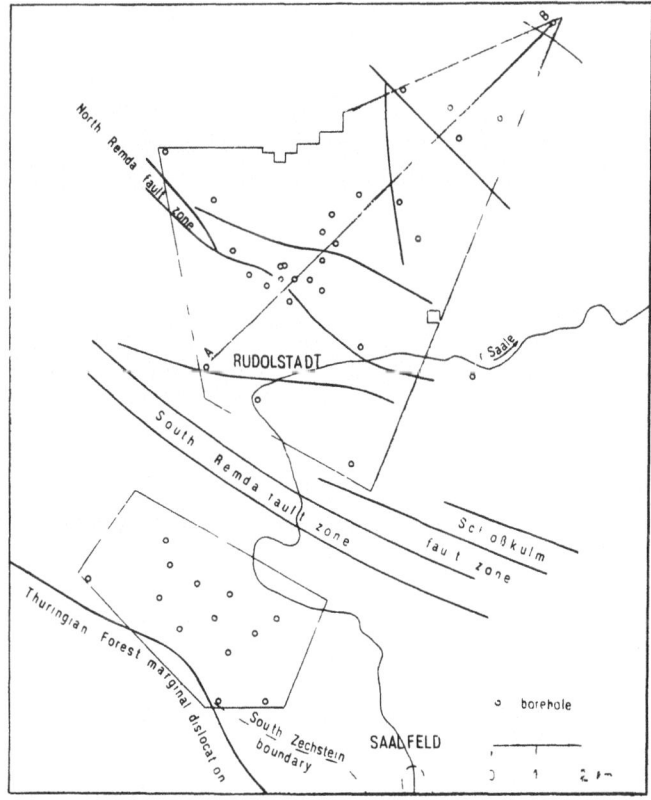

Figure 2. Structural situation within area investigated (after Seifert, 1972, Fig. 2).

Table 1. Normalized stratigraphic section

age		thickness	lithological character
Q		some meters	
— hiatus —			
T1		> up to	Middle Bunter
T1		> >400m	Lower Bunter
— hiatus —			
P2	z1	c. 10m	Lower Werra anhydrite (anhydrite, halite, mudstone)
	z1	4 - 25m	Lower Werra carbonate
	z1	0.2- 1.2m	"Kupferschiefer" (copper shale: marlstone up to carbonate)
	z1	0.6 - 1.2m	"Mutterfloez" (basal carbonate: gray dolostone)
	z1	0.5 - 3m	"Weissliegendes" (white underlying: gray, very sandy conglomerate)
P1	r3	10m	mixed conglomerates and pebbly sandstones
	r2	45m	conglomerates consisting of rounded quartzites
— discordance —			
P1...C2		40m	limestone conglomerates, well graded
— discordance —			
C1...O			folded Ordovician, Silurian, Devonian, Carboniferous

METAL CONTENTS WITHIN THE KUPFERSCHIEFER

The copper shale in the region of the Rudolstadt Basin is a dark gray to black, compact, laminated, coaly bituminous dolomitic marlstone-bearing nonferrous metal sulfides. Its thickness is about 0.2 up to 1.2 m, where the delimitation towards the capping beds in the vicinity of the basin margins is problematical.

Above the basement complex ridge in a greater distance from the margins of the Rotliegendes trough, the rock is developed as a lead-zinc-enriched marlstone with a galenite-sphalerite-paragenesis ± chalcopyrite. Unimportant copper enrichments which are proved by the local occurrence of the copper mineralization are connected with the vicinity of the margin faults of the Rudolstadt Basin.

In general, the regularities of the metal or the ore mineral distribution within the copper slate are known. A polystadial model of genesis was published by Rentzsch (1974). Tischendorf and Schwab (1989) summarized the different models of genesis and arranged the metallogeny of the Kupferschiefer type into the frame of their metallogenetic concept of the platform stockwork.

Using quantitative computer-aided simulation methods the connection between the copper sulfide contents and the paleotectonic (but not the paleogeographic) conditions is to be verified, although only a part of the Rudolstadt Basin is explored (Figs. 1 and 2). At the same time, the different behavior of the metals lead and zinc on the one hand with respect to copper on the other hand is to be analyzed quantitatively.

METHODOLOGY OF THE STATIC SIMULATION

A static basin model is to be understood as the computer-aided, three-dimensional quantitative description of the structure of a sedimentary basin. Of course, static models fundamentally differ from dynamical models of those used, for example, by Harff and others (1989) with the aim of reconstruction of paleosurfaces in the framework of the carbon hydrogen exploration or to solve other geological problems.

With the aid of traditional statistical analyses it is impossible to get a decisive knowledge as to static models. Thus, the correlation coefficients triangle matrix of some parameters only shows that the total nonferrous metal accretion practically depends only on the metal amount of lead and zinc, but not of copper. Also, the thickness of the Rotliegendes sediments does not correlate with the copper contents in a significant manner (Table 2). The use of the trend-surface analysis which is proved in copper shale regions with sufficient dense exploration pattern (Rentzsch and Thiergärtner, 1972; Thiergärtner and Abramovich, 1977) here is not suitable because of the limited number of exploration points. The regional subdivision of the Rudolstadt Basin by a classical cluster analysis with

Euclidic distance measure and flexible strategy after Lance and Williams
or the strategy after Ward (Table 3) including the following ordering of
the classes obtained to the exploration points (Fig. 3) in the methodology
published by Gradova (1975) can provide only limited results: three
cluster groups represent copper shale profiles without underlying
Rotliegendes with slight differing copper accretions; two classes reflect
profiles with underlying Rotliegendes sediments. It can be seen that the
maximum copper content does not occur in profiles with maximum
thicknesses of the Rotliegendes, but within the more marginal profiles of
the wedging out Rotliegendes. A continuous space simulation is not
possible.

Table 2. Results of linear correlation analysis between selected parameters

	thickness of Rotliegendes	copper accretion	lead accretion	zinc accretion	nonferrous metals
latitude	+0.43	+0.44	-0.19	-0.21	-0.15
longitude	+0.50	+0.26	-0.21	-0.12	+0.04
Rotliegendes		+0.23	-0.12	-0.08	+0.06
copper accr.			-0.10	-0.29	-0.03
lead accr.				+0.39	+0.76
zinc accr.					+0.86

significant at a 95% level correlation coefficients are underlined

A suitable static basin model can be generated by use of image-
processing methods consisting of interpolation and classification procedures.

Interpolation

Using the parameter values x(i) available from irregularly distributed exploration or data points i (Rotliegendes thickness, copper accretion, lead accretion), there was realized an interpolation towards an area-
covering quadratic grid (250m grid point distance in the nature, 4 pixels
in the image). The algorithm includes the following procedures:
   (1) A delimination of geologically interpretable partial areas by
       polygons.

Table 3. Results of the numerical cluster analysis

dendrogram	cluster group in Fig.3	number of points	thickness of Rotliegendes in m			copper accretion in kg/m²		
			mean	min	max	mean	min	max
		19	0	0	0	0.7	0.1	1.7
		5	0	0	0	2.7	2.2	3.5
		6	0	0	0	5.3	4.7	6.4
		3	11.9	2.4	22.8	16.2	15.7	17.0
		9	35.3	1.4	76.0	4.2	0.9	7.5
over all values in the area			7.5	0	87.9	3.3	0.1	17.0
over all mode in the area			0			0.5		

(2) The determination of the relevant data points i with respect to the calculation of interpolated values x(g) within grid points g by
- giving of a maximum search radius r(max) (r(max) = 1.25 km in the nature) and exclusion of all those points that correspond to the following relation:

$$((b(g) - b(i))^2 + (l(g) - l(i))^2)^{1/2} \quad > \quad r(max) \quad (1)$$

with   b(g) resp. l(g) coordinates of the grid point g
        b(i) resp. l(i) coordinates of the data point i;

- application of the Gabriel graph which secures that in each direction around the grid point only the nearest data points will

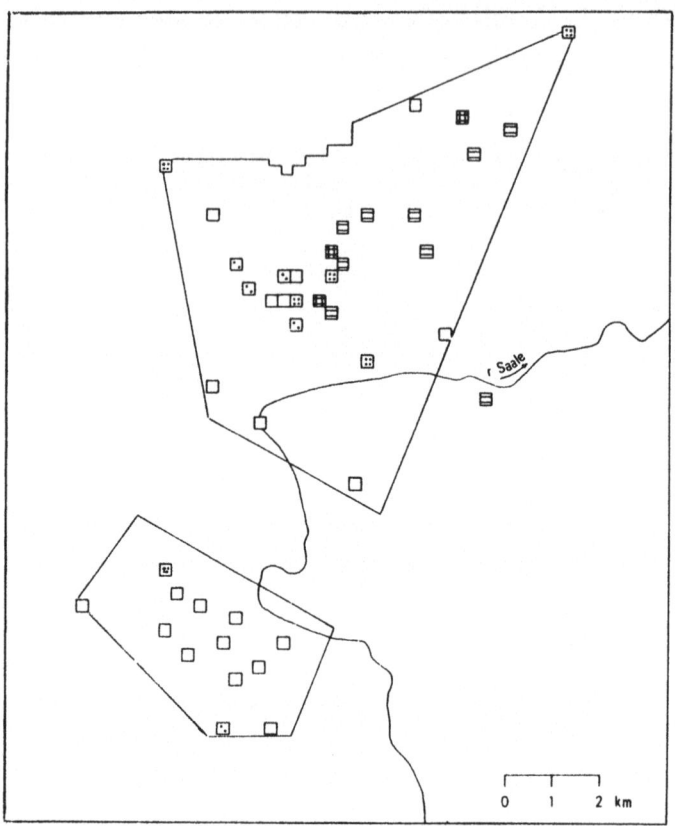

Figure 3. Areal distribution of the numerical cluster groups with respect to Table 3.

be taken into consideration by exclusion of a data point i than if the following relation is sufficient:

$$((b(g)-b(i))^2 + (l(g)-l(i))^2 > ((b(g)-b(j))^2 + l(g)-l(j))^2) \quad (2)$$

(At the same time, the calculation of values in usually not explored partial areas will be avoided.)

(3) The computation of the interpolated value x(g) in each grid point g by inverse distance weighting according to the formula

$$x(g) = \sum_{i=1}^{rel} (x(i)/r^c(i)) \; / \; \sum_{i=1}^{rel} (1/r^c(i)) \qquad (3)$$

with rel number of relevant data points
    x(i) data value within the point i
    r(i) geometrical distance between the point i and the grid point g
    c   weight of the weighting function (c=2).

(4) A transformation of the interpolated values x(g) into the range 0, ..., 254 of the gray-scale value by application of the formula

$$g(g) = x(g) * 254 / (x(max) - x(min)) \qquad (4)$$

with g(g) gray value within the grid point g
    x(g) interpolated value within the grid point g
    x(max) maximum value within the original data matrix
    x(min) minimum value within the original data matrix.

(5) Visualization of the interpolated values within the investigated area (Figs. 4, 5, 8) as well as along the NE-SW-striking cross section A - B (Fig. 7) which is situated within the northern polygonal region. For the purpose of a black-white-reproduction the interpolated values of the Rotliegendes thickness, of the copper accretion and the lead accretion were transformed into gray values by the following nonlinear coding scheme.

Gray-Scaling Value in Figures 4, 5, 8

gray value	thickness of Rotliegendes in m	copper accretion in kg/ m²	lead accretion in kg/ m²
white	0...2.0	—	—
bright gray	2.1...28.0	0...2.2	0...3.8
gray	28.1...45.5	2.3...4.7	3.9...10.9
dark gray	45.6...55.0	4.8...8.7	11.0...23.5
black	55.1...80	8.8...20	23.6...40

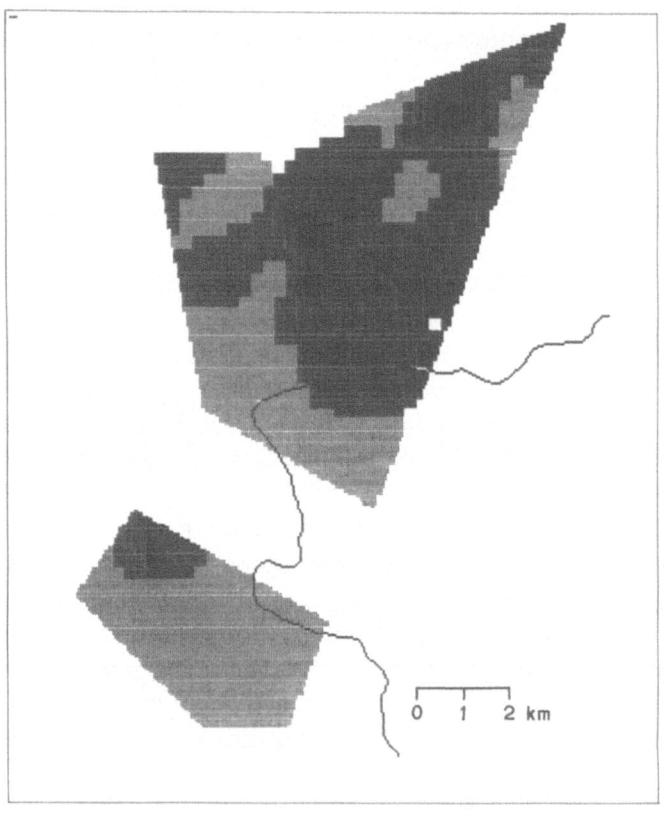

Figure 4. Areal distribution of copper accretion values.

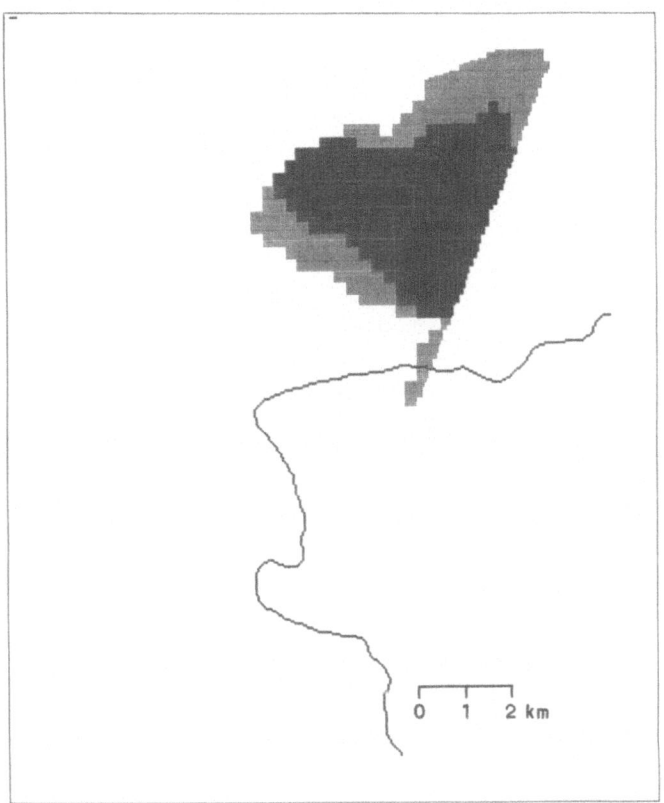

Figure 5. Areal distribution of Rotliegendes thickness values.

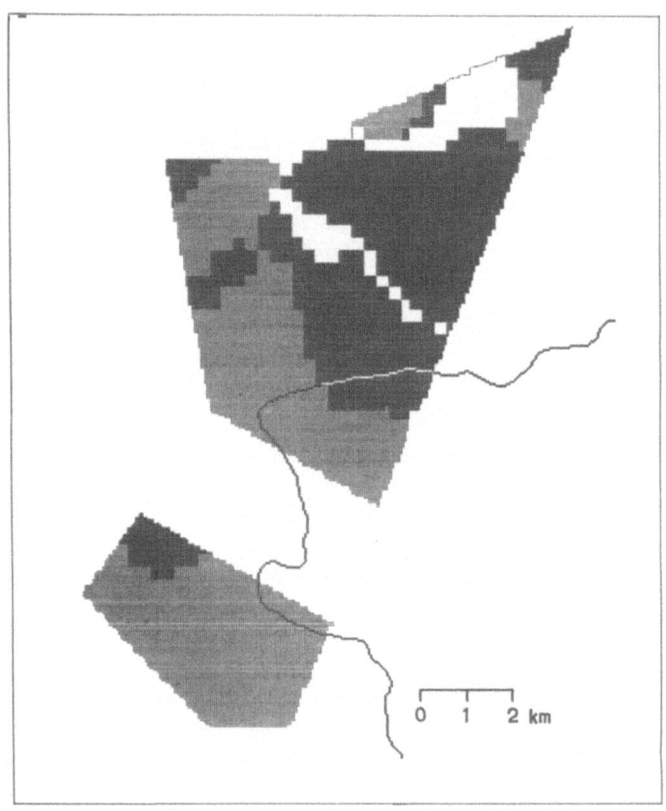

Figure 6. Areal distribution of image-processing cluster groups respective to Table 3.

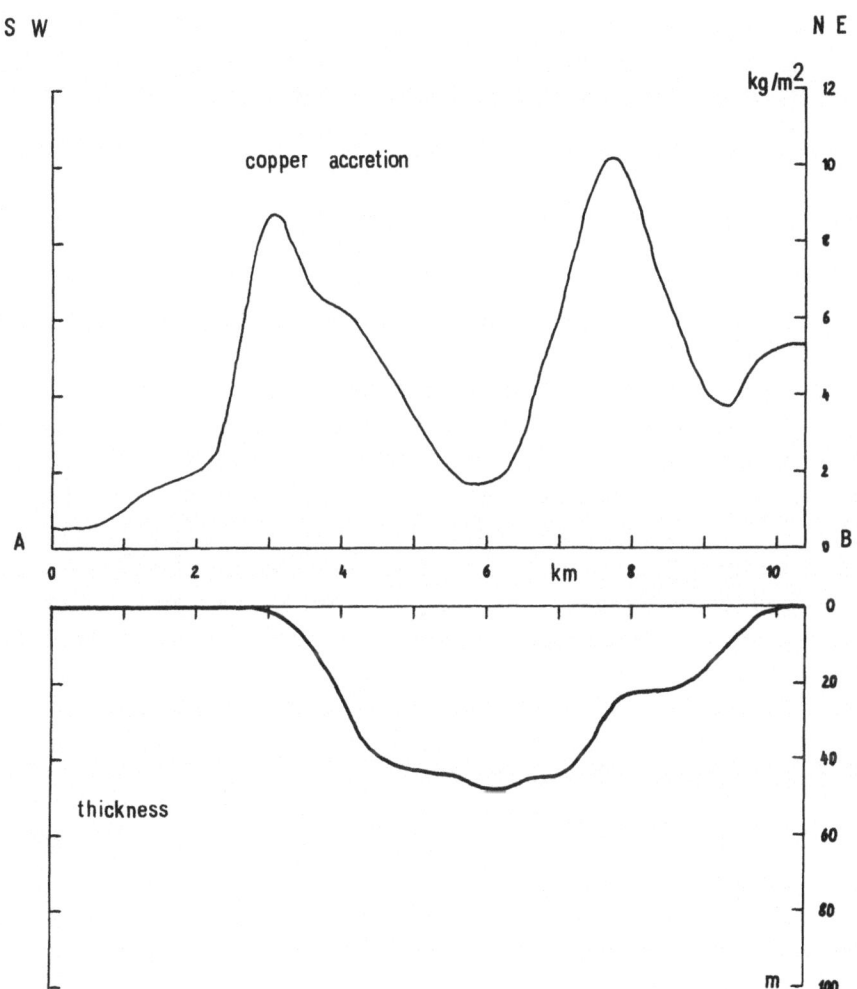

Figure 7. Cross section A - B through Rudolstadt Basin, upper part: copper accretion values, lower part: thickness of Rotliegendes.

In order to obtain the values along the cross section the interpolated values x(g) were smoothed according to the relation

$$x(g)^* = (x(g-1) + x(g) + x(g+1)/3 \qquad (5)$$

with x(g-1) as interpolated value in the neighbor grid point (g-1). This allows to get the regional tendencies by low-pass filtering.

Classification

On the basis of the parameters "Rotliegendes thickness" and "copper accretion" that was interpolated for each grid point and, consequently, for each pixel a nonsupervised multispectral classification with an algorithm after Sebesteyn and Eddie was carried out. It is the question of a nonhierarchical self-learning method of subdivision of a total set R of pixels in the investigated area into a number t of partial sets resp. clusters R(k), k= 1, ..., t. t is to be given. The clusters consists of each n(k) pixels. If p parameters would be included, a pixel p can be described by the vector x(q, p) = q of the interpolation values generated before, and a cluster k can be represented by a vector m(k, p) = m of the parameter averages and the vector s(k, p) = s of the parameter standard deviations between the n(k) pixels arranged around the cluster centrum. A concrete pixel q will be classified disjunctively to that cluster k* among all available clusters k, for which its standardized distance

$$d(q,k) = (x-m)/s \qquad (6)$$

with respect to the cluster centrum within the p-dimensional parameter space gets a minimum value and, at the same time, does not exceed a boundary value alpha:

$$d(q,k^*) = \min_{k=1,...,t} d(q,k) \qquad (7)$$

and

$$d(q,k^*) \leq alpha \qquad (8)$$

After the classifying, the properties of the hitherto existing cluster k^* that is expanded by 1 pixel now will be changed as follows (so-called teaching of the cluster), if not the restriction

$$beta < d(q,k)^* \leq alpha \qquad (9)$$

is true simultaneously:

number of pixels $n' = n(k^*) + 1$

vector of averages $m' = (n(k^*)/n(k^*)+1)*(m + x/n(k^*))$

vector of std. dev. $s' = (n(k^*)/n(k^*)+1)*(s^c+(m-x)^c)^{1/c}$

The projecting condition secures the neglect of the influence of such pixels that are situated in marginal zones of the parameter space around the cluster centrum on the cluster properties. The constant c regulates the velocity of the enlargement of the clusters while classifying new pixels. A pixel will locate a new cluster with the pixel number 1, the cluster centrum x and the variance 0, if the restriction

$$d(q,k^*) > alpha \qquad (10)$$

is true and if the pregiven number of clusters is not treated fully; unless it is necessary to change alpha and beta. To the cluster groups (t= 5) false color values according to the code table included in Table 4 were assigned. The areal distribution of the cluster classes can be seen in Figure 6 in the adequate scale.

Table 4. Results of clustering of 14337 pixels by image processing

color	gray	distribution of pixels	thickness of Rotliegendes in m		copper accretion in kg/m²	
value in Fig. 6			mean	std.dev.	mean	std. dev.
blue	bright gray	48.2 %	0.6	± 4.4	1.1	± 0.8
green	gray	22.3 %	1.9	± 3.8	4.5	± 1.4
white	white	9.6 %	24.2	±13.1	8.1	± 3.3
red	black	19.6 %	50.8	± 7.0	4.4	± 1.8

METALLOGENETIC INTERPRETATION

The results of the image analysis shows the following regularities in the Rudolstadt Basin:

(1) In general, the copper accretion attains low values about the northwestern part of the Rudolstadt Basin itself as well as about the regions transgressed at the beginning Zechstein time. The subdeposition of Rotliegendes sediments under the Kupferschiefer within the Rudolstadt Basin in the narrower region does not cause significantly higher copper concentrations and a dependence of the copper-bearing on the Rotliegendes thickness cannot be seen (Fig. 5 versus Fig. 4).

(2) More considerable copper enrichments, first of all, seem to be limited on the area near three boreholes only (Fig. 4). However, by the multispectral classification it is possible to prove that the zone of enrichment exactly occupies a narrow limited linear region at the north-northwest and the southwest margin of the Rotliegendes trough (white color in Fig. 6). This fact is confirmed with the aid of a cross section A - B (Fig. 7) through the Rudolstadt Rotliegendes Basin.

(3) The lead accretion corresponds well with the zinc one but not completely identical with it is determined by north - south and NW - SE-striking structure elements and does not follow the Rotliegendes basin or its boundaries (Fig. 8).

As results, the following regularities can be mentioned:

(1) Local occurrences of the copper-type Kupferschiefer marlstone are connected with the partial area of the marginal faults of the Rudolstadt Basin. Lead-zinc-bearing sedimentary rocks situated at the Zechstein Basin occur on the southeastern flank of the Schwarzburg Ridge beginning at the marginal Zechstein and outcrops up to the Rudolstadt Basin. The pure lead type shows a connection with a NW - SE-striking structure at the marginal Zechstein and with a north - south or a NE - SW structure near the southeastern margin of the Schwarzburg Ridge.

(2) The distribution of the mineralization types in the Zechstein Basin sediments is controlled by the paleotectonics of pre-Zechstein more than by the paleogeographical environment of the Kupferschiefer sea.

(3) The ascendent supplies of solutions into the Zechstein depression, first of all, took place   by NW-SE-striking marginal faults of the Rotliegendes trough that is situated across to the   general striking of the great basins and ridges during the Rotliegendes subformation. The Pb-Zn-enrichment of the Zechstein Basin striking parallel to the Schwarzburg Ridge point to preferred supplies of solutions within areas characterized by crossing NW-SE and NE-SW up to north-south-striking faults.

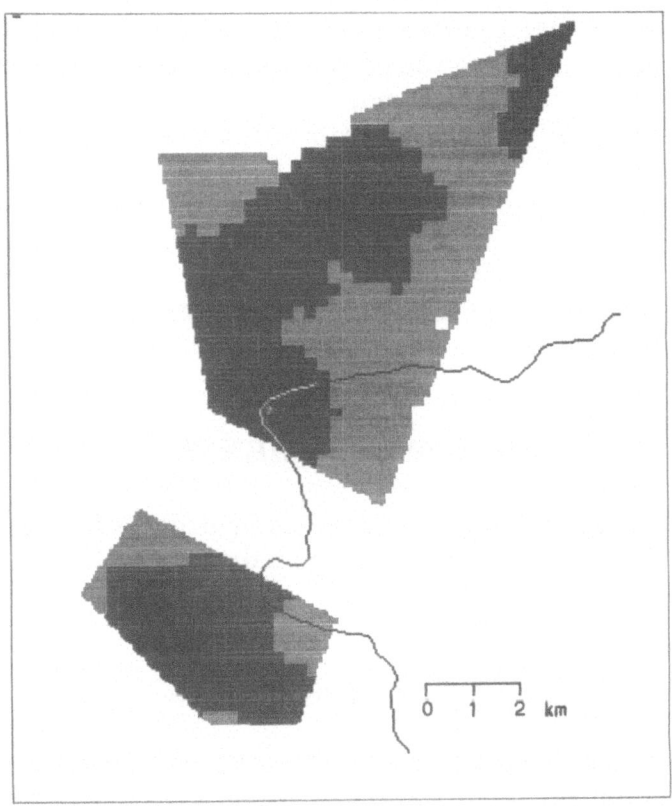

Figure 8. Areal distribution of lead accretion values.

CONCLUSIONS

It is possible to construct a static model of the ore mineralization in a sedimentary basin with relatively low exploration grade by image processing even if traditional statistical or classical classification methods fail to work. This possibility is founded by

(1) the undistorted area-covering interpolation on the basis of some exploration points only if suitable algorithms are selected;

(2) the power of monospectral, nonlinear splitting of the interpolated values in the gray value range using geological model ideas and hypotheses in a dialog working regime; and

(3) the geologically relevant control of the multispectral classification process.

The image processing and analysis prove the conclusion deduced from field observations that the metal distribution in the Rudolstadt Basin area is controlled more by paleotectonics of the pre-Zechstein then by the paleogeographical environment during deposition of the Kupferschiefer marlstone.

ACKNOWLEDGMENTS

We gratefully thank the director of the Central Geological Institute Berlin, Dr. J. Kamps, for supporting the analyses as well as U. Schmidt for his assistance in image processing.

REFERENCES

Gradova, T. A., 1977, Teoreticheskiye voprosy postanovki i resheniya zadach rayonirovaniya na EVM: Tome II VC SO AN SSSR, Novosibirsk. 27 p.

Harff, J., Schretzenmayr, S., Springer, J., and Eiserbeck, W., 1989, Mathematisch-numerische Modellierung regionaler bis lokaler Einheiten in sedimentären Becken an einem Beispiel für die Kohlenwasserstofferkundung im Rotliegenden: Z. geol. Wiss., Berlin, v. 17, no. 7, p. 747-759.

Hoppe, W., and Seidel, G., eds., 1974, Geologie von Thüringen: VEB Hermann Haack, Gotha/Leipzig, 1000 p.

Jungwirth, J., and Seidel, G., 1968, Die faziellen Änderungen der Zechsteintone in Thüringen: Jb. Geol., Berlin, v. 2/1966, p. 271-280.

Jungwirth, J., and Seifert, J., 1966, Zur Stratigraphie und Fazies des Zechsteins in Südwestthüringen: Geologie, Berlin, v. 15, no. 4/5, p. 421-433.

Rentzsch, J., 1974, The Kupferschiefer in comparison with the deposits of the Zambian copper-belt: Cent. Soc. Geol. Belg.- Gisements Strat. et Prov.cupr., Liege, p. 395-418.

Rentzsch, J., Tischendorf, G., Ungethüm, H., and Pilot, J., 1971a, Contributions to the geochemistry of the Kupferschiefer mineralization: Intern. Geochem. Congr., Abstracts, Moscow, Tome II, p. 382-383.

Rentzsch, J., Tischendorf, G., Ungethüm, H., and Pilot, J., 1971b, K geokhimii medistykh slantsev: I. Mezhdunar. geokhim. kongr., Moskva, v. 2, (1971) Tome IV, p. 172-184.

Rentzsch, J., and Thiergärtner, H., 1972, Vergleichende geologisch-mathematische Untersuchung der Buntmetallverteilung in Lagerstätten vom Typ Kupferschiefer: Z. angew. Geol., Berlin, v. 18, no. 12, p. 537-548.

Seifert, J., 1972, Das Perm am Südostrand des Thüringer Beckens: Jb. Geol., Berlin, v. 4/1968, p. 97-179.

Thiergärtner, H., and Abramovich, I.I., 1977, Geologische Bewertung verschiedener Flächentrendmodelle: Abhandl. d. ZGI, Berlin, v. 39, p. 135-142.

Tischendorf, G., and Schwab, G., 1989, Metallogenesis of the transition period between Hercynian and subsequent platform stage in central Europe: Z. geol. Wiss., Berlin, v. 17, no. 8, p. 815-842.

TOWARDS ASSESSMENT OF PLAYS CONTAINING MIGRATED PETROLEUM

David J. Forman, Alan L. Hinde, Steven J. Cadman,
Andrzej P. Radlinski,
Bureau of Mineral Resources, Australia
and John Morton
Department of Mines and Energy, South Australia

ABSTRACT

The undiscovered resources in the Patchawarra Formation in the Cooper Basin in South Australia have been assessed by a computer program using area of closure and resources per unit area of closure of the potential traps input as loglinear models. Compared to other forms of input, loglinear models help to describe dependent relationships that occur between these parameters and drilling or discovery sequence number and will allow the program to better simulate sequential drilling of the traps as is required for production forecasting.

Different results are obtained depending on how the assessment units are selected. The entire formation may be assessed as a single unit. The average value of the assessment is almost doubled, however, by assessing individual migration fairways within the formation. A better estimate of the accumulation sizes and a slightly higher assessment are obtained by assessing individual fractal plays within the assessment units. None of these assessments, however, takes account of migration between traps.

Additional information, such as the order in which the traps fill and the amounts that spill from one trap to another, is needed to assess plays that contain migrated petroleum. This additional information can be deduced using a structure contour map of the play, provided that the present structure is substantially the same as it was at the time of migration.

Computerized Basin Analysis, Edited by J. Harff
and D.F. Merriam, Plenum Press, New York, 1993

INTRODUCTION

No matter the method used, serious problems arise in assessment because of the way the assessment units are selected and because little or no account is taken of tertiary migration, that is migration from one trap to another. Using examples from the gas-prone Cooper Basin in South Australia, we show how the plays that are used normally in assessment may be broken into smaller assessment units and how the assessment of the undiscovered gas resources will change depending on how the units are selected. We also provide equations and suggest a framework by which plays that contain migrated petroleum may be assessed.

Selection Of Assessment Unit

The trap

The petroleum trap, which is the simplest assessment unit, is portrayed by a structure contour map of the contact between the reservoir and cap rocks. Traps may be simple with one culmination (Fig. 1) or complex (Forman and others, 1989) with two or more culminations (Fig. 2).

The simple trap has an area of closure (A_{clos}) which is outlined by a structure contour, fault, or stratigraphic pinchout and an area of drainage (A_{drain}) which is outlined by a drainage divide, fault, or stratigraphic pinchout, either separately or in combination. In anticlinal traps, A_{clos} and A_{drain} coincide at one point termed the spill-point, in fault traps they coincide along the line of the sealing fault or faults, and in isolated stratigraphic bodies they coincide along the stratigraphic pinchout. Whereas with one reservoir, the simple trap contains only one accumulation, a partly filled complex trap may contain a number of accumulations, depending on the geometry and the amount of petroleum.

The play

The structural and stratigraphic traps within the Permian Patchawarra and Toolachee Formations (Kantsler and others, 1983) of the Cooper Basin in South Australia have been referred to as individual plays (Forman, Hinde, and Cadman, in press). The Lower Permian Patchawarra Formation contains channel and point-bar sandstone reservoirs, sealed

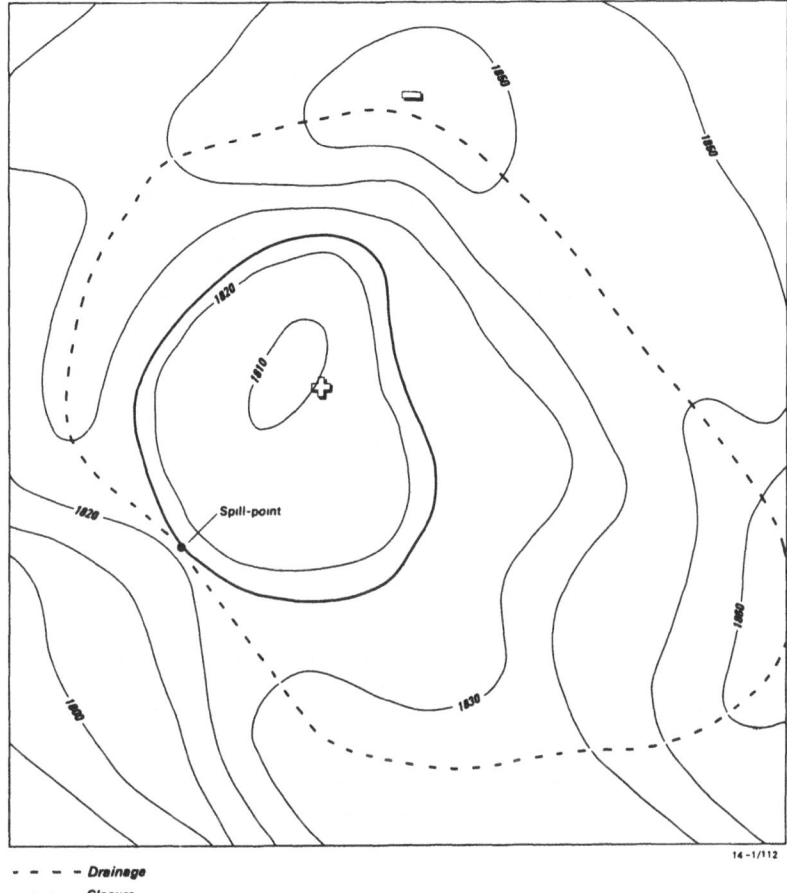

- - - - Drainage
———— Closure

Figure 1. Two-way time contour map of simple anticlinal trap showing lowest closing contour (thick line), drainage divide (dashed line), and spill-point.

by shale and sourced by up to 1300m (4250 ft) of interbedded siltstone, shale, and coal. The Upper Permian Toolachee Formation consists of up to 200m (650 ft) of coal measures sealed by shale of the Lower to Middle Triassic Nappamerri Formation.

The number and size of the petroleum accumulations in these Formations depend partly on the amounts of petroleum that have been generated in the drainage areas of the individual traps and partly on the amounts that have spilled between traps. Yet the Formations contain traps that are spread out along many migration fairways. We suggest, therefore, that the migration fairway itself is a better assessment unit,

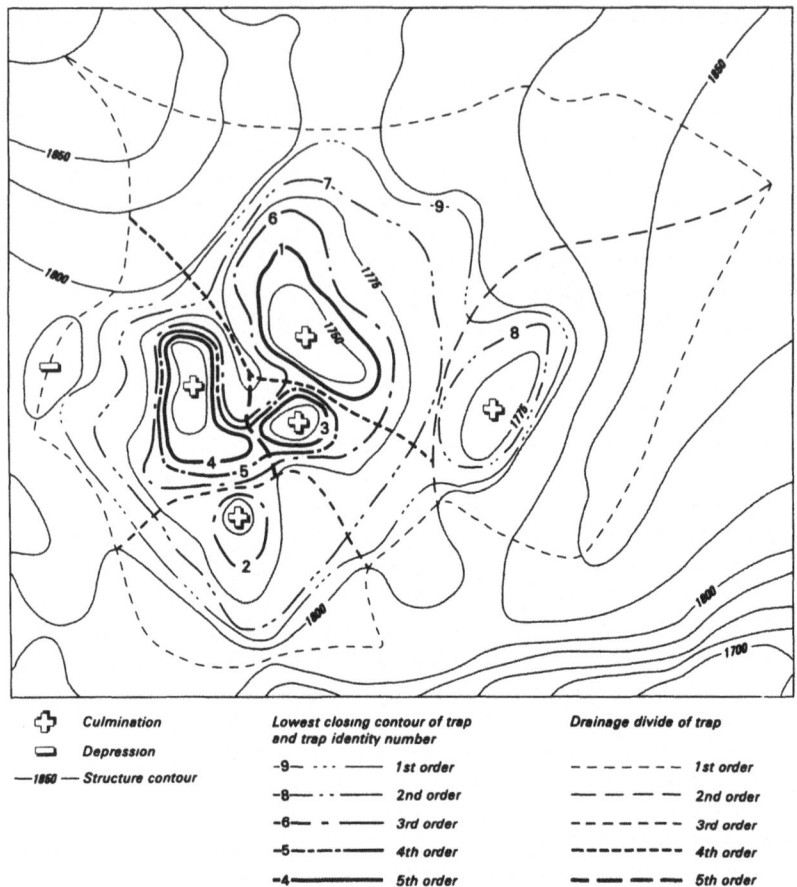

Figure 2. Two-way time contour map of complex trap showing lowest closing contours and drainage divides for nine possible traps, numbered 1-9, belonging to five orders.

particularly where it contains accumulations that are filled to their spill-points.

The migration fairway

Figure 3 is a map of a migration fairway within the Upper Permian Toolachee Formation. At its simplest, the migration fairway is a connected system of traps that lie within the same sequence of source, reser-

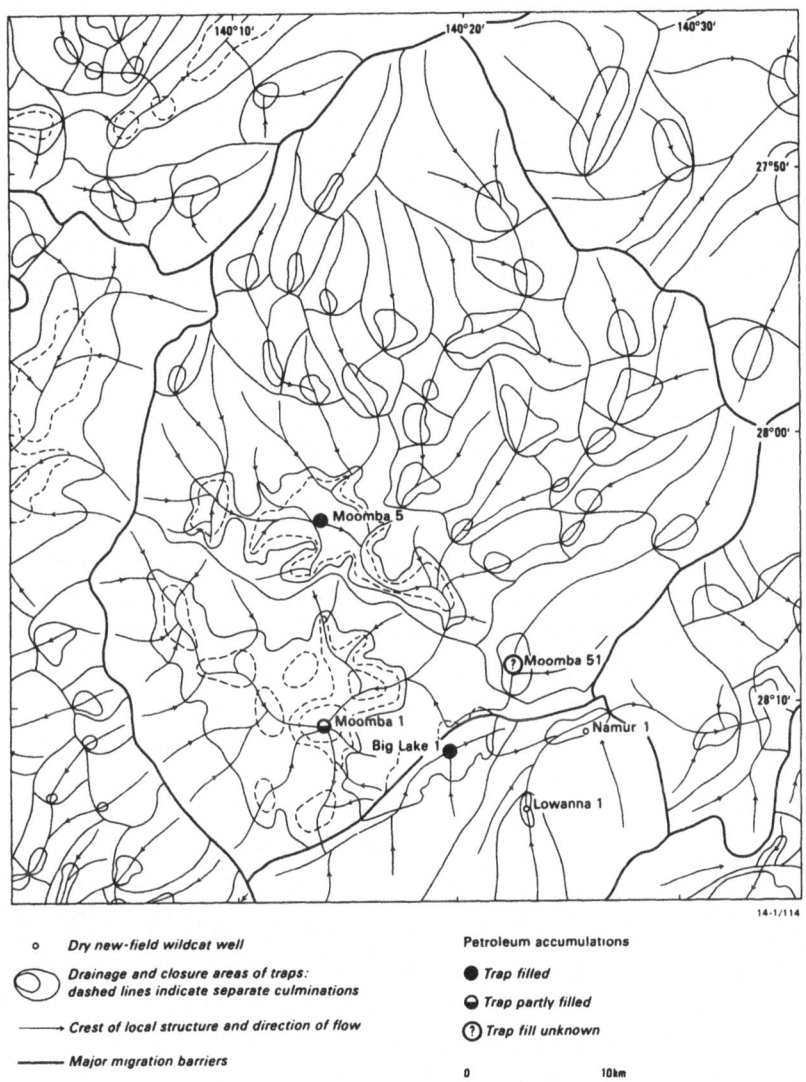

Figure 3. Map showing petroleum accumulations and lowest closing contours and drainage barriers of structural traps within Toolachee Formation in part of Cooper Basin in South Australia. Crests of traps are migration paths. Major drainage barriers, which are shown as thick lines, surround migration fairways.

voir, and caprocks and is separated from adjacent systems by barriers to lateral migration of petroleum, such as synclines, faults, or stratigraphic pinchouts. The example shown has been produced by drawing the lowest closing contours and the drainage barriers for each trap and the likely migration paths from one trap to another onto a structure contour map of the top of the reservoir sequence. The fairway then was outlined by emphasizing the regional drainage barriers.

A fairway is distinguished from a trap or complex trap by the fact that a closing contour can never be drawn around it or any grouping of traps within it. The traps always tilt upwards towards the one spill-point; in the fairway outlined the spill-point lies between Moomba 1 and Big Lake 1. There may be a number of spill-points from other fairways, however, but there are none in the example figured.

Although the fairway forms only part of a conventional petroleum play, it may be thought of as an assessment unit within which petroleum generation, migration, and entrapment are to some degree different to those in adjacent fairways, particularly if the present geometry is a reasonable match to the geometry at the time of migration. These differences occur partly because of differences in geometry and migration paths and partly because of differences in source richness, migration efficiency, and seal capacity.

The order

Figure 2 is a contour map of a complex anticlinal trap with five culminations. Altogether there are nine possible traps which are classified into orders according to the sequence in which they fill and coalesce. The lowest closing contour outlines the largest trap, trap 9, which forms no part of any other trap and is classified as first order. The two smaller traps, 7 and 8, that it incorporates are classified as second order. Similarly, traps 2 and 6 within trap 7 are classified as third order, and traps 5 and 1 within trap 6 are classified as fourth order. Traps 3 and 4, located within trap 5, are classified as fifth order.

There is an A_{drain}, outlined in Figure 2, corresponding to each A_{clos} and these also may be classified as first order, second order, and so on. Whereas, for example, the A_{clos} of the second-order traps must add up to a lower value than that of the A_{clos} of the first-order trap that contains them, their A_{drain} add up to the same value. It follows that DR, V_{Aclos}, TF, and the angle of tilt of the second-order traps will be higher, on average, than for the enclosing first-order trap. Hence, the traps belonging to each order of a play or fairway have their own characteristic range of parameters and should be treated as a separate assessment unit (Forman and others, 1989).

It is possible that the contact between the reservoir and cap rocks is a fractal surface (Mandelbrot, 1983) and that there is an infinite number of orders of traps. With higher orders of traps, however, A_{clos} approaches zero as DR approaches ∞ so that there is a practical limit to the number of orders that can be recognized. The number also depends on the overall tilt of the complex anticline. The number of orders will decrease from a maximum value at zero tilt, through to a value of one at the tilt angle at which the trap is essentially simple, to a value of zero when the tilt is too high and A_{clos} is too small for a trap to be detected.

Assessment Of Undiscovered Petroleum Resources

There are several ways of assessing the undiscovered recoverable petroleum resource (V) of a single petroleum trap. Forman and Hinde (1986) and Forman, Hinde, and Cadman (in press) used the equation:

$$V = A_{clos} \cdot V_{Aclos} \qquad (1)$$

where A_{clos} is the area of closure of the trap and V_{Aclos} is the recoverable resources per unit area of closure. Many others (Gehman, Baker, and White, 1981; Procter and Taylor, 1984) used the equation:

$$V = A_{clos} \cdot h \cdot \phi \cdot HS \cdot RF \cdot TF / HVF \qquad (2)$$

where A_{clos} is the area of closure of the trap, h is the thickness of the reservoir, ϕ is the average porosity of the reservoir, HS is the hydrocarbon saturation, RF is the recovery factor, TF is the fraction of the reservoir volume that contains hydrocarbons, and HVF is the hydrocarbon volume factor.

V_{Aclos} of the identified accumulations may be calculated by dividing the amount of petroleum in each trap (V) by A_{clos} and TF can be calculated by dividing V by the amount of petroleum that the trap would contain if filled to capacity (V_{cap}).

$$V_{Aclos} = V/A_{clos} \qquad (3)$$

and:

$$TF = V/V_{cap} \tag{4}$$

where V is taken to equal the proved + probable reserves of the petroleum industry plus some, unknown, part of the possible reserves and V_{cap} may equal the total of the proved + probable + possible reserves.

Equation (1) may be derived by assuming that each trap is simple (Fig. 1), that the trap is filled partly with petroleum that has originated in its immediate drainage area (A_{drain}) (Davis, 1987; Dembicki and Anderson, 1989), and that there has been no leakage through the cap rocks. Then, the amount of petroleum in the trap (Forman, Hinde, and Cadman, in press) is:

$$V = A_{drain} \cdot V_{Adrain} \tag{5}$$

where V_{Adrain} is the cumulative height in meters of recoverable gas and gas equivalent oil, measured at surface temperature and pressure, that has migrated into the trap from within the drainage area. Equation (5) can be converted to Equation (1), because $V_{Aclos} = V_{Adrain} \cdot DR$; where DR is the drainage ratio, defined as A_{drain}/A_{clos}. Similarly, Equation (5) may be converted to Equation (2), because $V_{Aclos} = V/A_{clos} = h \cdot \phi \cdot HS \cdot RF \cdot TF/HVF$.

Remembering the assumptions made, Equations (1) and (2) also may be used to assess the amount of petroleum in each trap within a complex trap. The assessor will need to have a structure contour map of the traps that also shows the locations of all the identified accumulations and must take care to calculate values of V_{Aclos}, V_{Acap}, or TF using values for the A_{clos} of the trap in which the accumulation occurs

There are problems, however, where migration has occurred between traps within the complex trap. We show that the number and size of the accumulations in the complex trap depend on the amounts of petroleum generated in the drainage areas of each culmination, the additional amounts that have spilled in from other culminations, and the sequence in which the accumulations coalesce.

Equations (1) and (2) form the basis for several methods for assessing the petroleum resources of a play. One method, based on Equation (1) (Forman and Hinde, 1986; Forman, Hinde, and Cadman, in press; Forman, Hinde, and Radlinski, 1992), uses the equation:

$$V_{ij} = A_{clos, ij} \cdot V_{Aclos, ij} \cdot c_{ij} \cdot d_{ij} \tag{6}$$

where Vij is the simulated volume of recoverable petroleum in trap j during iteration i, $A_{clos,ij}$ is the area of closure of the trap, $V_{Aclos,ij}$ is the resources per unit area of closure, c_{ij} is either 0 or 1 depending on whether V_{ij} is smaller or larger than the random value selected for iteration i from the distribution for the smallest size of accumulation to be included as a resource, d_{ij} is either 0 or 1 depending on whether a random number is greater or less than a random value selected for iteration i from the distribution for the success rate. The gas resources in the trap equal $V_{ij}.(1-P_{o,ij})$ and the oil resources equal $V_{ij}.P_{o,ij}$, where $P_{o,ij}$ is either 0, 1, or a random value, selected from a triangular distribution of possible values for the proportion of oil to oil plus gas, depending on whether a random number indicates that the accumulation contains gas, oil, or oil and gas.

The risked total undiscovered gas resources of a play (V_{tot}), simulated during an iteration (i), are determined using the equation

$$V_{tot,i} = E_i \sum_{j=1}^{n} V_{ij} . (1 - P_{o,ij}) \qquad (7)$$

where E_i is either 0 or 1 depending on whether a random number is greater or less than the existence risk and n is the number of traps. A similar equation is used for undiscovered oil.

A second method is based on Equation (2) (Miller, 1982; Procter and Taylor, 1984). It has many similarities to the first method, but uncertain parameters, such as A_{clos}, h, and TF, are input to the program as probability distributions instead of linear distributions.

Irrespective of the method, additional work is needed where petroleum has migrated from one trap to another within the play or fairway.

Loglinear models

The large number of probability distributions involved in the second method can not be used properly in a Monte Carlo simulation (Forman, Hinde, and Radlinski, 1992); they do not describe the dependent relationships that occur amongst the parameters and they do not change according to the size of their populations. Another disadvantage is that accumulation sizes are not output in discovery sequence suitable for discovery and production forecasting (Forman and Hinde, 1990).

Most of these problems are overcome by using loglinear models of A_{clos} and V_{Aclos} with drilling or discovery sequence number, as required in the first method. For instance, Equation (1) has fewer parameters than Equation (2) and therefore fewer dependent relationships. Although the models give only a partial description of these relationships, they do change with sequence number and they do allow the program to output accumulation sizes in sequence suitable for forecasting.

The bases for using a loglinear model of $\log A_{clos}$ versus drilling sequence number are the observations that the values of A_{clos} for the traps in most assessment units seem to form truncated lognormal distributions and that there is a tendency to drill the larger traps early. This tendency corresponds to a successive sampling model (Kaufman, 1986) in which the probability of drilling a trap next is proportional to its A_{clos} raised to the power lambda (λ).

Repeated iterations of a computer program that simulates drilling a population of traps according to this model (Forman and Hinde, 1985, 1986) yield plots of $\log A_{clos}$ versus drilling sequence number in which the average values are approximately linear, particularly when the standard deviation of the population is high and the value of lambda is low. The distributions of values of $\log A_{clos}$ for each drilling sequence number are approximately normal and each has a similar standard deviation. Also there is a dependent relationship among the average values of slope, intercept, standard deviation, and lambda which is given to within about five percent by the linear least-squares equations in conjunction with the empirical equation:

$$\bar{B} = B_{max}(1 - e^{-\sigma\lambda}), \quad \lambda > 0 \tag{8}$$

where $\bar{B}$ is the average slope for a given value of λ, B_{max} is the slope of the fitted line when $\lambda = \infty$, and σ is the standard deviation of the $\log A_{clos}$ of the traps that have been drilled and, in turn is dependent on the number of traps that have been drilled.

The similarity of the loglinear and successive sampling models, therefore, suggests that A_{clos} of the undrilled traps may be estimated by projection of a model of $\log A_{clos}$ versus drilling sequence number for the traps that already have been drilled. Wherever possible, A_{clos} estimated this way should be checked against measurements of the A_{clos} of the undrilled traps made directly from structure contour maps. Alternatively, the A_{clos} of the undrilled traps may be included in the model in an assumed order of drilling. In poorly explored plays, there may be no alternative but to build up the model by analogy with plays of similar structure.

Plots of $\log h$ versus drilling sequence number and of $\log V_{Aclos}$ and $\log$ TF versus discovery sequence number also are approximately linear and models of historic data may be projected to provide estimates of likely values for undiscovered accumulations. In areas with few or no discoveries, the loglinear models of V_{Aclos} or TF may be built up by analogy or from quantitative models, such as those outlined by Bishop, Gehman, and Young (1983) and Sluijk and Nederlof (1984).

A dependent relationship between any of these models and the model of A_{clos} versus drilling sequence number is indicated wherever both models have a nonzero slope. Ideally, a dependent relationship would be accounted for in the calculations by using the covariance of the lambda values. There is insufficient information available, however, to calculate this covariance and instead the lambda values are treated as independent of each other.

ASSESSMENT OF THE PATCHAWARRA FORMATION

Assessment As A Single Play

The Patchawarra Formation in South Australia was assessed initially as a single play. Assessment began with systematic collection of information about the area and geometry of the traps and the sizes of their petroleum accumulations. Some of these data then were sorted, plotted, and analyzed in order to examine the historic trend and to consider the future trend. The plots of A_{clos}, A_{drain}, and DR versus drilling sequence number and of V_{Aclos} and V_{Adrain} versus discovery sequence number are shown in Figures 4-6.

Figure 4 illustrates that the tendency to drill the traps with the largest A_{clos} early is stronger than the tendency to drill the traps with the largest A_{drain} early. This phenomenon is explained by the variable degree of tilt of the traps throughout the Patchawarra Formation. Tilting of a trap through a few degrees will have little effect on A_{drain}, but will substantially reduce A_{clos}; hence a proportion of the traps with small A_{clos} will have comparatively large A_{drain}.

Reducing A_{clos} by tilting also increases DR. The larger traps tend to be nearly horizontal with low DR and, because a proportion of them have been tilted, the smaller traps tend to have high average DR. This produces a dependent relationship between average DR and A_{clos} so that the loglinear plot of DR versus drilling sequence number (Fig. 5) has a positive slope. Figure 5 also shows variation in DR because of natural variation in the shape and type of traps. For instance, average DR is likely to be higher for anticlinal traps than for fault traps.

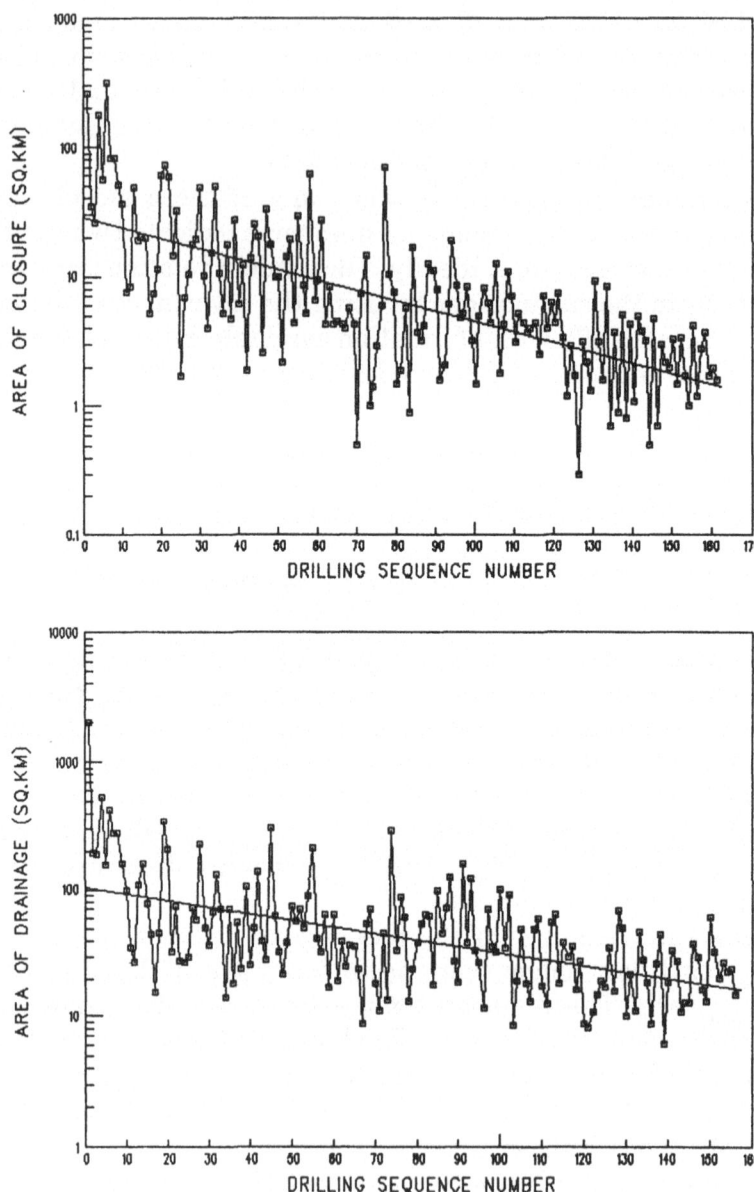

Figure 4. Areas of closure and drainage plotted on logarithmic scale against their drilling sequence numbers for structural traps within Patchawarra Formation. Straightlines have been fitted by method of least squares.

The different effect that tilting has on A_{clos} and A_{drain} of the structural traps is illustrated further by the divergent slopes of the loglinear plots of V/A_{clos} and V/A_{drain} with discovery sequence number (Fig.

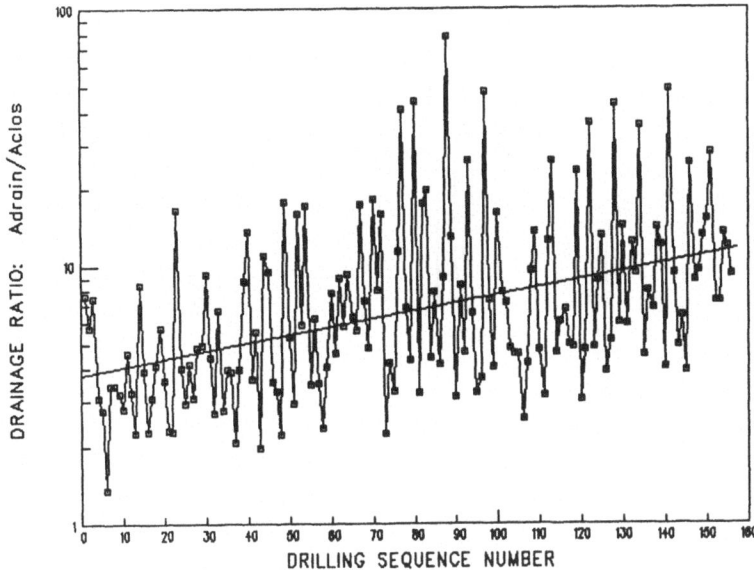

Figure 5. Drainage ratios (A_{drain}/A_{clos}) plotted on logarithmic scale against their drilling sequence numbers for structural traps within Patchawarra Formation. Straightline has been fitted by method of least squares.

6). Average log V/A_{drain} declines slightly with discovery sequence number, whereas log V/A_{clos} inclines.

The average value of V/A_{clos} can only continue increasing with discovery sequence number until the traps are filled to their spill-points. Beyond this, it should decrease, because there is a well-known positive correlation between A_{clos} and h (Masters and others, 1986). Average TF ($TF = V_{Aclos}/V_{Acap}$) should vary in a similar fashion.

The undiscovered recoverable resources of sales gas within structural traps in the Patchawarra Formation within the Cooper Basin in South Australia were assessed (Fig. 7, curve a) using a projection of the log A_{clos} model of the historic data, a model of log V_{Aclos} with horizontal slope based on historic data, a triangular distribution of success rates based on historic patterns, and an existence risk of one .

Assessment Of Orders

Of 161 tests of structural traps within the Patchawarra Formation in the Cooper Basin in South Australia, 102 tested first-order traps, 13 tested second order, three tested third order, 32 were joint tests of the first two orders, six were joint tests of the first three orders, one was a joint test of the first five orders, two were a joint test of second and third orders, and

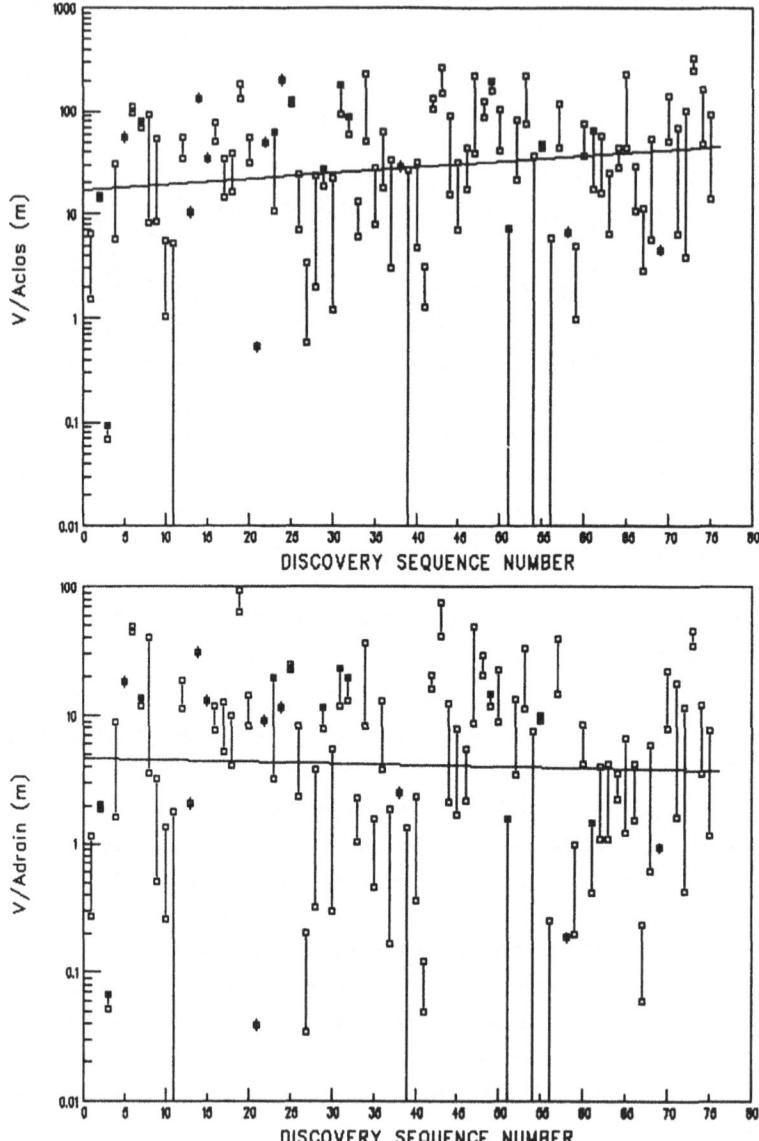

Figure 6. V/A$_{clos}$ and V/A$_{drain}$ versus discovery sequence number for accumulations within Patchawarra Formation. Lower rectangles show proved + probable reserves and upper rectangles show proved + probable + possible reserves. Solid rectangles show accumulations known to be filled to their spill-points. Straightlines have been fitted by method of least squares.

two were a joint test of the third and fourth orders. Plots showing historic trends of A$_{clos}$, A$_{drain}$, and DR versus drilling sequence number, V/A$_{clos}$ versus discovery sequence number, and discovery sequence number

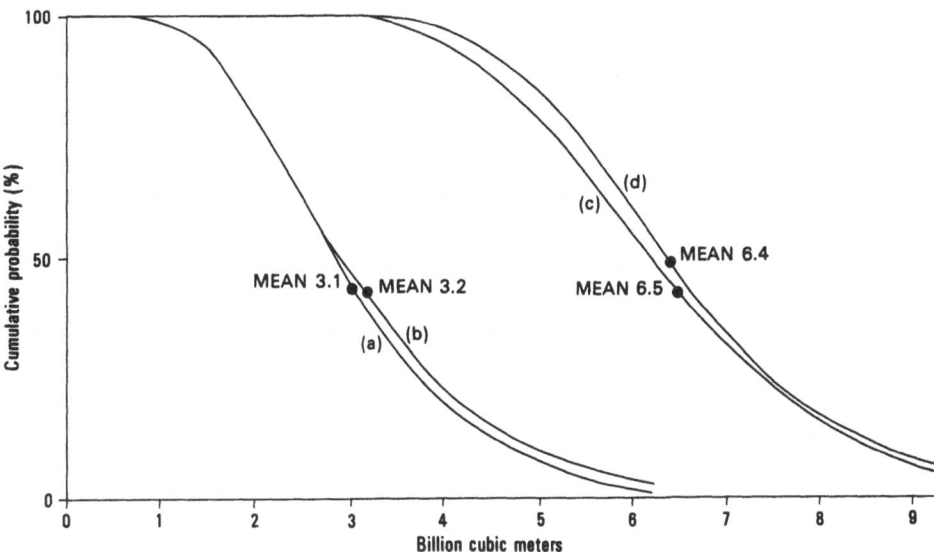

Figure 7. Cumulative probability distributions showing four assessments of undiscovered resources of recoverable sales gas within structural traps in Patchawarra Formation in Cooper Basin in South Australia as at 24/10/89: A, Patchawarra Formation; B, Patchawarra Formation broken into orders; C, Patchawarra Formation broken into migration fairways; D, Patchawarra Formation broken into fairways and into orders.

versus drilling sequence number were prepared for each order of traps. Each order then was assessed separately using projections of the loglinear plots of A_{clos} and V_{Aclos} and Equations (6) and (7). The assessments were added together to give a total for the Patchawarra Formation. The total (Fig. 7, curve b) is only a little higher than the result that was obtained ignoring the complex nature of the traps. Differences occur, however, in the number and size of the accumulations that were estimated.

Assessment Of Migration Fairways

The first step was to examine the historic trends in A_{clos}, A_{drain}, and A_{drain}/A_{clos} versus drilling sequence number, V/A_{clos} versus discovery sequence number, and discovery sequence number versus drilling sequence number for each of the migration fairways in the Patchawarra Formation. Where available, the trends in A_{clos} of the traps that remain to be drilled were examined. Then loglinear models of Aclos in order of drilling and V_{Aclos} in order of discovery were selected as representative of the undrilled traps in each fairway and their likely success rates were predicted. Each fairway was assessed separately and the results added

together assuming that they are independent of each other to give a total
for the Patchawarra Formation in South Australia (Fig. 7, curve c).

The assessments are not strictly comparable, because assessments
shown in curves a and b in Figure 7 are essentially projections of historic
data whereas assessment shown in curve c included additional geological
information and subjective judgement. In some fairways there was little
information and the model of V_{Aclos} and a triangular distribution of the
success rate were determined using available geological information and
analogy. Nevertheless, the assessment of the undiscovered resources is
almost doubled by subdividing the Patchawarra Formation into migration fairways before assessment. This occurs because the large amount of
data (Figs. 4-6) for the Patchawarra Formation overall creates the
impression that the future potential is well known. In fact, there are
comparatively few data for the individual fairways and the potential is
comparatively poorly known.

Assessment Of Migration Fairways And Orders

Each order of traps within each fairway was assessed as a separate
assessment unit and then added together assuming that they are independent of each other. The new total for the undiscovered gas resources
of the structural traps within the Patchawarra Formation of South
Australia is shown in curve d of Figure 7.

Once again, the assessment is only a little higher than the result
obtained ignoring the complex nature of the traps. The potential of the
higher orders is low because of the maturity of exploration in the area.
Examination of results for individual fairways, however, show assessments either significantly higher or lower than the assessments of the
same fairways made after they have been broken into individual orders.
Thus assessments made ignoring orders are based, incorrectly, on averaged
data and the corresponding accumulation size distributions will be in
error also.

ASSESSMENT OF MIGRATED PETROLEUM

Additional information, other than estimates of A_{clos}, V_{Aclos}, success
rate, and existence risk, is needed to assess complex traps and plays in
which petroleum has migrated between traps. It is essential to know the
value of V_{Adrain} at which each trap fills, the order in which the traps fill,
and the order in which the traps are likely to be drilled.

Migration Within The Complex Trap

Imagine charging the five culminations of the complex anticline (Fig. 2) with gas that has been generated uniformly within the drainage area and migrates uniformly into the reservoir rock; that is each trap within the complex receives the same resource of gas per unit area of drainage (V_{Adrain}). Assuming no migration from adjacent, first-order traps and no leakage through the cap rock, V_{Aclos} of each trap is given by the equation

$$V_j / A_{clos,j} = V_{Adrain} \cdot DR_{j+[k]} - (\sum_{[k]} V_{cap,k}) / A_{clos,j} \quad (9)$$

where V_j is the volume of petroleum in trap j, $A_{clos,j}$ is the area of closure of trap j, V_{Adrain} is the resources per unit area of drainage assumed uniform throughout the complex trap, $DR_{j+[k]} = (A_{drain,j} + \sum_{[k]} A_{drain,k}) / A_{clos,j}$ is the overall drainage ratio of trap j and the set [k] of all traps spilling into it at that particular value of V_{Adrain}. $A_{drain,k}$ is the area of drainage of one of the traps in set [k], and $V_{cap,k}$ is the volume of petroleum in one of the traps in set [k] when it is filled to spill-point. The equation simply states that the resources per unit area of closure in trap j is equal to the amount of petroleum, if any, that has been generated in the drainage area of trap j, plus the amount that has been generated in the drainage areas of the traps that spill into trap j, minus the amount reservoired in the traps that spill into trap j, divided by the area of closure of trap j.

When V_{Aclos} is plotted against V_{Adrain} (Fig. 8), each trap will plot as one or more straightlines; the first term in Equation (9) gives the slope and the last term gives the intercept. The culminations fill independently in the beginning and their respective sets [k] are empty. Thus they are represented by straightlines with slope DR_j intersecting the origin. The top of each straightline shows the value of $VA_{clos,j}$ at which the trap overflows. This maximum value is termed the capacity factor (V_{Acap}) and in this example it is assumed to be the same for each trap. An additional straightline with zero slope and an intercept equal to V_{Acap} has been added to Figure 8 to show traps filled to spill-point.

Traps 1 and 2 will fill first, because they have the highest DR_j. Thereafter, gas generated in their drainage areas will spill into traps 3 and 4, which have an abrupt increase in their drainage areas and, hence, in their drainage ratios ($DR_{3+[k]}$ and $DR_{4+[k]}$, where [k] = [1,2]) and rates of fill. Trap 3 will fill first and then spill into trap 4, causing a further abrupt increase in its rate of fill. The accumulations within traps three and four

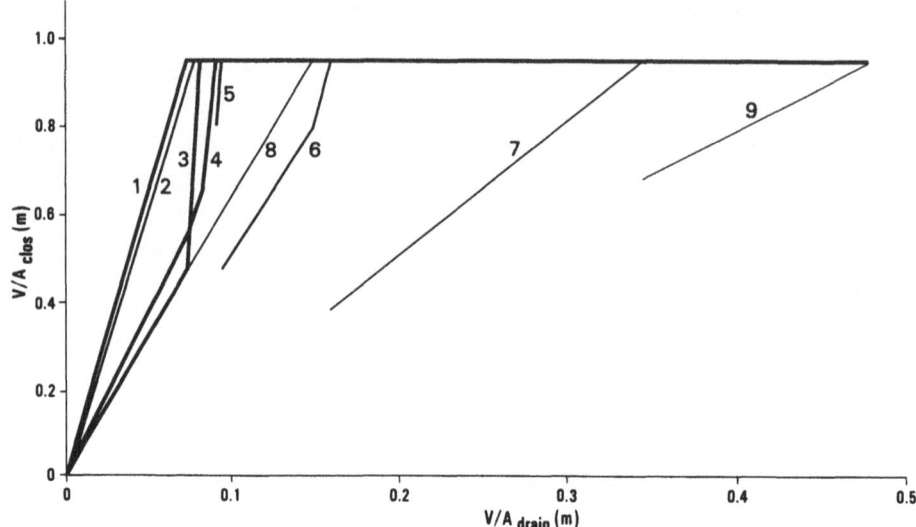

Figure 8. Resources per unit area of closure (V/Aclos) plotted against resources per unit area of drainage (V/A_drain) for traps 1 - 9 shown in Figure 2.

will coalesce to form an accumulation within trap five, which, in turn, fills and then coalesces with an accumulation within trap 1 to begin forming an accumulation within trap 6.

The plot for trap 8 shows a steady linear filling to its spill-point. Trap 6 which is a third-order structure receives gas from its own drainage area and it also receives gas spilled from trap 2. As it fills further, it also receives gas spilled from trap 8. At the value of VA_{drain} at which the accumulation in trap 6 coalesces with the accumulation in trap 2 to form an accumulation in trap 7, there will be a discontinuity in which the value of V_{Aclos} will reach a maximum, representing trap 6 filled to capacity, and will jump to a lower value representing trap 7 partly filled. The trap will fill to spill-point and V_{Aclos} will again reach a maximum value as it receives gas both from within its own drainage area and also spilled over from the drainage area of trap 8. At the spill-point, the accumulations within traps 7 and 8 coalesce and trap 9, the last part of the structure, will begin filling. Because trap 9 is isolated, the straight line representing it slopes towards the origin.

Hence, Figure 8 shows two types of discontinuity when single phase hydrocarbons are introduced into the structure. One, a discontinuity in the effective drainage area which occurs when one accumulation spills into another, produces a change in $DR_{j+[k]}$ and hence a change in slope of the linear plot. The other, a discontinuity in A_{clos} which occurs because A_{clos} changes from one order of trap to the next, causes a change in the value of V_{Aclos} and hence a discontinuity in the linear plot. The tertiary migration paths also seem to undergo a series of discontinuities as each individual trap fills and ceases to be an effective trap for additional petroleum.

Assuming uniform generation and migration of petroleum and that all traps have the same capacity factor, Figure 8 may be used to indicate which traps are filled, which are partly filled, and the value of V_{Aclos} for each trap. For instance, using a value of 12 cms of gas for V_{Adrain}, the figure indicates that traps 1,2,3, and 4 should be filled and traps 8 and 6 should be partly filled. Similar diagrams may be drawn to illustrate other situations, such as the sequence of events when the complex structure is filled with a mixture of oil and gas, oil followed by gas, or gas followed by oil.

The information contained in the filling sequence diagram may be compiled into a table showing an identity number for each trap, the identity numbers (where applicable) of the two traps that coalesced to form it, the identity number of the simple trap into which it spills, and values of A_{clos}, A_{drain}, and V_{Acap}. A recursive algorithm then may be used to identify the trap, simple or complex, into which each trap spills. The computer may calculate $DR_{j+[k]}$, the values of V_{Adrain} at which each trap overflows, and the corresponding set [k]. After a random value of V_{Adrain} has been selected from the input distribution, a program can use Equation 9 to calculate the sizes of the accumulations within the complex trap.

Number of wells and order of drilling

The likely order of drilling the five new-field wildcat wells that could be needed to test all the culminations in the complex trap in Figure 2 can be determined on the assumption that the explorer will wish to test the largest traps first. The method assumes uniform reservoir characteristics throughout and can be illustrated by arranging the resources estimated in each of the traps within the complex trap into a drilling sequence diagram (Fig. 9).

Figure 9 shows that trap 7 and associated higher order traps could contain more gas than trap 8, trap 6 and associated higher order traps could contain more gas than trap 2, and traps 3, 4, and 5, combined, could contain more gas than trap 1. The first well therefore should be drilled on trap 4, the larger of the fifth order traps. A well drilled on trap 4, would

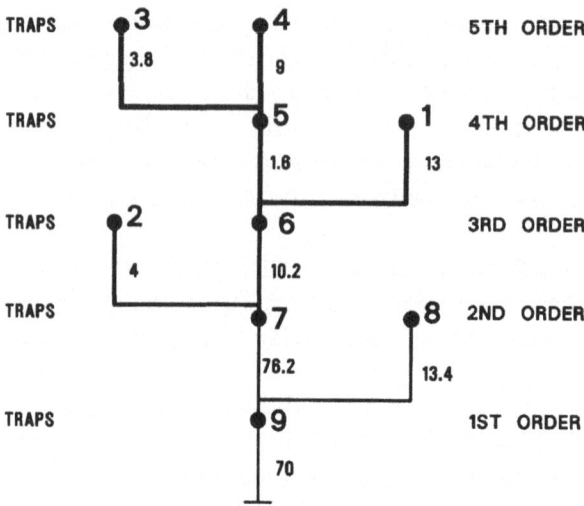

Figure 9. Drilling sequence diagram showing potential in-place gas resources (in billions of cubic meters) that could be identified in each of nine traps by drilling five culminations shown in Figure 2.

identify any gas in traps 5, 6, 7, and 9. If the first-order trap, trap 9, contains gas, all the second, third, fourth, and fifth order traps must be filled and the complex anticline needs no further testing. If, on the other hand, trap 9 proved water-wet, but gas was identified within trap 7, one more well would be needed to test trap 8. If traps 7 and 9 proved to be water-wet, but gas occurred in trap 6, two more wells would be needed to test traps 8 and 2, in that order. If traps 6,7, and 9 are water-wet but gas occurs in trap 5, three more wells would be needed to test traps 8, 1, and 2. Finally, if gas is identified only in trap 4, up to four more wells would be needed to test traps 8, 1, 2, and 3.

Success rate

The new-field wildcat success rate for the simulated drilling of a complex trap may be thought of as made up of a dependent probability that gas has been generated in each of the culminations and an independent probability that trap and seal are adequate to retain the gas. An additional probability, that the gas accumulation is larger than the smallest size of accumulation to be included as a resource, can be calculated by the computer program.

When random numbers selected for generation and trap probabilities simulate a discovery, the filling sequence model can be used to determine the trap in which the accumulation occurs. If the accumulation occurs in a first-order trap no further testing will be required. If the discovery is restricted to one of the culminations, however, it is necessary to test the other culminations and the probability of simulating a discovery in each of them is the probability that trap and seal are adequate. When the random number selected for generation indicates that the complex trap is dry, no further discoveries will be simulated in that computer iteration, even if the random number selected for trap and seal indicate that they are adequate.

More complex situations can be modeled where the generation and trap probabilities are independent or partly dependent throughout the complex trap.

Migration Within The Fairway

Petroleum can migrate by spilling from filled traps in the mature part of a fairway and form major accumulations up dip, even in the immature part of a fairway. Lateral tertiary migration of this type could account for part of the gas within accumulations in the Toolachee Formation at Moomba 5 and Moomba 1 (Fig. 3). Assuming no leakage from or into the system, the resources per unit area of closure of a trap j $(V_j/A_{clos,j})$ at the end of the migration stream is given by Equation 9.

More generally, $V_j/A_{clos,j}$ for a trap at the end of the migration stream is given by the equation

$$V_j / A_{clos,j} = (V_{Adrain,j} \cdot A_{drain,j} + \sum_{[k]} V_{Adrain,k} \cdot A_{drain,k})/A_{clos,j}$$

$$- (\sum_{[k]} V_{cap,k})/A_{clos,j} \qquad (10)$$

where the resources per unit area of drainage (V_{Adrain}) varies throughout the fairway. This equation is no longer a simple linear plot of $V_j/A_{clos,j}$ versus V_{Adrain} but a multidimensional plot of $V_j/A_{clos,j}$ versus $V_{Adrain,j}$ and the independent variables $V_{Adrain,k}$. The determination of $V_j/A_{clos,j}$ as a function of time, now requires simultaneous estimates of $V_{Adrain,j}$ and $V_{Adrain,k}$ as functions of time.

In the example shown in Figure 3, the accumulation at Moomba 5 is filled to its spill-point (Fairburn, 1989) and the accumulation at Moomba 1 is at least partly filled. It is likely, therefore, that a considerable resource of undiscovered gas occurs down-dip in the northeastern part of the fairway in traps that are mostly filled to spill-point. Regional consideration, however, indicates that the reservoirs in these traps are likely to be tight and the gas may be uneconomic to recover.

Migration Through Leaky Caprocks

The assumption that some, or even most, traps are filled with petroleum to a leak-point (rather than a spill-point) complicates the interpretation of the genesis of the accumulations in terms of the preceding models and makes selection of analogues and interpretation of data trends particularly difficult, indicating a vital area for further study. There are at least two possible ways in which the cap rocks may leak; the leak may occur either at a specific location in the cap rock, such as along a fault or the leak may be more diffuse and may occur more or less uniformly across the area of the caprock.

Studies of petroleum traps (Nederlof, 1981; Sluijk and Nederlof, 1984) indicate that there is a correlation between the thickness of the cap rocks and the buoyancy pressures exerted on them by the underlying hydrocarbon columns. Further study is needed to see if a correlation can be established for Australian caprocks.

CONCLUSIONS

Provided that the traps are simple and that they are not filled to spill-point, the undiscovered petroleum resources of a play can be assessed using Equations (6) and (7). A better assessment is obtained by assessing each migration fairway within the play separately. Areas of closure and drainage and likely migration paths should be drawn onto structure contour maps and regional drainage barriers should be outlined. Where a number of complex traps is involved, the assessment can be improved further by assessing each order separately and then by combining the assessments. The assessor must have a clear perception of the likely complexity of the traps and of the origin of the petroleum if values of A_{clos} and V_{Aclos} are to be determined by analogy or by quantitative modeling.

Complex traps and fairways that might contain migrated petroleum can be assessed using additional information, such as the value of V_{Adrain}

at which each trap fills, the sequence in which the traps fill, and the sequence in which the traps are likely to be drilled. Success rate in complex traps should be thought of as made up of a dependent probability for the generation of petroleum and an independent probability that trap and seal are adequate. Resources per unit area of closure may be estimated using Equation 9 or 10; sophisticated quantitative modeling would be required to provide the input for Equation 10. The value of V_{Aclos} and the complexity of the traps also will affect the number of wells required to identify the undiscovered resources.

ACKNOWLEDGMENTS

The authors thank SANTOS Ltd for supplying the geological maps used in the compilation of Figures 1-6 and the Department of Mines and Energy, South Australia for supplying the estimates of identified resource that were used in preparing Figure 6 and the information about the location and extent of the identified resources shown in Figure 3. Figures 1-3 and 7-9 were drawn by the Bureau of Mineral Resources, Geology and Geophysics (BMR) drafting office.

REFERENCES

Bishop, R.S., Gehman, H.M., Jr., and Young, A., 1983, Concepts for estimating hydrocarbon accumulation and dispersion: Am. Assoc. Petroleum Geologists Bull., v. 67, no. 3, p. 337-348.

Davis, R.W., 1987, Analysis of hydrodynamic factors in petroleum migration and entrapment: Am. Assoc. Petroleum Geologists Bull., v. 71, no. 6, p. 643-649.

Dembicki, H., Jr., and Anderson, M.J., 1989, Secondary migration of oil: experiments supporting efficient movement of separate, buoyant oil phase along limited conduits: Am. Assoc. Petroleum Geologists Bull., v. 73, no. 8, p. 1018-1021.

Fairburn, W.A., 1989, The geometry of Toolachee Unit 'C' fluvial sand trends, Moomba field, Permian Cooper Basin, South Australia, in O'Neil, B.J., ed., The Cooper and Eromanga Basins Australia: Proc. Petroleum Exploration Society of Australia, Soc. Petroleum Engineers, Australian Soc. Exploration Geophysicists (SA Branches), Adelaide, p. 239-250.

Forman, D.J., Cadman, S.J., Hinde, A.L., and Radlinski, A.P., 1989, Assessment of petroleum plays containing complex traps: 28th Intern. Geological Congress, Abstracts v. 1, Washington, D.C., p. 501.

Forman, D.J., and Hinde, A.L., 1985, Improved statistical method for assessment of undiscovered petroleum resources: Am. Assoc. Petroleum Geologists Bull., v. 69, no. 1, p. 106-118.

Forman, D.J., and Hinde, A.L., 1986, Examination of the creaming methods of assessment applied to the Gippsland Basin, offshore Australia, in Rice, D.D., ed., Oil and gas assessment - methods and applications: Am. Assoc. Petroleum Geologists Studies in Geology # 21, p. 101-110.

Forman, D.J., and Hinde, A.L., in press, Computer-assisted estimation of discovery and production of crude oil from undiscovered accumulations, in Gaal, G., and Merriam, D.F., eds., Computer applications in resource estimation - prediction and assessment for metals and petroleum: Pergamon Press, Oxford, p. 253-271.

Forman, D.J., Hinde, A.L., and Cadman, S.J., 1990, Use of closure area and resources per unit area for assessing the undiscovered gas resources of the Cooper Basin, South Australia, in Sinding-Larsen, R., and others, eds., Proc. IUGS Research Conference on Petroleum Resource Modeling and Forecasting, Loen, Norway 1988: Intern. Union Geological Sciences Publ. Series.

Forman, D.J., A.L. Hinde, and Radlinski, A.P., 1992, Assessment of undiscovered petroleum resources by the Bureau of Mineral Resources, Australia: Energy Sources, v. 14, p. 183-203.

Gehman, H.M., Baker, R.A., and White, D.A., 1981, Assessment methodology - an industry viewpoint, in Proc. Seminar on Assessment of Undiscovered Oil and Gas, Kuala Lumpur, Malaysia: United Nations ESCAP, CCOP Tech. Publ., 10, p. 113-121.

Kantsler, A.J., Prudence, T.J.C., Cook, A.C., and Zwigulis, M., 1983, Hydrocarbon habitat of the Cooper/Eromanga Basin, Australia: The APEA Journal, v. 23, p. 75-92.

Kaufman, G.M., 1986, Finite population sampling methods for oil and gas estimation, in Rice, D.D., ed., Oil and gas assessment - methods and applications: Am. Assoc. Petroleum Geologists Studies in Geology # 21, p. 43-53.

Mandelbrot, B.B., 1983, The fractal geometry of nature: W.H. Freeman and Co., New York, 468p.

Masters, C., Robinson, K., Procter, R., and Taylor, G., 1986, Petroleum resource assessment: East Asia Workshop Comm. Coordination of Joint Prospecting for Mineral Resources in Asian Offshore Areas CCOP and Intern. Union Geological Sciences, 79p.

Miller, B.M., 1982, Application of exploration play-analysis techniques to the assessment of conventional petroleum resources by the USGS: Jour. Petroleum Technology, v. 34, p. 55-64.

Nederlof, M.H., 1981, The use of quantitative petroleum geology in evaluating exploration prospects: Science and Technology Newsletter, v. 49, 3p.

Procter, R.M., and Taylor, G.C., 1984, Evaluation of oil and gas potential of an offshore west coast Canada play an example of Geological Survey of Canada methodology, in Masters, C.D., ed., Petroleum resource assessment : Intern. Union Geological Sciences Publ. 17, p. 39-62.

Sluijk, D., and Nederlof, M.H., 1984, Worldwide geological experience as a systematic basis for prospect appraisal, in Demaison, G., and Murris, R.J., eds., Petroleum geochemistry and basin evaluation: Am. Assoc. Petroleum Geologists Mem. 35, p. 15-26.

ARTIFICIAL INTELLIGENCE FOR BASIN ANALYSIS

G.J. Peschel, H.H. Poppitz, and M. Mokosch
Ernst-Moritz-Arndt-University of Greifswald, Department of Geosciences, Greifswald,Germany

ABSTRACT

The swift proliferation of the use of artificial intelligence for basin analysis may be characterized by the fact that there are about a hundred expert systems and knowledge-based programs on this topic published at the last half decade. They concern the following three groups of application:

(1) General problems and auxiliary methods of basin analysis. There are 27 expert systems for geographical and cartographical foundations, photogeology and remote sensing, classification of sedimentary basins, selection of the exploration strategy and prognostication of deposits, and also for the statistical data analysis.

(2) Exploration and evaluation of deposits. This topic concerns 43 systems used for field-geophysical and geochemical survey, well-logging interpretation, classification, and identification of minerals, rock types and raw materials, paleontology, geochronology and paleoclimatology, facies analysis and identification of depositional environments, stratigraphic correlation and for structural modeling of deposits, and resource appraisal by geostatistic and other methods.

Computerized Basin Analysis, Edited by J. Harff
and D.F. Merriam, Plenum Press, New York, 1993

(3) Special exploration systems and advising tools. There are 24 systems for oil exploration, hydrogeology and environment protection, engineering geology, and for seismic risk evaluation.

The available features of 94 essential artificial intelligence programs were acquired by the authors and installed in form of a database on a computer diskette. The paper reviews the up-to-date state of development, completed by selected results of the Ernst-Moritz-Arndt-University of Greifswald.

The application of mathematical methods and computers to geosciences has affected the field of sedimentology and the exploration of deposits within sedimentary basins. Sophisticated programs aid the exhausting use of all the available information and also the representation of the results. But the limits of the impetus of conventional data processing to geosciences already are visible. Formerly the geologist must have experience and skill to interpret the observed facts. Todays must have these also and a comprehensive knowledge too on how to use the huge number of mathematical methods and programs.

The reason of this dilemma is that geological research and exploration, because of their complicated subjects, need many procedures, such as, for instance, for comparisions, decisions, and for the assertion and examination of hypotheses which not could be realized only by conventional data processing. This is clearly the field of artificial intelligence which is trying to provide methods for knowledge processing and even for knowledge-based data processing.

DEVELOPMENT STATE OF ARTIFICIAL INTELLIGENCE IN GEOSCIENCES

During the last half decade we have noted a swift profileration on artificial intelligence applications to the geosciences.

Special conferences and seminars with essential contributions on the development of expert systems and other intelligent programs have been the

- Joint U.S. Geological Survey and Geological Survey of Canada workshop on mineral resource assessment, Leesburg, VA, U.S., September, 1985 (McCammon and Cheslow, 1986),

- International Conference: Quantitative analysis of mineral and energy resources, Lucca, Italy, June/July 1986 (Koch and others, 1988),
- 3rd International conference on geoscience information, Adelaide, Australia, June, 1986 (Bichteler, 1986),
- International Mining Pribram and Geochautauqua Conference, Pribram, Czechoslovakia, September, 1987 (Nemec, 1987),
- CAIPEP, the first Conference on Artificial Intelligence in Petroleum Exploration and Production, Tulsa, OK, U.S., August 1988 (Wong, Startzam, and Kuo, 1988).

The literature contains descriptions of several systems in various stages of perfection for practical use. Some are available already commercially but others are yet prototypes and a lot of them are only an idea or conception. Special reviews have been published by Ripple and Ulshoefer, (1987) and by Robinson and Frank (1987) for geographic problems, by Fabbri, Fung, and Yatabe (1988) for remote sensing, by Horvath, Carayannó, Povlos, and McCormack (1985), Brouwer (1986) and by Chiaruttini, Roberto, and Saitta (1989) for seismic signal processing, by Fang, Shultz, and Chen (1986) for petroleum exploration, by Usery, and others (1988) for geological engineering, and by Tabesh (1989) for merging qualitative and quantitative geological databases.

A first systematical review on artificial intelligence in the geosciences was given by Yatabe and Fabbri (1988) at the CODATA Conference in Wroclaw, Poland in June 1987. They have analyzed and compared 33 expert systems by a set of 11 questions, concerning the field of application, the type of knowledge representation and inference, computer, language, etc.

Starting from this source material, the authors of this paper studied a set of references by surveying the Scientific and Technical Network which contains 70 databases and especially the databases GeoRef, COMPENDEX, and INSPEC. Furthermore the AIDOS database of the Eastern European countries, the information bulletin "Referatifnij shurnal" of the USSR and several other information bulletins of geosciences, information science, geography, cartography, remote sensing, mathematics, and statistics were analyzed. The papers and summaries were analyzed utilizing the 21 characteristics of each system (listed in Table 1).

Because of the incomplete description of the characteristics in some of the systems, it was not possible to use all of them. Thus, some of the incompletely described systems have been omitted from this study.

Table 1. Characteristics of expert systems for geological basin analysis

1 Name of the expert system
2 Authors and references
3 Scientific discipline
4 Field of application
5 Input
6 Output
7 Components
8 Knowledge base
9 Inference
10 Probability methods
11 Classification methods
12 Fuzzy measures
13 Hardware
14 Operating system
15 Program language
16 Interaction mode
17 Data base interface
18 Number of modules, rules etc.
19 Producer, Developing institute etc.
20 Commercial application
21 Examples of application

Summarizing this work, we can say there are 130 serious systems for geosciences including mining problems and tasks of the technical utilization of raw materials. 94 of them are important for the analysis of sedimentary basins. Their characteristics have been accumulated in form of a diskette database.

GENERAL PROBLEMS AND AUXILIARY METHODS; GEOGRAPHICAL AND CARTOGRAPHICAL FOUNDATIONS

Expert system-techniques have been investigated by Albert (1988), Summers and McDonald (1988), Ripple and Uelshoefer (1987), Fisher and others (1988), and Usery and others (1988) to implement a set of rules for geological map production. A Knowledge-Based Geographic Information System which is termed KBGIS has been developed by Albert 1988. It allows the design of a rule base for both GIS processing and for geological applications. The rule bases are implemented in the Goldworks expert-system development shell interfaced to the Earth Resources Data Analysis System (ERDAS) a raster-based GIS for input and output. The database is managed by ARC/INFO. The KBGIS-II system (Robinson and Frank, 1987) is used for the preparation of decisions about geographical problems and for geographical data management. Other expert systems for the representation of geographical knowledge are MERCATOR (Davis, 1986) and GEOMYCIN (Fisher and others, 1988). The TAX (Argialas and Narasimhan, 1988) is an prototype for terrain analysis.

An unnamed intelligent front end for the Land Analysis System has been described by Doescher, Scholz, and Quirk (1988). Furthermore HYDRA (Gude, 1988) aids the automatic digitalization and interpretation of cartographic patterns and AUTONAR (Fisher and others, 1988) the automatic arrangement of patterns inclusive names etc. on maps.

PHOTOGEOLOGY AND REMOTE SENSING

A review by Fabbri, Fung, and Yatabe (1988) analyzes intelligent information systems that handle symbolic and spatial data in the field of remote sensing. Applications range from land-cover analysis, forest clear-cut monitoring, map assisted photointerpretation, and cartographic feature extraction to LANDSAT TM (Thematic Mapper) image classification.

In particular, RESHELL is an expert-system shell developed by Goodenough and others (1987) for advising the extraction of resource information from remotely sensed data by digital image processing. Matsuyama and Hwang (1985) have developed SIGMA an expert system for image understanding in photointerpretation with a geometric reasoning system. The LAS, described by DoescherScholz, and Quirk (1988) is a Land Analysis System, containing 200 PROLOG modules for remote-sensed images.

Other expert systems for remote-sensing interpretation are LES (Perkins, Laffey, and Nguyen, 1985) and the multipurpose system NASA G.S. (Fisher and others, 1988). An unnamed educational expert-system approach to the interpretation of aerial photographs for coastal landforms has been described by Kozai (1987).

Furthermore there are some expert systems for the classification of rock types from remote-sensed data which we will consider next.

CLASSIFICATION OF SEDIMENTARY BASINS

A comprehensive research work for the development of expert systems for classifying world sedimentary basins has been executed by Miller (1986A, 1986B). The result was the operational prototype expert-system muPETROL which provides for classifying sedimentary basins as a first step in developing an integrated expert-systems approach to estimating undiscovered worldwide petroleum resources (Miller, 1987A, 1987B).

The basin classification is based on nine sedimentary basin models using Klemme's (1975, 1981) recognition criteria for a world sedimentary basin classification system. Each model is defined by a rule-based system and embodies the geological concepts of plate tectonics modified by regional tectonics and lithologic and depositional sequences. The muPETROL expert system is constructed on an IBM-XT by using muLISP and contains inference nets with about 200 rules and nearly 400 nodes. An integrated tutorial model has about 45 rules and 60 nodes. The rules include matching the basin to be classified with a database containing nearly 800 world basins already classified.

The general structural form of the inference network is constructed as follows:

(1) Establish the nature of the crustal basin basement (i.e. continental or oceanic crust).

(2) Determine the tectonic location relative to past plate movement involved in the formation of the basin (divergent, convergent, etc.).

(3) Determine the geographic location of the regional or basin area relative to global geography.

(4) Describe the structural evolution of the basin by identifying the characteristics of the basin-modifying or basin-forming tectonics.

(5) Describe the stratigraphic evolution of the basin by characterizing the depositional cycles relative to the tectonic periods.

The basin models are documented on computer diskettes (Miller, 1987C).

SELECTION OF THE EXPLORATION STRATEGY AND PROGNOSTICATION OF DEPOSITS

The first intelligent program for the prognostication of deposits was FINDER (Singer, 1985). The rule-based Pascal program for locating mineral deposits uses Bayesian statistics and a powerful area-of-influence method of differently shaped and orientated targets, selected by geological, geochemical, and geophysical characteristics. It provides estimates of the possible number of deposits for each target in the form of contour maps.

A similar philosophy obviously also influenced by Agterberg and others (1972) probability mapping method is the expert system EVALUATOR for the estimate of mineral endowment proposed by Koch (1986).

Probably the most discussed expert systems in geology are PROS-PECTOR (Duda, 1980) and PROSPECTOR II (Mc Cammon 1987, 1989). They have been developed to assist geologists in searching for mineral deposits.

PROSPECTOR uses probability weighted inference within seman-tic networks. The reasoning process of the experienced geologist is emulated by evaluating the likelihood that the target area corresponds to one or more models of the knowledge base. Starting from the given facts acquired by interaction the Bayesian posterior probability of a hypotheti-cal and more complex characteristic of the interesting area will be evaluated and used as the a prior probability of this feature to compute the posterior probability of a deposit model. PROSPECTOR is combined with an image-processing system producing image files up to 128 x 128 pixels for the construction of simplified digital maps of a prospective area. The emulation of the PROSPECTOR expert system with a raster GIS has been discussed by Katz (1988).

PROSPECTOR II differs from PROSPECTOR mainly by the knowledge representation. In contrast to the rule-based PROSPECTOR this expert system uses an object-orientated knowledge representation by frames. Another important difference is that PROSPECTOR II has been expanded to use the graphics display capability of the Interlisp-D programming environment. Thereby the interaction can be supported by comfortable window handling and by bit map knowledge representation.

muPROSPECTOR (McCammon, 1983, 1986) is a simplified version of the PROSPECTOR expert system using the programming language muLISP. It was adapted so that it could serve as base for some other systems, such as muPETROL (Miller, 1987a) and EVALUATOR (Koch, 1986) already mentioned.

Another forecast expert system is GENESIS developed by Bugaets, Vostroknutov, and Vostroknutova (1987), and Bugaets and Vostroknutov (1989). It contains approximately 1000 rules which mimic the heuristic knowledge of skilled geologists, as well as their intuition and facility of drawing the correct forecast conclusions based on incomplete data. At the same time, considerable experience of formal mathematical model applications for several problem classes is accumulated.

Furthermore there are systems for strategic planning of the search and exploration work of deposits, such as ASOD-PROGNOS described by Beljashov and others (1987) and the Exploration Planning Assistant (Quick and Schuyler, 1988). The latter supports user organizations less experienced in exploration planning the sophistications needed to create a viable exploration plan.

STATISTICAL DATA ANALYSIS

The statistical analysis of the data acquired by geological, geo-chemical, and geophysical exploration mostly is executed by statistical program packages such as, for example, SPSS. But such programs are unintelligent in that they do not use the data to check if the assumptions underlying the statistical procedure being used are satisfied. Furthermore such programs contain a large number of different tests and methodical variants so that the user must be an experienced expert of statistics to select procedures best fitted for the given data and the problem to be solved.

Intelligent front ends and expert systems for the statistical data analysis could be important for overcoming these impediments. But unfortunately they are mostly unknown in geosciences. Reviews about statistical expert systems were made by Gale (1986), Hand (1984, 1985, 1986), Haux (1986), Nelder (1988), and Streitberg (1988). We will consider some examples here.

PANOS is an expert system for PArametric and NOnparametric Statistics (Wittkowski, 1988). It is a rule-based assistant for selecting a statistical data model and formulating the desired test problem.

GLIMPS is an intelligent front end of the statistical program package GLIM (General Linear Interactive Modeling). It has been developed by Nelder and Wolstenholme (1986) as a PROLOG program for matching the best-fitted linear model to the given data whereby the user interaction can be executed on several levels depending from a prior knowledge and user experience. If the specifications for data analysis are determined, the FORTRAN program GLIM will be activated and served by the actual parameters.

DINDE, developed by Oldford and Peters (1985), is a frame-orientated system for statistical data analysis. It is characterized by a sophisticated window management supporting an easy survey on data structures, available statistical methods, and helping tools.

EQUANT is a rule-based intelligent front end to the cluster analysis module of the statistical program package BMDP (Havranek and Soudsky, 1988). The knowledge base contains approximately 500 rules. EQUANT is implemented in PL/1 for IBM 370 mainfraimes. The user interaction is supported by TSO and allows answers with several degrees of certainty.

The REgression Expert REX, developed by Gale and Pregibon (1984) for matching regression models to given data, is a frame orientated system written in LISP and aided by an comfortable window handling and graphics. A more developed successor of REX has been named STUDENT.

Statistical expert systems mostly were made by statisticians for statisticians. So, for instance, the user of EQUANT must answer 49 questions. That will be possible if a 'normal' statistician needs the help of a cluster analysis expert. But for geosciences we need such systems being more intelligent and deciding the right methodology of data analysis from properties of the data. Furthermore they must have a knowledge base containing improved statistical models for determined geological objects and tasks too.

A prototype of such systems is OPTREX being developed by Peschel and Kolyschkow. OPTREX should have a geological database, that is to say a collection of well-known models of the mutual relation of geological properties. Furthermore there will be a inference net to determine models for unknown relations by different criteria such as minimum estimate deviation, maximum simplicity of the model, the type of the inverse relation, the type of the statistical distribution of a regarded pair of characteristics etc. An essential part of this planned system is OPTRIS, the available FORTRAN-77 program for OPtimal Regression AnalysIS (Kolyschkow and Peschel, 1985).

EXPLORATION AND EVALUATION OF DEPOSITS; FIELD GEOPHYSICAL AND GEOCHEMICAL SURVEY

According to the importance of geophysical and geochemical survey for the analysis of sedimentary basins there are some expert systems for planning and interpretation of field experiments. A Geophysics Advisor Expert System has been described by Olhoeft (1988). Other systems for the optimization of geophysical field methods and parameters are DIAPASON for the selection of refraction seismic registration arrangements (Jebrak and Loesener, 1987) and an expert system for determining vibroseis field parameters was described by Penas, Hadsell, and Stout (1986).

The difficult problem of segmentation of seismic sections for facies and stratigraphic analysis has been investigated by Simaan, Zhang, and Love (1987) and by Defigueiredo and Shaw (1987) by image processing, pattern recognition, and spectral methods. An intelligent LISP-based program which is termed RESOLVER was developed by Love and Simaan (1985). It supports the segmentation of stacked seismic sections into zones of common texture by heuristic rules. These rules concerning the relative depth, sequence, and shape of sedimentary units as well as the neighbors of individual pixels applied for the classification of the pixel gray tones of seismic sections whereby a fuzzy membership conception is used.

An expert system for the recognition and classification of geochemical anomalies was developed by Bonnefoy and others (1989). It is termed SERGE and contains roughly 150 rules concerning the quantitative characteristics of geochemical anomalies and their geological interpretation.

WELL-LOG INTERPRETATION

Well logging probably furnishes the greatest information quantity for basin analysis. Therefore it is a broad field of artificial intelligence application and in this section we will consider only methods for the petrophysical and lithological interpretation of wireline logs. Problems of stratigraphical correlation, of facies analysis and identification of depositional environments (see DIPMETER ADVISOR), and problems of the construction of geological sections will be discussed next.

Expert systems for the petrophysical and lithological interpretation of well-log data are GPEX (Smith, Saleken, and Rockel, 1986), ELAS (Apte and Weiss, 1985), LITHO (Bonnet, Harry, and Ganascia, 1982), HESPER (Peveraro and Lee, 1988) and two unnamed systems described by Skandsen and Lile (1987) and by Kassenaar and Dusseault (1988). These systems mostly assist the segmentation or zonation of well logs (Chen and Fang, 1986), the qualitative lithological interpretation, and the evaluation of reservoir properties for oil exploration. Most systems are rule based. GPEX is object orientated by frames. Interesting results of comparing an expert system to human expertise in well-log analysis and interpretation were published by Einstein and Edwards (1988). The zonation of well logs also will be assisted by the PROSPECTOR expert system mentioned previously. Furthermore there are systems providing both results of geological as well of technical interest, for example a real-time expert system for fault detection and failure prevention described by Chung (1988).

CLASSIFICATION AND IDENTIFICATION OF MINERALS, ROCK TYPES, AND RAW MATERIALS

In contrast with the systems just considered the following ones mostly use information observed from mineral and rock samples. Artificial intelligence approaches have been described by Takeda (1986) for mineralogical database management and by Fang and Stanley (1986) for mineral identification via thin section and X-ray powder diffraction.

MINID (Reeves, 1989) is an expert system for optical identification of minerals in thin sections utilizing up to 18 characteristics such as color, symmetry, etc. The mineral dictionary covers 255 usual designations. A peculiar feature of this system is the ability for crystallographic pattern recognition. Based on that Diemer and others (1989) handled with teaching mineral identification skills using digitized video images. Other systems for mineral identification in samples and thin and polished sections are XMIN (Donahoe, Green, and Fang, 1989), MINIDENT (Donahoe, Green, and Fang, 1989), MICA (Hart, McQueen, and Newmarch, 1988), and KBS (West, 1984). The expert system GENEX described by Meech (1988) is intended for the solution of mining problems also contains a module for mineral diagnostics.

The identification of rock types is a more complicated problem. The rule-based systems PETRO-AP and PETRO-HP described by Bkouche-

Palen (1986) assist the rock-type diagnosis by using characteristics such as color, texture, mineral content, etc. but they are estimated by the creator to be only tutorial systems. This also is true for INTAL (Hawkes, 1985) an expert system for the identification of igneous rocks in the hand specimen. The classification of rock types based on remote-sensed data will be aided by the expert system GEMS (Papacharalampos and Koch, 1986). This rule-based system uses gray-tone values scanned by LANDSAT pictures. An artificial intelligence approach to soil classification is described by McCracken and Cate (1986).

The classification and identification of raw materials by artificial intelligence methods is yet a poorly developed matter. There is only BURUGOL (Peschel, 1987, 1990), a rule-based expert system for the classification of lignites. BURUGOL contains a database of up to 200 properties of 48 important lignite deposits of Hungary, Bulgaria, Germany, Mongolia, Poland, Soviet Union, and Czechoslovakia and a knowledge base comprehending roughly 50 models of the mutual relations between the properties of the raw material and 300 rules about the possible utilization and refinement of lignites. With the help of BURUGOL in 1986-87, a new classification and codification system of low-rank coals were worked out by an expert commission of the seven states mentioned, which was suggested also should be used by the Economic Commission for Europe.

PALEONTOLOGY, GEOCHRONOLOGY, AND PALEOCLIMATOLOGY

The first expert system for the identification of fossils was described by Alexander (1985) and by Brough and Alexander (1986). It is named FOSSIL and characterized by a frame-orientated knowledge base written in microPROLOG. Another expert system is EXPAL (Conrad and Beightol, 1988) which determines taxa of marine fossils by morphological descriptors. An extensive database (256 Kbytes) contains also stratigraphical positions and references. Problems of the application of artificial intelligence principles to paleontological taxonomy also are discussed by Wang and Zhang (1988).

An artificial intelligence datafile management program for a radiocarbon database has been described by Kalin und Long (1989).

There also is an unnamed paleoclimatic expert system described by Scotese (1987) that predicts coastal upwelling.

FACIES ANALYSIS AND IDENTIFICATION OF DEPOSITIONAL ENVIRONMENTS

Although the DIPMETER ADVISOR (Smith and others, 1985) firstly is an expert system for geophysical well-log interpretation, it is one of the most important tools for the analysis of the sedimentary facies and environment identification. It was developed at Schlumberger and is commercially available. This expert system assists the structural and facies interpretation of the dip values of thin formation layers measured by three high reliable resistivity sondes in boreholes. Depositional sequences of determined environments but also structural deformation such as folds, faults, etc. produce a typical dip pattern. DIPMETER ADVISOR can help to identify such characteristic dip patterns during an advanced and user-friendly graphic interaction.

Another expert system for determining clastic depositional enviroments is XEOD (Shultz and others, 1986, 1988). It contains 58 facies types characterized by frames of up to 166 properties of terrigenous clastic sediments and approximately 1000 rules. The inference machine is supported by fuzzy measures of certainty.

GEOX described by Chiou (1985) assists the classification of lithological facies types based on remote-sensed data. Furthermore are unnamed expert systems described by Krystinik (1986) for the identification of foreshore depositional environments and by Mabile and others (1989) for formation recognition.

A tutorial PROLOG expert system has been made at the Ernst-Moritz-Arndt-University of Greifswald for the facies interpretation of lignites. Out of some lithological characteristics of the coals and also based on the sequence of the facies types can be identified the lithofacies type, biofacies type, swamp environment, and some hints for the utilization.

STRATIGRAPHIC CORRELATION

The problem of stratigraphic correlation using geostatistics and artificial intelligence was discussed first by Hohn and Fontana (1986) and by several papers of Kuo (1986), Startzman and Kuo (1986, 1987), Kuo and Startzman (1987), and Startzman, Kuo, and Wong (1987). These papers describe an expert system for field-scale stratigraphic correlation based on well logs.

Another artificial intelligence approach to lithostratigraphic correlation using geophysical well logs was made by Olea and Davis (1986). From that paper followed the CORRELATOR (Olea, 1988). This is a program system consisting of 24 conventional modules joined with an expert system. The methodology is characterized by computing a weighted correlation coefficient for two logs for each reference and matching a borehole pair. One use is for the evaluation of the normalized shale similiarity between the two boreholes to be correlated. The other use is for the computation of the correlation function between a shale layer of the reference borehole and some neighboring shale layers of the matching borehole. The results of this computation will be interpreted by the expert system containing 23 antecedents and 11 hypotheses in 19 production rules, such such as that correlations do not cross the other correlations of a bundle of correlated beds.

The well-correlation system GEOLOGIX described by Budding and others (1989) combines three knowledge-based systems. The first one determines the facies type of the different units (coal, shale, channel, mouthbar, etc.). The second system is the correlation advisor and the third one the stratigraphical advisor. The latter will give explanations for thickness differences of units in different wells, thus providing a geological model of the area. The authors have emphasized that a correlation between approximately 20 wells can be made within a day by GEOLOGIX.

Unfortunately there are missing expert systems or semantic networks for general stratigraphic foundations to assist the geologist at the comparision of standard scales of different areas especially for the least units with local names well known only in a small circle of workers. Such a system written in PROLOG, is curently under development at the Ernst-Moritz-Arndt-University of Greifswald, considering such predicates as older, of the same age, younger, not older, not younger, synonymous, and contained.

STRUCTURAL MODELING OF DEPOSITS AND RESOURCE APPRAISAL BY GEOSTATISTICAL AND OTHER METHODS

Because structural modeling mostly is beginning with the design of cross sections, we first must point to the GROG program described by

Hamburger Bessis, and Parrod (1989). It is based on a method of recognition of geological objects which is not characterized by recognizing their geometrical features but their historical features. Using artificial intelligence methods a formalism for Representation of Objects in Geology (FROG) is implemented at the GROG program. Its essential goal is the symbolic representation of a cross-section picking and structuring the graph rising by that into a topochronological tree.This methodology constitutes a frame to set up the scenario of basin formation. The chronology of sequence formation can be derived and each subgraph associated to a sequence is characterized.

Two- and three-dimensional structural modeling may be joined with the application of geostatistical methods. To assist the sophisticated use of such methods also by less experienced geologists an intelligent interface to the geostatistical program package VARIO3 was developed by Dimitrakopoulos (1987) and by Dimitrakopoulos and David (1987). The system is named BOU-1 and integrates expert knowledge for geostatistical problem solving with numerical data processing for variogram computation and modeling.

A similar system, for structural modeling is DES, described by Sarkar, O'Leary, and Mill, (1988) which joins a knowledge base and a decisions system with algorithms for numerical evaluation of distribution and amount of raw materials by geostatistical methods. Another but unnamed expert system for structural modeling has been described by Kleyn and Helwig (1987).

Also important for this topic is the GRIDDING ADVISOR (Maslyn, 1987), an PROLOG expert system which assists the selection of the basic grid for interpolation by geostatistical or other methods. By 20 rules that can be selected, to determine which grid from 6 grid types is best suited for given data and a given task.

Expert systems for the evaluation of deposits and resource appraisal are GEOVALUATOR (Koch and Papacharalampos, 1988), EMACS (Janakiraman, 1986), and GEXSYS (Sarkar, O'Leary, and Mill, 1988). GEOVALUATOR accepts two-dimensional digital images as input and has been used for the resource appraisal of kaolin deposits in Georgia, U.S.A. EMACS is suited for the geostatistical ore reserve estimation.

GEOCAD, discussed by Cheimanoff, Deliac, and Mallet (1989), is an alternative CAD and artificial intelligence tool that helps moving from geological resources to mineable reserves. Another artificial intelligence approach to the integration of mineral data by image processing has been treated of by Fabbri and Kasvand (1988).

SPECIAL EXPLORATION SYSTEMS AND ADVISING TOOLS; OIL EXPLORATION

Most of the expert systems regarded here are used for oil exploration, but mostly in generally sense for the solution of applied geological problems. In addition to that there are several expert systems, immediately orientated and special equipped to the search, exploration and evaluation of oil and gas deposits.

One of the first has been the rule-based EXPLOR (Maslyn, 1986) which is written in BASIC. EXPLOR contours structural elevation, structural dip, thickness, and water-saturation values derived from well-log data. Structural modeling is executed by trend-surface analysis. By a special form of spatial reasoning structural traps are interpolated and classified.

Other expert systems generally for oil exploration are muEXPLOR (Hanley and Merriam, 1986), OILXPERT (Fang, Shultz, and Chen, 1986) and unnamed systems described by Beaumont (1987) and by Quincy and Kubichek (1987). More specialized for determined trap types are the expert system for carbonate reservoirs, presented by Stout (1986) and the expert system RIF (Sejful-Mulyukov and Nemirovskij, 1987) for the prognostication of buried reefs.

Furthermore there are some expert systems for solving problems of reservoir mechanics. ELFIN (Martin-Clouaire, 1984) and SIMMAIS (Jebrak and Loesener, 1987) assist the investigation of the avenue of migration of hydrocarbons within an area. WSULOG, presented by Linehan and Sutterlin (1986) aids the evaluation of saturation values and the identification of the reservoir type out of a set of 10 such types. Dharan, Turek, and Vogel (1989) have described a fluid properties measurement expert system and Guerillot, Blanc, and Madre (1989) an intelligent interface for reservoir simulators.

An extensive research program aiming at vigorous productivity increases by expert systems and neural networks has been executed by the Arco Oil and Gas Corporation, Texas, U.S.A. (Smutz, 1989). For example, an expert system has been developed for the selection of a correlation in predicting pressure loss for multiphase flow. Neural networks should be used for enhanced pattern recognition (Stoisits, 1989).

HYDROGEOLOGY AND ENVIRONMENT PROTECTION

Because of some similarities to the problems of reservoir mechanics, here are attached some information about hydrogeological expert systems. There are HYDRO for the evalution of flow modeling parameters (mentioned in a review of Jebrak and Loesener, 1987), an unnamed database and expert system for groundwater modeling discussed by Newell and Bedient (1987) and HYDROLAB (Poyet and Detay, 1989). The latter aids the evaluation of groundwater resources and has been used successfully in North Cameroon.

Closely associated with hydrogeological problems are the problems of groundwater protection. They are supported by such expert systems as have described by McClymont and Schwarz (1986) for water contamination hydrology, GEOTOX (Mikroudis, Fang, and Wilson, 1986) for the evaluation of dump locations for special garbage, and ALTRISK presented by Skala (1988). The latter aids the evaluation of the risk that groundwater can be contaminated by old dumps.

ENGINEERING GEOLOGY AND SEISMIC RISK EVALUATION

The development of expert systems for engineering - geological applications has been concentrated up to now on problems of the slope stability. There are two unnamed expert systems, described by Brown (1988) and by Grivas and Reagan (1988) and the XPENT presented by Faure and others (1988).

A further topic touching at engineering geological problems is seismic risk evaluation. For that was developed an expert system which is specialized for sites in California (Dong, Lamarre, and Boissnade, 1986; Lamarre, Boissonade, and Shah, 1987; Lamarre, 1988). An expert system for earthquake intensity evaluation termed EIE has been presented by Liu and Wang (1986).

REFERENCES

Agterberg, F.P., Chung, C.F. Fabbri, A.G., Kelly, A.M., and Springer,J.S., 1972, Geomathematical evaluation of copper and zinc potential of the Abitibi area, Ontario and Quebec: Geol. Survey Canada, p. 71-41.

Albert, T.M., 1988, Knowledge-based geographic information systems (KBGIS) new analytic and data management tools: Jour. Math. Geology, v. 20, no. 8, p. 1021-1035.

Alexander, I.F., 1985, FOSSIL: an expert system for paleontology: Tertiary Research, v. 7, no. 1, p. 1-11.

Apte, C.V., and Weiss, S.M., 1985, An approach to expert control of interactive software systems (abst.): Trans. Patt. Anal. and Machine Intell., v. 7, no. 6, p. 591.

Argialas, D.P., and Narasimhan, R., 1988, TAX; prototype expert system for terrain analysis: Jour. Aerospace Engineering, v. 1, no. 3, p. 151-170.

Beaumont, D.F., 1987, Knowledge-based expert system to evaluate petroleum exploration opportunities (abst.): Am. Assoc. Petroleum Geologists Bull., v. 71, no. 5, p. 529.

Beljashov, D.N., Vinogradov, S.N., Savcenko, I.S., and Golubeva, V.A., 1987, Intellektualnye funkcii ASOD- PROGNOZ, in Nemec, V., ed., The Mining Pribram: Pribram, Czecheslovakia, v. M2, p. 527-530.

Bichteler, J.H., 1986, Expert systems for geoscience information processing: Proc. 3rd Intern. Conf. Geoscience Information, v.1, Glenside, Australia, p. 179-191.

Bkouche-Palen, E., 1986, Systeme expert dans la classification des roches: Science de la Terre, Informatique Geologique, v. 25, p. 57-82.

Bonnefoy, D., Jebrak, M., Rousset, M.C., and Zeegers, H., 1989, SERGE: an expert system to recognize geochemical anomalies: Jour. Geochem. Explor., v. 32 no. 1-3, p. 343-344.

Bonnet, A., Harry, J., and Ganascia, J.G., 1982, LITHO, un systeme expert inferant la geologie de sous-sol: Technique et Science Informatique, v. 1, no. 5, p. 393-402.

Brough, D.R., and Alexander, I.F., 1986, The Fossil expert system: Expert Systems, v. 3, no. 2, p. 76-83.

Brouwer, J.H., 1986, Expert systems in seismic interpretation: Europ. Assoc. Exploration Geophysicists , 48th Meeting, Technical Programme and Abstracts of Papers, Ostend, p. 40-41.

Brown, D.J.,1988, An expert system for slope stability assessment: unpubl. doctoral dissertation, Univ. Nottingham, Nottingham, UK, 279 p.

Budding, M.C., van der Pas, P., Rieuwerts, H., and ten Tusscher, A.B.G.M., 1989, Geologix: a well correlation system (abst.): Intern. Geol. Congress, Washington, v. 1, p. 210.

Bugaets, A.N., Vostroknutov, E.P., and Vostroknutova, A.I., 1987, Methods of artificial intelligence in the research of geological prognostication (abst.), in Nemec, V., ed., The Mining Pribram: Pribram, Czecheslovakia, v. M2, p. 513-520.

Bugaets, A.N., and Vostroknutov, A.I., 1989, Geological forecasting and artificial intelligence methodology (abst.): Intern. Geol. Congress, Washington, v. 1, p. 210.

Cheimanoff, N.M., Deliac, E.P., and Mallet, J.L., 1989, GEOCAD: an alternative CAD and artificial intelligence tool that helps moving from geological resources to mineable reserves, in Weiss, A., ed, Application of computers and operation research in the mineral industry: Society of Mining Engineers, New York, p. 471-478.

Chen, H., and Fang, J.H., 1986, A heuristic search method for optimal zonation of well logs: Jour. Math. Geology, v. 18, no.5, p. 489-500.

Chiaruttini, C., Roberto, V., and Saitta, F., 1989, Artificial intelligence techniques in seismic signal interpretation: Geophys. Journal, v. 98, no. 2, p. 223-232.

Chiou, W.C., 1985, NASA image-based geological expert system development project for hyperspectral image analysis: Applied Optics, v. 25 no. 14, p. 2085-2091.

Chung, D.T., 1988, Realtime expert system for fault detection and failure prevention and application: unpubl. doctoral dissertation, Univ. Maryland, College Park, MD, U.S.A, 227 p.

Conrad, M.A., and Beightol, D.S., 1988, Expert systems identify fossils and manage large paleontological databases: Geobyte, v. 3, no. 1, p. 42-46.

Davis, E., 1986, Representing and acquiring geographic knowledge: Pitman Publ., London, and Morgan Kaufmann Publ., Los Altos, California, 223 p.

Defigueiredo, R.J.P., and Shaw, S.W., 1987, Spectral and artificial intelligence methods for seismic stratigraphic analysis, in Helbig, K., and Treitel, S., eds., Handbook of geophysical exploration, Section I, Seismic exploration: Geophys. Press, London, p. 426-445.

Dharan, M.B., Turek, E.A., and Vogel, J.L., 1989, The fluid properties measurement expert system: Proc. Petroleum Computer Conference, Soc. Petroleum Engineers, Richardson, Texas, p. 131-134.

Diemer, J.A., Frakes, W.B., Gandel, P.B., and Fox, C.J., 1989, Teaching mineral-identification skills using an expert system computer program incorporating digitized video images: Jour. Geol. Education, v. 37, no. 2, p. 121-127.

Dimitrakopoulos, R., and David, M., 1987, Expert (system) technology for geostatistical applications (abst.): Eos, Trans. Am. Geophys. Union, v. 68, no. 44, p. 1267.

Dimitrakopoulos, R., 1987, The explicit knowledge formalism as perspective for geostatistical operations: unpubl. predoctoral thesis, Ecole Polytechnique, Montreal, p. 46-61.

Doescher, S.W., Scholz, D.K., and Quirk, B.K., 1988, An expert- system front end for the Land Analysis System, in Wiltshire, D.A., ed., Selected papers in the applied computer science: U. S. Geol. Survey Bull. B 1841, p. A1-A11.

Donahoe, J.L., Green, N.L., and Fang, J.H., 1989, An expert system for identification of minerals in thin section: Jour. Geol. Education, v. 37, no. 1, p. 4-6.

Dong, W., Lamarre, M., and Boissonade, A.C., 1986, Expert system for seismic risk evaluation: Proc. 8th European conference on earthquake engineering, v. 1, p. 2.4/23-30.

Duda, R.O., 1980, The PROSPECTOR system for mineral exploration: Final Report, Stanford Research Inst. Intern., Stanford, California, 120 p.

Einstein, E.E., and Edwards, K.W., 1988, Comparison of an expert system to human experts in well log analysis and interpretation: Proc. 1988 Society of Petroleum Engineers Annual Technical Conf. and Exhibition, Houston Texas, p. 253-262.

Fabbri, A.G., Fung, K.B., and Yatabe, S.M., 1988, The usage of artificial intelligence in remote sensing; a review of applications and current research, in Chung,C.F.,Fabbri, A.G.,and Sinding-Larsen,R., eds., Quantitative analysis of mineral and energy resources: D. Reidel Publ. Comp., Dordrecht, p. 489-512.

Fabbri, A.G., and Kasvand, T. ,1988, Automated integration of mineral resource data by image processing and artificial intelligence, in Chung, C.F., Fabbri, A.G., and Sinding-Larsen,R., eds., Quantitative analysis of mineral and energy resources: D. Reidel Publ. Co., Dordrecht, p. 215-239.

Fang, J.H., and Stanley, D.A., 1986, Mineral identification via thin section and X-ray powder diffraction; an artificial intelligence approach (abst.): Geol. Soc. America, 99th Ann. meeting, San Antonio, Texas, p. 597.

Fang, J.H., Shultz, A.W., and Chen, H.C. ,1986, Expert systems and petroleum exploration; an overview: Geobyte, v. 1, no.1, p. 6-11.

Faure, R.M., Leroueil, S., Rajot, J.P., Larochelle, P., Seve, G., and Tavenas, F., 1988, XPENT, systeme expert en stabilite des pentes: Proc. Intern. Symposium on Landslides, v.5, p. 625-629.

Fisher, P.F., Mackaness, W.A., Peacegood, G., and Wilkinson, C.G., 1988, Artificial intelligence and expert systems in geodata processing: Progr. Phys. Geogr., v. 12, no.3, p. 371-388.

Gale, W., and Pregibon, D., 1984, REX: an expert system for regression analysis: Proc. COMPSTAT 84 congress, Wien.

Gale, W., 1986, Artificial intelligence and statistics: Addison-Wesley Publ. Co., New York, 322p.

Goodenough, D.G., Goldberg, M., Plunkett, G., and Zelek, J., 1987, An expert system for remote sensing: Trans. Geoscience and Remote Sensing, v. GE-25, no.3, p. 349-359.

Grivas, D.A., and Reagan, J.C., 1988, An expert system for the evolution and treatment of earth slope instability: Proc. Intern. Symposium on Landslides,v. 5, p. 649- 654.

Gude, T., 1988, Automatic digitization and interpretation of cartographic drawings: Geol. Jahrbuch, v. A104, p. 57-62.

Guerillot, D.R., Blanc, G.A., and Madre, V., 1989, Intelligent interfaces for reservoirs simulators: Proc. Petroleum Computer Conference, Soc. Petroleum Engineers , Richardson,Texas, p. 135-144.

Hamburger, J., Bessis, F., and Parrod, Y., 1989, Geological object recognition and representation (abst.): Intern. Geol. Congress, Washington, v. 2, p. 15-16.

Hand, G., 1984, Statistical expert systems; Design: The Statistician, v. 33, p. 351-369.

Hand, G., 1985, Statistical expert systems; Necessary attributes: Jour. Applied Statistics, v. 12, no.1, p. 19-28.

Hand, G.,1986, Expert systems in statistics: Knowledge Engineering Review, no.1, p. 2-10.

Hanley, J.T., and Merriam, D.F., eds., 1986, Microcomputer applications in geology: Pergamon Press, Oxford, 258 p.

Hart, A.B., McQueen, K.G., and Newmarch, J.D.,1988, A computer program which uses an expert system approach to identifying minerals: Jour. Geol. Education, v. 36, no.1, p. 30-33.

Haux, R., ed.,1986, Expert systems in statistics: Gustav Fischer Verlag, Stuttgart, 215 p.

Havranek, T., and Soudsky, O., 1988, Using an expert system shell for setting statistical parameters: Computational Statistics Quarterly, no. 3, p. 11-28.

Hawkes, D.D., 1985, INTAL; an expert system for the identification of igneous rocks in the hand specimen: Geol. Journal, v. 20, no. 4, p. 367-375.

Hohn, M.E., and Fontana, M.V., 1986, Geostatistics and artificial intelligence applied to stratigraphic correlation (abst.): Am. Assoc. Petroleum Geologists Bull., v. 70, no.5, p. 601.

Horvath, P., Carayannopoulos, N.L., and McCormack, M.D., 1985, Effective seismic data processing using knowledge- based systems (abst.): Europ. Assoc. Explor. Geophysicists, 47th Meeting, Technical Programme and Abstracts of Papers, Budapest, p. 143-144.

Janakiraman, C., 1986, A knowledge-based consultant for geostatistical ore reserve estimation (abst.): Proc. CIM Symposium, Montreal, p. 375.

Jebrak, M., and Loesener, C., 1987, Systemes-expert et geologie: Geochronique, no. 22, p. 12- 15.

Kalin, R.M., and Long, A., 1989, Radiocarbon data base; Q&A; an artificial intelligence data file management program: Radiocarbon, v. 31, no 1, p. 1-6.

Kassenaar, J.D.C., and Dusseault, M.B., 1988, Expert system analysis of log data: Proc. Second Intern. Symposium on Borehole Geophysics for Minerals, Geotechnics, and Groundwater Applications, Houston, Texas, p. 119-130.

Katz, S.S., 1988, Emulating the prospector expert system with a raster GIS, in Thomas, H.F., ed., GIS: integrating technology and geoscience applications: Nat. Acad. Sci., Washington, p. 27-28.

Klemme, H.D., 1975, Giant oil fields related to their geologic setting - a possible guide to exploration: Bull. Can. Petroleum Geologists, v. 23, no.1, p. 30-36.

Klemme, H.D., 1981, Field size distribution related to basin characteristics: The Oil and Gas Jour., v. 81, no. 26, p. 168-176.

Kleyn, M., and Helwig, J.A., 1987, Structural modeling using an expert system (abst.): Am. Assoc. Petroleum Geologists Bull., v. 71, no. 5, p. 578.

Koch, G.S., and Papacharalampos, D., 1988, GEOVALUATOR, an expert system for resource appraisal; a demonstration prototype for kaolin in Georgia, U.S.A., in Chung, C.F., Fabbri, A.G., and Sinding-Larsen, R., eds., Quantitative analysis of mineral and energy resources: D. Reidel. Publ. Co., Dordrecht, The Netherlands, p 513 - 527.

Koch, G.S., Campbell, A.N., Fabbri, A.G., Lanyon, W., Pereira, H.G., Shulman, M.J., Sinding- Larsen, K.S., Stokke, P.R., Toister, A., and

Watney, W.L., 1988, Workshop on mineral and energy resource expert system development, in Chung, C.F., Fabbri, A.G., and Sinding-Larsen, R., eds., Quantitative analysis of mineral and energy resources: D. Reidel Publ. Co., Dordrecht, The Netherlands, p. 707-713.

Koch, G.S., Jr., 1986, EVALUATOR; a proposed expert system to estimate mineral endowment (abst.), in Sheahan, P., ed., Computer Applications in Mineral Exploration: Computer Appl. Miner. Explor., Oakville, Ontario, Canada, p. 44.

Kolyschkow, P., and Peschel, G., 1985, Untersuchung der Zusammenhaenge zwischen den Eigenschaften geologischer Objekte mit Hilfe nichtlinearer Regressionsbeziehungen: Z. angew. Geol., v. 31, no.7, p. 175-178.

Kozai, K., 1987, An educational aid to interpret aerial photographs for coastal landforms; an expert system approach: Am. Soc. Photogramm. and Remote Sensing Technical Papers, v. 4, p. 80-86.

Krystinik, K.B., 1986, An expert system for the identification of foreshore depositional environments: U. S. Geol. Survey Open-File Report No. OF 86-0513, 13 p.

Kuo, T.B., and Startzman, R.A., 1987, Field-scale stratigraphic correlation using artificial intelligence: Geobyte, v. 2, no. 2, p. 30-35.

Kuo, T.., 1986, Well log correlation using artificial intelligence: unpubl. doctoral dissertation, Texas A&M Univ., College Station, Texas, 150 p.

Lamarre, M., Boissonade, A.C., and Shah, H.C., 1987, Development of an expert system for seismic hazard evaluation (abst.): Seis. Research Letters, v. 58, no .1, p. 22.

Lamarre, M., 1988, Seismic hazard evaluation for sites in California; development of an expert system: Stanford Univ., John A. Blume Earthquake Engineering Center, Rept. No. 85, 170 p.

Linehan, J.M., and Sutterlin, P.G., 1986, WSULOG, microcomputer-based well-log evaluation for carbonate reservoirs in Kansas: Computers & Geosciences, v. 12, no. 4B, p. 499-517.

Liu, X., and Wang, P., 1986, Expert system for earthquake intensity evaluation (EIE): Earthquake Engineering and Engineering Vibration, v. 6, no. 3, p. 27-34 (Chinese with English abstract).

Love, P., and Simaan, M., 1985, Segmentation of a seismic section using image processing and artificial intelligence techniques: Pattern Recognition, v. 18, no. 6, p. 409-420.

Mabile, C.M., Hamelin, J.P.A., Amaudric du Chaffaut, B., and La Fonta, J.G.M., 1989, An expert system helps in formation recognition: Proc. Petroleum Computer Conference, Soc. Petroleum Engineers, Richardson, Texas, p. 117-122.

Martin-Clouaire, R., 1984, ELFIN - Une Approache Systeme Expert et Theorie des Possibilites Appliquees en Geologie Petrolier: unpubl. dissertation, Universite Paul Sabatier de Toulouse, France, 180 p.

Maslyn, R.M., 1986, EXPLOR AND PROSPECTOR - Expert systems for oil and gas and mineral exploration, in Hanley, J.T., and Merriam, D.F., eds., Microcomputer applications in geology: Pergamon Press, Oxford, p. 89-103.

Maslyn, R.M., 1987, Gridding Advisor - an expert system for selecting gridding algorithms: Geobyte, v. 2, no. 4, p. 42-43.

Matsuyama, T., and Hwang, V., 1985, SIGMA - a framework for image understanding of bottom-up and top-down analysis: Proc. Intern. Joint Conference on Artificial Intelligence, p. 1-8.

McCammon, R.B., 1983, Operation manual for the muPROSPECTOR consultant system: U. S. Geol. Survey Open-File Report No. OF 83-804, 29 p.

McCammon, R.B., and Cheslow, R., 1986, Seminar on expert systems: U.S. Geol. Survey Circ. 0980, p. 329-330.

McCammon, R.B., 1986, The muProspector mineral consultant system: U.S. Geol. Survey Bull. 1697, 35 p.

McCammon, R.B., 1987, Artificial intelligence applications using PROSPECTOR: U. S. Geol. Survey Open-File Report No. OF 87-0314, 40 p.

McCammon, R.B., 1989, Prospector II: Proc. Annual AI Systems in Government, Conf., p. 88-92.

McClymont, G.L., and Schwartz, F.W., 1986, An application of expert system technology to contaminant hydrology (abst.): Eos, Trans. Am. Geophys. Union, v. 67, no. 44, p. 941.

McCracken, R.J., and Cate, R.B., 1986, Artificial intelligence, cognitive science, and measurement theory applied in soil classification: Soil Science Soc. America Jour., v. 50, no. 3, p. 557-51.

Meech, J.A., 1988, Using GENEX to develop expert systems, in Computer Applications in Mineral Industry: Proc. 1st Canadian Conf., Quebec, Brookfield Publ., Rotterdam, p. 555-562.

Mikroudis, G.K., Fang, H., and Wilson, J.L., 1986, Development of GEOTOX expert system for assessment of hazardous waste sites: Intern. Symposium on Environmental Geotechnology, Envo Publ. Co. U.S.A., v. 1, p. 223-232.

Miller, B.M., 1986a, Building an expert system helps classify sedimentary basins and assess petroleum resources: Geobyte, v. 1, no. 2 , p. 44-48.

Miller, B.M., 1986b, Development of expert system for classifying sedimentary basins as aid to assessing petroleum resources (abst.): Am. Assoc. Petroleum Geologists Bull., v. 70, no. 5, p. 621.

Miller, B.M., 1987a, The muPETROL expert system for classifying world sedimentary basins: U.S. Geol. Survey Bull. B 1810, 87 p.

Miller, B.M., 1987b, An expert system for sedimentary basin analysis for assessment of mineral and energy resources (abst.): U.S. Geol. Survey, Rept. No. C0995, p. 45-46.

Miller, B.M., 1987c, Sedimentary basin models documented on computer diskettes for USGS Bulletin 1810 for the muPETROL expert system for classifying world sedimentary basins: U. S. Geol. Survey Open-File Report, OF 87-0404, 5 p., diskette.

Nelder, J., and Wolstenholme, D., 1986, A front end for GLIM, in Haux, R., ed., Expert systems in statistics: G. Fischer Verlag, Stuttgart.

Nelder, J., 1988, The role of expert systems in statistics, in Faulbaum, F and Uelinger, H., eds., Fortschritte in der Statistik, Software 1: G. Fischer Verlag, Stuttgart.

Nemec, V., ed., 1987, The Mining Pribram in the science and technology 1987: Mathematical methods in geology, Pribram, Czechoslovakia, 612 p.

Newell, C.J., and Bedient, P.B., 1987, Development and application of a ground water modeling database and expert system: Proc. NWWA/ API Conference on Petroleum Hydrocarbons and Organic Chemicals in Ground Water; Prevention, Detection, and Restauration: Nat. Water Well Assoc., Dublin, Ohio, p. 559-578.

Oldford, R., and Peters, S., 1985, Implementation and study of statistical strategy (abst.): Workshop on AI Statistics, Princeton, New Jersey, U.S.A., p. 14.

Olea, R.A., and Davis, J.C., 1986, An artificial intelligence approach to lithostratigraphic correlation using geophysical well logs: Soc Petroleum Engineers, Richardson, Texas, separate print no.SPE 15603, 12 p.

Olea, R.A., 1988, CORRELATOR - an interactive computer system for lithostratigraphic correlation of wireline logs: Kansas Geol. Survey, Petrophysical Series No. 4, 85 p.

Olhoeft, G.R., 1988, Geophysics Advisor Expert System: U. S. Geol. Survey Open-File Report No. OF 88-0399-B, 1 diskette.

Papacharalampos, D., and Koch, G.S., 1986, Automating rock-type classification from satellite data; developing an expert system:

Application of computers and operations research in the mineral industry, v. 19, p. 809-815.

Penas, C., Hadsell, F., and Stout, J., 1986, An expert system for determining vibroseis field parameters: Geobyte, v. 1, no. 5, p. 42-44.

Perkins, W.A., Laffey, T.J., and Nguyen, T.A., 1985, Rule-based interpreting of aerial photographs using LES: Appl. Artificial Intelligence, v. 2, p. 138-146.

Peschel, G.J., 1987, BURUGOL - an expert system for the exploration and evaluation of lignites (abst.), in Nemec, V., ed., The Mining. Pribram: Pribram, Czechoslovakia, v. M1, p. 26.

Peschel, G.J., 1990, BURUGOL - an expert system for the exploration and evaluation of lignites: Z. angew. Geologie, v. 36, no. 4, p. 147-150.

Peveraro, R.C.A., and Lee, J.A., 1988, HESPER; an expert system for petrophysical formation evaluation: Proc. European Petroleum Conference, London, p. 61-370.

Poyet, P., and Detay, M., 1989, HYDROLAB; An example of a new generation of compact expert systems: Computers & Geosciences, v. 15, no. 3, p. 255-267.

Quick, A.N., and Schuyler, J.R., 1988, Expert system for strategic planning: The Oil and Gas Jour., v. 86, no. 8, p. 75-79.

Quincy, E.A., and Kubichek, R.F., 1987, Expert system for establishing hydrocarbon prospects using statistical pattern recognition input, in Helbig, K., and Treitel, S., eds., Handbook of geophysical exploration, Section I, Seismic Exploration: Geophys. Press, London, p. 446-471.

Reeves, M., 1989, MINID; a BASIC program to assist in the optical identification of minerals in thin section: Computers & Geosciences, v. 15, no. 1, p. 121-133.

Ripple, W.J., and Ulshoefer, V.S., 1987, Expert systems and spatial data models for efficient geographic data handling: Photogrammetric Engineering and Remote Sensing, v. 53, no. 10, p. 1431-1433.

Robinson, V.B., and Frank, A.U., 1987, Expert systems for geographic information systems: Photogrammetric Engineering and Remote Sensing, v. 53, no. 10, p. 1435-1441.

Sarkar, B.C., O'Leary, J., and Mill, A.J.B., 1988, An integrated approach to geostatistical evaluation: Mining Magazine, v. 159, no. 3, p. 199-207.

Scotese, C.R., 1987, Paleoclimatic expert system that predicts coastal upwelling (abst.): Am. Assoc. Petroleum Geologists Bull., v. 71, no. 5, p. 611.

Sejful'-Muljukov, R.R., and Nemirovskij, E.A., 1987, RIF -Expert system for prognostication of buried reefs (abst.), *in* Nemec, V., ed., The Mining Pribram: Pribram, Czechoslovakia, v. M, p. 525-526.

Shultz, A.W., Fang, J.H., Burston, M.R., Chen, H.C., and Beasley, M., 1986, Expert system for determining clastic depositional environments (abst.): Am. Assoc. Petroleum Geologists Bull., v. 70, no. 5, p. 647-648.

Shultz, A.W., Fang, J.H., Burston, M.R., Chen, H.C., and Reynold, S., 1988, XEOD; An expert system for determining clastic depositional environments: Geobyte, v. 3, no. 2, p. 22-32.

Simaan, M., Zhang, Z., and Love, P.L., 1987, Artificial Intelligence and expert systems for seismic data; Image processing and knowledge-based methods for segmentation of a seismic section based on signal character, *in* Helbig,K., and Treitel,S., eds., Handbook of geophysical exploration, Section I, Seismic Exploration: Geopys. Press, London, p. 389-424.

Singer, D.A.,1985, Preliminary version of FINDER, a Pascal program for locating mineral deposits: U. S. Geol. Survey Open-File Report No. OF 85-590, 24 p.

Skala, W., 1988, Expertensysteme in der Umweltgeologie: Prospekt, Freie Universitaet Berlin, 6 p.

Skandsen, J., and Lile, O.B.,1987, An expert system for interpretation of lithology from wireline log data (abst.): Geoexploration, v. 24, no. 3, p. 265.

Smith, B.J., Saleken, L., and Rockel, E.,1986, GPEX; Geophysical expert system: Proc. Canadian Industrial Computer Systems Conference, Ecole Polytechnique, Montreal, Quebec, p. 16-1 - 16-8.

Smith, R., Young, R., Lafue, G.,and Winston,H., 1985, AI in engineering: Schlumberger-Doll Research Note, 5 p.

Smith, R.G., 1984, On the development of commercial expert system: AI Magazine, v.5, no. 3, p. 61-77.

Smutz, J.,1989, Arco foresees productivity increases with AI technologies: The Oil and Gas Jour., v. 87, no. 2, p. 42-48.

Startzman, R.A., and Kuo, T.,1987, An artificial intelligence approach to well log correlation: The Log Analyst, v. 28, no.2, p. 175-183.

Startzman, R.A., Kuo, T.B., and Wong, S.A., 1987, Field scale well log correlation using an expert system (abst.): Am. Assoc. Petroleum Geologists Bull.,v. 71, no. 5, p. 618.

Startzman, R.A., and Kuo, T., 1986, A rule-based system for well log correlation, *in* Petroleum industry application of microcomputers: Soc. Petroleum Engineers, Richardson, Texas, p. 113-124.

Stoisits, R.F., 1989, AI is vital part of production computing: The Oil and Gas Jour., v. 87, no. 2, p.48-50.

Stout, J.L., 1986, An expert system for carbonate reservoirs: Am. Assoc. Petroleum Geologists Bull., v. 70, no. 8, p. 1057-1058.

Streitberg, B., 1988, Expertensysteme in der Statistik - Experten oder Ignoranten?: Fortschritte in der Statistik, Software 1, G. Fischer Verlag, Stuttgart, 412 p..

Summers, E.G., and MacDonald, R.A., 1988, Experiments with microcomputer-based artifificial intelligence environments: Jour. Math. Geology, v. 20, no. 8, p. 1037-1047.

Tabesh, E., 1989, Artificial intelligence techniques for merging qualitative and quantitative geological data bases: Proc. Geoscience Information Society, v. 19, p. 169- 175.

Takeda, H., 1986, An approach for artificial intelligence data bases in mineralogy: Jour. Mineralogical Society of Japan (Japanese with English abstract), v. 17, no. 6, p. 277-288.

Usery, E.L., Altheide, P., Deister, R.R.P., and Barr, D.J., 1988, Knowledge-based GIS techniques applied to geological engineering: Photogrammetric Engineering and Remote Sensing, v. 54, no. 11, p. 1623-1628.

Wang, Y., and Zhang, Y., 1988, Application of artificial intelligence principle to paleontologic taxonomy: Acta Palaeontologica Sinica (Chinese with English abstract), v. 27, no. 4, p. 521-524.

West, J., 1984, Towards an expert system for identification of minerals in thin section: Jour. Math. Geology, v. 17, no. 7, p. 743-753.

Wittkowski, K., 1988, Intelligente Benutzerschnittstellen für statistische Auswertungen: Fortschritte in der Statistik, Software 1, G. Fischer Verlag, Stuttgart, p. 47-63.

Wong, S.A., Startzman, R.A., and Kuo, T., 1988, CAIPEP; report on the first conference on artificial intelligence in petroleum exploration and production: Geobyte, v. 3, no. 3, p. 10- 16.

Yatabe, S.M., and Fabbri, A.G., 1988, Artificial intelligence in the geosciences; a review: Science de la Terre, Informatique Geologique, v. 21, pt. 1, p. 37-67.

CONTRIBUTORS

Bachu, S., Alberta Research Council, P.O. Box 8330, Postal Station F, Edmunton, Alberta, T6H 5X2, Canada

Bayer, U., Institute of Petroleum and Organic Geochemistry (ICH-5), KFA Jülich, P.O. Box 1913, D-5170 Jülich, F.R.G.

Berthold, A., University of Leipzig, Section of Physics, Institute of Geophysics, Geology and Meteorology, Talstr. 35, Leipzig, 7010, Germany

Bidstrup, Torben, Geological Survey of Denmark, Thoravej 8, DK-2400 København NV, Denmark

Bohling, Geoff, Kansas Geological Survey, 1930 Constant Ave., Lawrence, KS 66047 USA

Cadman, Steven J., Bureau of Mineral Resources, Geology and Geophysics, P.O. Box 378, Canberra City, A.C.T. 2601, Australia

Cao, S., Alberta Research Council, P.O. Box 8330, Postal Station F, Edmonton, Alberta, T6H 5X2, Canada

Davis, John C., Kansas Geological Survey, 1930 Constant Ave., Lawrence, KS 66047 USA

DeConto, Robert M., Cooperative Institute for Research in Environmental Sciences (CIRES), Department of Geology, and Museum, University of Colorado, Boulder, CO 80309 USA

Dietrich, H., Department of Geology, University of Griefswald, F.-L. Jahn-Str. 17a, Greifswald 0-2200, Germany

Dowd, P.A., Department of Mining and Mineral Engineering, University of Leeds, Leeds LS2 9JT, England, United Kingdom

Eiserbeck, W., Erdöl-Erdgas-Geotechnologie Gommern GmbH, 0-3300 Gommern, Magdeburger Chaussee, Germany

Forman, David J., Bureau of Mineral Resources, Geology and Geophysics, P.O. Box 378, Canberra City, A.C.T. 2601, Australia

Fuhr, Beth A., Stratigraphic Studies Group, Wichita State University, Wichita, KS 67208; current address: Pintail Petroleum, 412 E. Douglas Avenue, Suite 1, Wichita, KS 67202 USA

Hanemann, K.-D., Geological Survey of Saxony, 9200-Freiberg, Halsbrücker Str. 31 A, Germany

Hansen, Kirsten, Institute for Petrology, Copenhagen University, Østervoldgade 10, DK-1350 København K, Denmark

Harbaugh, John W., Department of Applied Earth Sciences, Stanford University, Stanford, CA 94305-2225 USA

Harff, J., Central Institute for Physics of the Earth, Telegrafenberg, O-1561 Potsdam, Germany; current address: Institute for Baltic Research, Seestr. 15, O-2530 Warnemuende, Germany

Hay, William W., Cooperative Institute for Research in Environmental Sciences (CIRES), Department of Geology, and Museum, University of Colorado, Boulder, CO 80309 USA; current address: GEOMAR, Wischhofstr. 1-3, D-2300 Kiel 14, Germany

Herzfeld, Ute C., Geologic Research Division, Scripps Institution of Oceanography, La Jolla, CA 92093 USA

Hinde, Alan L., Bureau of Mineral Resources, Geology and Geophysics, P.O. Box 378, Canberra City, A.C.T. 2601, Australia

Hoth, P., Central Institute for Physics of the Earth, Telegrafenberg, O-1561 Potsdam, Germany

Huang, L., Department of Geology, Syracuse University, Syracuse, NY 13224 USA

Jensen, P.K., Risø National Laboratory, DK-4000 Roskilde, Denmark

Kunzendorf, Helmar, Risø National Laboratory, DK-4000 Roskilde, Denmark

Lytviak, A., Alberta Research Council, P.O. Box 8330, Postal Station F, Edmonton, Alberta, T6H 5X2, Canada

Medevev, S.E., NPO "Sojuznefteotdacha", Kommunisticheskaya, 29 Ufa, 450076, USSR

Menschner, K., University of Leipzig, Section of Physics, Institute of Geophysics, Geology and Meteorology, Talstr. 35, Leipzig, 7010, Germany

Merriam, D.F., Stratigraphic Studies Group, Wichita State University, Wichita, KS 67208; current address: Kansas Geological Survey, The University of Kansas, Lawrence, KS 66047 USA

Mokosch, M., Ernst-Moritz-Arndt-University of Greifswald, Department of Geosciences, F.-L. Jahn-Str. 17 A, 0-2200 Greifswald, Germany

Morton, John, Department of Mines and Energy, South Australia, 191 Greenhill Road, Parkside 5063, Australia

Ondrak, R., Institute of Petroleum and Organic Geochemistry (ICH-5), KFA Jülich, P.O. Box 1913, D-5170 Jülich, Germany

Peschel, G.J., Ernst-Moritz-Arndt-University of Greifswald, Department of Geosciences, F.-L. Jahn-Str. 17 A, 0-2200 Greifswald, Germany

Poppitz, H.H., Ernst-Moritz-Arndt-University of Greifswald, Department of Geosciences, F.-L. Jahn-Str. 17 A, 0-2200 Greifswald, Germany

Radlinski, A. P., Bureau of Mineral Resources, Geology and Geophysics, P.O. Box 378, Canberra City, A.C.T. 2601, Australia

Rentzsch, J., Central Geological Institute, Invalidenstrasse 44, 1040 Berlin, Germany

Richers, D., Department of Geology, Syracuse University, Syracuse, NY 13224 USA

Robinson, J.E., Department of Geology, Syracuse University, Syracuse, NY 13224 USA

Schwab, G., Central Institute for Physics of the Earth, Telegrafenberg A51, O-1561 Potsdam, Germany

Shaw, C.A., Cooperative Institute for Research in Environmental Sciences (CIRES), Department of Geology, and Museum, University of Colorado, Boulder, CO 80309 USA; current address: Exxon Production Research Company, 3120 Buffalo Speedway, Houston, TX 77252-2189 USA

Springer, J., Central Institute for Physics of the Earth, Telegrafenberg A51, O-1561 Potsdam, Germany

Thiergärtner, H., Central Geological Institute, Invalidenstrasse 44, 1040 Berlin, Germany

Watney, W. Lynn, Kansas Geological Survey, 1930 Constant Ave., Lawrence, KS 66047 USA

Wold, Christopher N., Cooperative Institute for Research in Environmental Sciences (CIRES), Department of Geology, and Museum, University of Colorado, Boulder, CO 80309, USA; current address: GEOMAR, Wischhofstr. 1-3, D-2300 Kiel 14, Germany

Wong, J.C., Kansas Geological Survey, 1930 Constant Ave., Lawrence, KS 66047 USA

Zeissler, K.-O., Geological Survey of Saxony, 9200-Freiberg, Halsbrücker Str. 31 A, Germany

INDEX

The manufacturer's authorised representative in the EU is Springer
Nature Customer Service Centre GmbH, Europaplatz 3, 69115 Heidelberg,
Germany. If you have any concerns regarding our products, please
contact ProductSafety@springernature.com

Printed and bound by CPI Group (UK) Ltd, Croydon, CR0 4YY
29/04/2026
02099472-0014